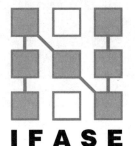

第八届结构工程新进展论坛文集
Industry-Acadamia Forum on Advances in Structural Engineering（2018）

可持续结构与材料

Sustainable Structures and Materials

肖建庄　主编

Editors in Chief：Xiao Jianzhuang

中国建筑工业出版社
CHINA ARCHITECTURE & BUILDING PRESS

图书在版编目（CIP）数据

可持续结构与材料/肖建庄主编．—北京：中国建筑工业出版社，2018.8
（第八届结构工程新进展论坛文集）
ISBN 978-7-112-22378-7

Ⅰ.①可… Ⅱ.①肖… Ⅲ.①工程材料-结构性能-文集 Ⅳ.①TB303-53

中国版本图书馆CIP数据核字（2018）第137679号

本书为"第八届结构工程新进展论坛"特邀报告人论文集。本届论坛主题为"可持续结构与材料"，合为以下10个主要议题：再生混凝土材料及结构；海砂海水混凝土材料及结构；基于复合材料的可持续结构；基于自然材料（木、竹、自然纤维等）的可持续结构；高性能结构钢及高性能钢结构；高性能混凝土及高性能混凝土结构；绿色建造（预制装配、可拆装技术、3D打印、BIM技术）；碳足迹及生命周期评价；可持续新型结构（组合结构、混合结构）抗灾理论与实践；结构可持续性设计及评价。本论文集选编了部分特邀报告人的论文，涵盖了大多数论坛议题。

责任编辑：赵梦梅　刘婷婷
责任校对：姜小莲

第八届结构工程新进展论坛文集
可持续结构与材料
肖建庄　主编

*

中国建筑工业出版社出版、发行（北京海淀三里河路9号）
各地新华书店、建筑书店经销
北京红光制版公司制版
天津翔远印刷有限公司印刷

*

开本：787×1092毫米　1/16　印张：17¼　字数：421千字
2018年8月第一版　　2018年8月第一次印刷
定价：**68.00**元
ISBN 978-7-112-22378-7
　　（32266）

版权所有　翻印必究
如有印装质量问题，可寄本社退换
（邮政编码 100037）

鸣 谢

本次论坛得到以下单位的支持和资助

同济大学建筑设计研究院（集团）有限公司

华建集团华东建筑设计研究总院

上海建工集团股份有限公司

上海通正铝合金结构工程技术有限公司

上海同济绿建土建结构预制装配化工程技术有限公司

前 言 Preface

"结构工程新进展论坛"自 2006 年首次举办以来，十余年间已经打造成为土木建筑行业内一个颇有影响的学术交流平台。论坛旨在促进我国结构工程界对学术成果和工程经验的总结及交流，汇集国内外结构工程各方向的最新科研信息，提高学科交叉与学术水平，推动我国建筑行业科技创新发展。

论坛原则上以两年一个主题的形式轮流举办，前七届的主题分别为：
- 新型结构材料与体系（第一届，2006，北京）
- 结构防灾、监测与控制（第二届，2008，大连）
- 钢结构研究和应用的新进展（第三届，2009，上海）
- 混凝土结构与材料新进展（第四届，2010，南京）
- 钢结构（第五届，2012，深圳）
- 结构抗震、减震技术与设计方法（第六届，2014，合肥）
- 工业建筑及特种结构（第七届，2016，西安）

"结构工程新进展论坛"已作为结构工程领域重要的学术会议在国内外产生了重要影响，历届论坛都吸引了众多专家学者、工程设计人员、青年学生等参会。第八届论坛由中国建筑工业出版社、同济大学《建筑钢结构进展》编辑部、香港理工大学《结构工程进展》（Advances in Structural Engineering）编委会联合主办，由同济大学土木工程学院建筑工程系承办，于 2018 年 9 月 7 日～9 日在同济大学召开。

本次论坛的主题是"可持续结构与材料"。"促进可持续的建筑业活动"是联合国《21 世纪议程》中的重要内容，工程建设的可持续发展也是当前国际上建设领域科技发展热点和研究的前沿。在本次论坛中我们荣幸地邀请到了 30 余位特邀报告人，他们的报告主题涵盖了近年来与"可持续结构和材料"相关的最新学术思想、研究成果、设计方法、施工技术以及相应新型可持续材料及结构体系的应用；阐述了在这些领域内的最新发展动态；同时也向与会者提供了一个与专家互动并获取宝贵经验的机会。

感谢特邀报告人，他们不仅在大会上作了精彩的主题报告，而且还奉献了精心准备的论文，使得本书顺利出版。

感谢参加本次论坛的所有代表，正是大家的积极参与，才使得本次论坛能够顺利进行。还要特别感谢为本书出版辛勤工作的同济大学李征老师和刘玉姝老师。

感谢中华人民共和国住房和城乡建设部执业资格注册中心、中国建筑工业出版社、同济大学《建筑钢结构进展》编辑部、香港理工大学《结构工程进展》编辑部对本次论坛的指导、支持和帮助。

感谢同济大学建筑设计研究院（集团）有限公司、华建集团华东建筑设计研究总院、上海建工集团股份有限公司、上海通正铝合金结构工程技术有限公司、上海同济绿建土建结构预制装配化工程技术有限公司对本次论坛成功举办的资助和支持。

目 录 Contents

1. 工程结构整体可靠性分析研究进展/李 杰 ········· 1
2. Making Recycled Aggregate Suitable For Structural Concrete/
   Surendra P. Shah　Yuxi Zhao　Weilai Zeng　Hongru Zhang ········· 17
3. An overview on the use of 2D nanomaterials in concrete/
   Ezzatollah Shamsaei　Felipe Basquiroto de Souza　Xupei Yao Shujian Chen
   Wenhui Duan ········· 50
4. Sustainable Concrete For The Next Century：Multi-recycled Aggregate Concrete/
   Jorge de Brito　Luís Evangelista　Rui V. Silva ········· 62
5. Properties of Concrete With Recycled Construction and Demolition Waste：
   A Research Experience in Belgium/Zengfeng Zhao　Luc Courard　Frédéric
   Michel　Simon Delvoie　Mohamed ElKarim Bouarroudj　Charlotte
   Colman　Jianzhuang Xiao ········· 79
6. 地聚合物混凝土收缩控制剂研究/李柱国　冈田朋友　桥爪进 ········· 91
7. 基于IMU的混凝土结构隐蔽工程质量实时监控系统/李 恒　杨新聪 ········· 102
8. 钢管再生混凝土构件力学性能研究（摘要）/韩林海　吕晚晴 ········· 128
9. 预应力正交胶合木剪力墙抗侧性能分析/何敏娟　孙晓峰　李 征 ········· 131
10. BIM-Based Bridge Performance Assessment/Weixiang Shi Jian Chai Peng
    Wang Xiangyu Wang ········· 154
11. 土木工程结构生命周期可持续量化评价/
    王元丰　章玉容　王京京　石程程　刘胤杉　梅生启　周硕文 ········· 166
12. 现代竹结构的进展/肖 岩　单 波　李 智 ········· 197
13. 组合混凝土结构基本概念和原理/肖建庄　张青天　丁 陶 ········· 207
14. 铝合金结构在我国的应用和发展/张其林 ········· 221
15. Mechanics of Biological Materials：A Civil Engineering Approach/
    Lihai Zhang ········· 232
16. Crumb Rubberized Concrete (CRC) Application in Structural Engineering/
    Yan Zhuge　Xing Ma　Danda Li　Jianzhuang Xiao ········· 240

第八届结构工程新进展论坛简介 ········· 260
第八届论坛特邀报告论文作者简介 ········· 264

1 工程结构整体可靠性分析研究进展

李 杰

(同济大学土木工程学院)

摘 要：本文系统回顾了工程结构可靠性近百年的研究历程，剖析了自20世纪60年代中期以来工程结构整体可靠性研究的基本思想与方法。在此基础上，论述了工程结构整体可靠性分析的物理综合法。这一方法，综合了结构受力力学行为的物理力学机制、从材料损伤到结构破坏的物理失效准则、随机性在物理系统中的传播规律，从而，基本完整地建立了工程结构可靠性分析新的理论体系。文中较为系统地介绍了这一理论的基本构架和主要方程，择要给出了若干实际工程的应用案例。

Advances in Global Reliability Analysis of Engineering Structures

Jie Li

(School of Civil Engineering, Tongji University)

Abstract: The present paper systematically retrospect the research developments of engineering reliability in the past century. Basic ideas and methodologies for global reliability research of engineering structures since the mid-1960s have been studied. On this basis, a physically-based synthesis method for the global reliability analysis of engineering structures is proposed. This method integrates the physical mechanisms of mechanical behavior of structures, the physical failure criteria from material damage to structural failure, and the propagation law of randomness inherent in physical systems. Therefore, a new theoretical system for the global reliability analysis of engineering structures is completely established. The fundamental framework and essential equations of this theoretical system are introduced systematically. For illustrative purposes, typical case studies on practical applications are presented briefly.

1. 前言

土木工程，是人类文明发展的重要标志之一。近代科学的兴起，则为人类理性地、而不仅仅是依凭经验地建造土木工程结构提供了基础。始自伽利略的材料力学传统与始自柯西的固体力学传统，使人们对工程结构的受力力学行为有了定量的认识与分析工具。据此，工程结构设计开始有了科学意义上的保证。19世纪初，设计安全系数的观念开始在分析基础上得以经验性的确立。在本质上，这种经验的结构安全系数是对工程中各种不确定性影响的一种综合估计。试图对这种总体的、定性的综合估计给出科学上的解释与技术上的定量修正，催生了工程结构可靠度理论。20 世纪初，基本确立了采用概率论反映客观随机性、度量结构可靠性的研究发展路线。经过近百年的发展，今天的工程师，已经可以基本清晰地认识到来自结构作用、结构性质、施工影响等方面的随机性、并采用分项系数的方法从设计角度来避免这些随机性的不利影响。然而，自 20 世纪 80 年代分项系数设计法在工程中普及以来，工程师关于结构整体安全性与可靠性的观念却越来越模糊了：按照基于二阶矩的近似概率设计准则设计工程结构，无法计算并定量评估结构的整体安全性！

如果说这种重大缺陷对于在正常使用条件下的结构尚不足以造成重大威胁的话，在灾害性动力作用下（如地震、强风、爆炸等等）、结构的整体安全性仍不能被定量评估，就构成了人们不可回避的基本问题。在过去 20 年间，本文作者和他的研究梯队对这一基本问题进行了持续、深入的研究，基本完整地建立起来结构整体可靠度分析理论，并示范性地应用于工程实践。本文，将比较系统地介绍这一研究进展的主要背景、基本成果和典型工程应用实例。

2. 工程结构可靠性研究历史的简要回顾

结构可靠性研究的核心，是解决存在随机性条件下结构安全性的科学度量问题。早在 1911 年，匈牙利学者卡钦奇就提出用统计数学分别研究结构荷载和材料强度的概率分布的思想[1]。1926 年，德国学者 Mayer 出版了名为《结构安全性》的专著[2]，第一次系统论述了采用概率论研究结构安全性的学术设想。20 世纪 40 年代后期，在美国学者 Freudenthal 和苏联学者斯特列津斯基、尔然尼钦等人的倡导下，工程结构可靠性问题开始得到学术界与工程界的普遍重视，并逐步成为土木工程研究中的核心与热点问题[3,4]。

事实上，在 20 世纪中期，由于文化的隔阂，关于结构可靠度的研究探索是在苏联和西方世界分别独立进行的。一般认为：苏联学者尔然尼钦于 1947 年最早提出了利用荷载效应与强度的均值及方差计算结构可靠度的二阶矩法（本质上属于中心点法），同年，苏联学者斯特列津斯基提出了将结构安全系数分为荷载系数和强度系数的方法[1,4]。在西方世界，美国学者 Frendenthal 提出了结构可靠度分析的全概率法。用今天的科学语言表述，即

$$P_s = P(z \geqslant 0) = \int \cdots \int_{z \geqslant 0} f_x(x_1, x_2, \cdots, x_n) \mathrm{d}x_1 \mathrm{d}x_2 \cdots \mathrm{d}x_n \tag{1}$$

式中，$Z = g(x_1, x_2, \cdots, x_n)$ 为结构功能函数；$x_1, x_2, \cdots, x_n$ 为结构作用与结构参数中的基本随机变量；$f(x_1, x_2, \cdots, x_n)$ 为基本随机变量的联合分布函数。

这一表述，在一般意义上将结构可靠度定义为结构功能函数大于零的概率，并通过对基本随机变量的联合概率密度函数在安全域内的多维积分定量表述结构可靠度。直至今日，这一基本表述仍具有基础性地位。事实上，对上述多维积分的近似求解，构成了经典结构可靠度理论研究的出发点和归宿点。

一个最直接而简单的近似是假定基本随机变量服从正态分布，而功能函数可以表述为基本随机变量的线性函数，由此，美国学者 Cornell 于 20 世纪 60 年代末提出了一次二阶

矩的基本方法和可靠度指标的概念[5]。在这一研究中，Cornell 引入了结构抗力综合变量 R 和结构荷载效应综合变量 S 的概念，这一创造性处理，使问题的理论表达得到极大简化。在一般意义上，结构功能函数可以表述为：

$$Z = R - S \tag{2}$$

而结构可靠度指标可以简单表述为：

$$\beta = \frac{\mu_z}{\sigma_z} \tag{3}$$

式中，μ_z、σ_z 分别为 Z 的均值与标准差。

由于假定基本随机变量为正态分布，上述结构可靠度指标与结构安全概率之间的关系可以精确地表述为：

$$P_s = \Phi(\beta) \tag{4}$$

式中：$\Phi(\cdot)$ 为标准正态分布的分布函数。

上述基本表达式，为将结构可靠性理论应用于实际工程打开了第一个缺口。形成了在西方科学世界被普遍接受的结构可靠度分析中心点法的基础。

在试图将中心点法推广到非线性功能函数和非正态随机变量时，丹麦的 Ditlevesen 等多位研究者发现：中心点法不能保证可靠度指标的不变性。即当功能函数采用力学本质等价的不同表达式时，可靠度指标 β 值不唯一[6]。由此，引发了一系列研究[7-9]，并最终形成了被国际上逐渐公认的验算点法（也称 JC 法）。这一方法，结合 Lind 在 1971 年所完成的分项系数表达法[10]，成为被普遍接受的工程结构构件可靠分析的基本方法，也成为 20 世纪 80 年代以来世界各国制订工程设计规范的理论基础。

依据这一理论，结构设计的基本表达式被表述为：[11]

$$\gamma_R R_k \geqslant \sum_{i \geqslant 1} \gamma_{G_i} S_{G_{ik}} + \gamma_{Q1} S_{Q_{1k}} + \sum_{j > 1} \gamma_{Q_j} \psi_{C_j} S_{Q_{jk}} \tag{5}$$

式中，γ_R 是结构抗力分项系数，R_k 是结构抗力标准值，γ_{G_i} 是第 i 个永久作用分项系数，$S_{G_{ik}}$ 是第 i 个永久作用标准值的效应，γ_{Q_j} 是第 j 个可变作用分项系数，$S_{Q_{jk}}$ 是第 j 个可变作用标准值的效应，ψ_{C_j} 是第 j 个可变作用的组合系数。

在结构设计中，采用分项系数的方式分别表述结构抗力随机性的影响和荷载效应随机性的影响，虽然看似精细而深刻，却在很大程度上混淆了工程师关于结构安全的观念。结构的整体安全性如何？如果用概率来度量这种安全性，应该如何计算？不同因素的随机性对结构整体安全性有何影响？这些问题，是任何一个真正关心结构安全的工程师和严肃的科学家所无法回避的基本问题！

3. 传统结构整体可靠性研究思想的剖析

试图对结构整体安全性给出概率意义上的度量，从 20 世纪中期就展开了持续努力。早在 1966 年，Freudenthal 就创造性地将结构系统比拟为串联系统，采用系统可靠度的研究思想，给出了系统失效概率的上界[12]。1975 年，洪华生先生和他的学生一起，发展了概率网络评估技术，初步提出了结构体系中失效相关问题的解决方案[13]。1979 年，

Ditlevesen 提出了结构体系可靠度的窄界限公式[14]。至此，采用串-并联系统的基本思想分析结构整体可靠度有了一个初步可行的雏形。但这里最为关键的，是如何将结构系统、尤其是超静定结构系统等效为系统可靠度分析的串-并联（或更为复杂的）系统。引用结构塑性分析中机构法构造结构失效模式，会带来组合爆炸问题。为此，在20世纪80年代，一批学者致力于结构主要失效模式研究，发展了一系列筛选结构主要失效模式、形成等效系统分析模型的方法[15-17]。这些方法，形成了结构整体可靠度分析的基本传统，并一直延续至今天[18,19]。但是，由于这些研究中根深蒂固的现象学研究传统，问题并未真正得到解决。

事实上，任何一种以搜索、确定结构主要失效模式的传统结构整体可靠度分析方法，都是从"破坏后果"出发来考虑问题的，因而在本质不能反映结构非线性发展的真实物理过程。以传统结构整体可靠度分析中的分枝限界法为例，对图1所示的简单框架结构，其基本分析过程是[20]：

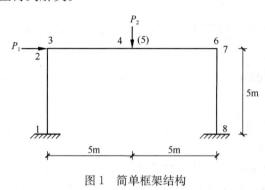

图 1 简单框架结构

（1）按照给定的荷载、依线性结构分析法计算荷载效应；

（2）根据荷载效应按照 R-S 模式列出典型截面（如图1中1～8截面）功能函数，求解截面失效概率；

（3）按照具有最大失效概率者先失效的原则、确定失效截面的位置；

（4）在失效截面，以屈服弯矩代替实际弯矩、计算不平衡节点力并反加在结构上；

（5）按加入塑性铰后的结构形成修正刚度矩阵；

（6）按新的结构形态计算荷载效应；

（7）根据新的荷载效应按 R-S 模式列出典型截面功能函数，再次计算各截面失效概率；

（8）以第（3）步确定的失效截面为起点，按并联系统方式计算失效路径产生概率；

（9）按最大产生概率原则确定失效路径；

（10）按照上述步骤（3）～（9）反复搜索失效截面、失效路径与失效模式（以图1中截面7失效为起点，给出的失效路径示例于图2），直到找到所有主要失效模式。

为了减少上述搜索工作的复杂性，Thoft-Christensen 和 Murotsu 分别引入了分枝限界操作。其实质，是预先规定一个认为此后可以忽略的失效概率，当上述第（3）步和第（9）步的最大失效概率或最大产生概率分别小于规定限值时，不再执行多余的搜索。显然，这不仅给分析过程带来了很大的主观性，也不符合概率的基本原则。在真实世界中，小概率事件并不是不会发生的事件，忽略小概率事件虽然可以给计算带来简单性，也不会对整体结构可靠度分析结果带来太大的影响，但若小概率事件所对应的失效模式不被发现，其带来的后果依然可能是灾难性的。

经过上述繁复的分析，结构系统最终可以用如图3所示的串-并联系统加以表示，其中，每一串联系统中的并联单元表示失效路径，而每一串联单元表示一个失效机构（失效模式）。

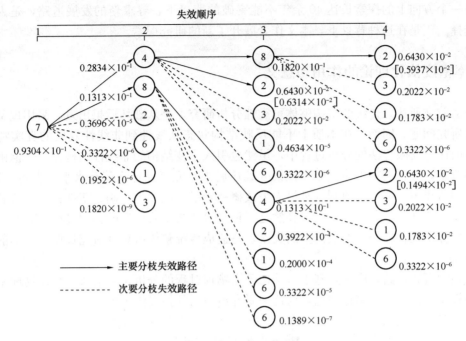

图 2 分枝限界法搜索失效路径

由于失效机构的产生概率（失效概率）已由前述分析获得，故所有失效机构构成的串联系统的失效概率可由下式计算：

$$P_f = P(\bigcup_{i=1}^{M} A_i)$$
$$= \sum_{j=1}^{M} (-1)^{j-1} \sum_{1 \leqslant i_1 < i_2 \cdots < i_j = j} P(\bigcap_{l=1}^{j} A_l) \quad (6)$$

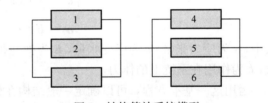

图 3 结构等效系统模型

显然，由于不同失效模式之间失效相关，采用上式计算结构整体失效概率具有高度复杂性。事实上，当 $j > 2$ 时，$P(\bigcap_{l=1}^{j} A_l)$ 的计算必然涉及高维积分！而忽略高阶相关、引入窄界限公式计算结构整体失效概率，其精度甚为可疑。

事实上，对于一般的框架结构，失效机构数可以用下式计算：

$$M = 2^{n-l} - 1 \tag{7}$$

式中，n 为潜在塑性铰数目，l 为结构超静定次数。

上式表明，框架结构的失效机构数呈指数级增长。因此，试图通过搜索主要失效模式来降低计算工作量，无异于杯水车薪。

失效模式相关和失效机构数按指数级增长（组合失效），构成了传统结构整体可靠度分析研究中几乎不可逾越的两大障碍！

进一步注意到：传统结构整体可靠度的研究，至今没有超出杆系结构的范围，也没有超出理想弹塑性所限制的范围，对于板、墙、块体等构成的复杂结构以及一般的物理非线性问题，这一方法基本无能为力。这一背景，使我们不得不怀疑传统结构整体可靠度研究的道路正确性。

在一个方向上的探索长达 40 余年不能突破的背景下，寻求新的发展道路，是人们必然的选择。正是在这种背景下，本文作者展开了新的研究探索。

4. 概率密度演化理论的作用与意义

正如前文所述，传统的结构整体可靠性分析研究，是从破坏后果出发、采用现象学的方式来研究问题。因而，在本质上不能反映结构状态从线性到非线性发展的真实物理过程。在前述分支限界法的分析过程中，虽然也引入了变结构的非线性结构分析，但由于随机性的背景，分析者并不能判断塑性铰（破坏后果）究竟在什么荷载水平上发生，因此，只能使用"概率可能"的原则（具最大失效概率者最可能发生）确定塑性铰的位置和失效路径可能。这样，失效、失效路径必然是一种经过"概率修饰"的物理过程，而不是真实的物理过程。这种在分析方法论上的失误，是造成传统整体可靠度研究长期停滞不前的根本原因。

正确的研究道路，应该是基于对结构受力物理过程的分析，考察随机性在物理系统中的传播规律。概率密度演化理论，为这一研究道路的开拓提供了可能。

以一般的弹性动力系统为例，众所周知，弹性力学基本方程是：

$$\left. \begin{array}{l} \nabla \cdot \boldsymbol{\sigma} + \boldsymbol{b} = \rho \ddot{\boldsymbol{u}} + \eta \dot{\boldsymbol{u}} \\ \boldsymbol{\varepsilon} = \dfrac{1}{2}(\nabla \boldsymbol{u} + \nabla^T \boldsymbol{u}) \\ \boldsymbol{\sigma} = \boldsymbol{E} : \boldsymbol{\varepsilon} \end{array} \right\} \quad (8)$$

式中，$\boldsymbol{\sigma}$ 为应力张量，$\boldsymbol{\varepsilon}$ 为应变张量，$\boldsymbol{E}$ 为弹性张量，$\boldsymbol{u}$ 为位移场，ρ 为材料密度，η 为阻尼系数，$\boldsymbol{b}$ 为作用在系统上的体力。

运用这一基本方程，可以确定一般结构在外部和自身重力作用下的变形与运动状态。问题在于：如果结构的外部作用（如地震动）或结构基本力学性质（如 η、$\boldsymbol{E}$）具有随机性，结构响应应如何分析？换句话说：当 $\ddot{u}_g$ 具有随机性，或基本的弹性性质因材料制作或形成过程中具有不可控制性而形成随机量 $\boldsymbol{E}$，这些随机性是如何经过上述物理系统的作用而演化为结构响应 $\boldsymbol{u}$（或 $\boldsymbol{\sigma}$、$\boldsymbol{\varepsilon}$）的随机性呢？显然，这是一个随机性在物理系统（也是工程系统）中的传播问题。

概率守恒原理告诉我们：随机源所决定的概率测度，在数学和物理变换中守恒[21]。换句话说：物理规律并不因系统中含有随机性而改变！本文作者，把这一规律称之为随机系统中的客观不变性原理❶。利用概率守恒原理的随机事件描述，可以导出与式（8）对应的广义概率密度演化方程：

$$\frac{\partial p_{U\Theta}(\boldsymbol{u},\boldsymbol{\theta},t)}{\partial t} + \sum_{j=1}^{N} \dot{u}_j(\boldsymbol{\theta},t) \frac{\partial p_{U\Theta}(\boldsymbol{u},\boldsymbol{\theta},t)}{\partial u_j} = 0 \quad (9)$$

式中，$p_{U\Theta}(\boldsymbol{u},\boldsymbol{\theta},t)$ 是随机变量 $\boldsymbol{U}$ 与 $\boldsymbol{\Theta}$ 的联合概率密度函数。

显然，上述方程可以容易地退化为关于指定点位移 u_j 的一维概率密度演化方程：

$$\frac{\partial p_{U_j\Theta}(u_j,\boldsymbol{\theta},t)}{\partial t} + \dot{u}_j(\boldsymbol{\theta},t) \frac{\partial p_{U_j\Theta}(u_j,\boldsymbol{\theta},t)}{\partial u_j} = 0 \quad (10)$$

---

❶ 这一提法，是本文作者于 2011 年 4 月 11 日在中国科学院力学研究所讲学时首次阐发的。

代入 t_0 时刻的初始条件，联立求解式（8）与式（10），不难获得 U_j 与 $\mathbf{Q}$ 的联合概率分布，而结构位移反应 U_j 的概率密度随时间的变化过程，可由关于 θ 的积分获得，即

$$p_{u_j}(u_j,t) = \int_\Omega p_{u_j\theta}(u_j,\theta,t)\mathrm{d}\theta \tag{11}$$

如何通过对基本随机变量所构成的概率空间进行合理剖分，以精确高效地求解概率密度演化方程、进行上述积分，我们进行了大量的研究[22-25]，一个新近的进展可见文献[26]。

事实上，广义概率密度演化方程的列式是具有非常宽广的适应性的。例如，对于如式（8）的弹性动力学基本方程，如果我们关心的重点是应力状态，则可以列出关于应力的概率密度演化方程：

$$\frac{\partial p_{\sigma\Theta}(\sigma,\boldsymbol{\theta},t)}{\partial t} + \sum_{i=1}^{M}\dot{\sigma}_i(\boldsymbol{\theta},t)\frac{\partial p_{\sigma\Theta}(\sigma,\boldsymbol{\theta},t)}{\partial \sigma_i} = 0 \tag{12}$$

式中，M 为所考察应力点数。

据此，可以方便地给出结构应力状态的概率分布密度及其演化过程。显然，给定强度失效准则，可以非常容易地求取结构可靠度。

考察随机性在物理系统中如何传播是十分有趣的事情。不妨以式（12）为例，当退化到一维概率密度方程时，有（不妨去除脚标）：

$$\frac{\partial p_{\sigma Q}(\sigma,\boldsymbol{\theta},t)}{\partial t} + \dot{\sigma}(\boldsymbol{\theta},t)\frac{\partial p_{\sigma Q}(\sigma,\boldsymbol{\theta},t)}{\partial \sigma} = 0 \tag{13}$$

将上式移项表述为：

$$\frac{\partial p_{\sigma\Theta}(\sigma,\boldsymbol{\theta},t)}{\partial t} = -\dot{\sigma}(\theta,t)\frac{\partial p_{\sigma\Theta}(\sigma,\boldsymbol{\theta},t)}{\partial \sigma} \tag{14}$$

对于一个弹性动力系统，我们关心的是：初始随机源 $\boldsymbol{\Theta}$ 所具有的随机性，是如何经过物理系统作用、转化（传播）为结构响应的随机性的。概率密度演化理论是从目标物理量（这里是 σ）与本源随机性（这里是 $\boldsymbol{\Theta}$）的联合概率分布（$p_{\sigma\Theta}$）来描述这一问题的。方程（14）十分清楚地告诉我们：联合概率分布 $p_{\sigma\Theta}$ 关于时间的变化率与关于应力的变化率在每一时刻均成比例，比例系数是 $\dot{\sigma}(t)$。而 $\dot{\sigma}(t)$ 恰恰代表了应力状态在时刻 t 的综合变化率。这就十分清楚地告诉我们，是物理状态的变化[$\dot{\sigma}(t)$]促成了概率密度 $p_{\sigma\Theta}$[因之 $p_\sigma(t)$]的演化。

物理规律如何作用于随机系统，并因之促进了随机系统的状态变化，概率密度演化理论给出了旗帜鲜明的回答！

将上述诠释联系于概率密度演化方程的求解，不难发现：由于形如式（10）和式（13）的偏微分方程，解为一特征线，从而导致了每一样本所携带的概率随时间变化不断地被重新分配到响应状态量的不同值域之中，进而导致各状态量概率分布密度的变化（概率重分配）。换句话说，物理方程的耦合作用导致各状态量相互作用，从而形成初始概率测度在不同值域内的重新分配（样本相互作用机制！）。这样，就清晰地解释了初始随机性是如何在物理系统变化过程中得到了传播、演化的机理。

值得附带指出的是，在广义概率密度演化方程中，关于时间变量应该作广义的理解。

事实上，这牵涉对动力系统的理解。本文作者认为：任何存在状态变化的系统都应该视为动力系统。在物理上，这类系统总可以表现为一类含有广义时间的偏微分方程。当这类系统中存在随机因素影响时，也总是可以给出相应的广义概率密度演化方程。例如，对于前述弹性动力系统，当不考虑外部动力作用、仅考虑静力加载机制时，存在弹性静力学基本方程：

$$\left.\begin{array}{l}\nabla \cdot \boldsymbol{\sigma} + \boldsymbol{b} = \boldsymbol{0} \\ \boldsymbol{\varepsilon} = \dfrac{1}{2}(\nabla \boldsymbol{u} + \nabla^{\mathrm{T}} \boldsymbol{u}) \\ \boldsymbol{\sigma} = \boldsymbol{E} : \boldsymbol{\varepsilon}\end{array}\right\} \quad (15)$$

引入比例加载机制、并取基本的加载物理量（如结构顶点位移）为广义时间 τ，则关于应力状态的广义概率密度演化方程可以表达为：

$$\frac{\partial p_{\sigma Q}(\boldsymbol{\sigma},\boldsymbol{\theta},\tau)}{\partial \tau} + \sum_{i=1}^{m} \dot{\sigma}_i(\boldsymbol{\theta},\tau) \frac{\partial p_{\sigma Q}(\boldsymbol{\sigma},\boldsymbol{\theta},\tau)}{\partial \dot{\sigma}_i} = 0 \quad (16)$$

显然，这种广义的理解与表达方式，加深了我们对客观世界的认识，也有利于研究与分析一般物理与工程系统。

5. 等价极值原理

应用概率密度演化理论，可以方便地对求出随机物理系统任意状态量的概率分布及其演化过程，这为从精确概率分析角度计算结构可靠度奠定了基础。但是，对于工程结构的整体可靠度分析，还要解决不同状态量失效相关的问题。

文献[27]较为完整地解决了这一问题。经过严格的数学推导，文献[27]断言，无论对于串联系统、并联系统、还是串-并联系统（这些系统，基本囊括了传统结构整体可靠度分析中所应用的主要系统模型），均可以找到多元功能函数的等价极值分布，并据之求取系统可靠度。事实上，求系统可靠度，等价于求这个等价极值函数大于规定值的概率。例如，对于串联系统，系统可靠度为：

$$R = \Pr\{\bigcap_{j=1}^{m} g_j(\theta) > 0\} \quad (17)$$

式中，$g_j(\theta)$ 为多元功能函数，m 为串联系统中单元的个数。

文献[27]从数学上严格证明，存在等价极值事件：

$$Z_{\text{ext}} = \min_{1 \leqslant j \leqslant m} g_j(\theta) \quad (18)$$

使得

$$\Pr\{\bigcap_{j=1}^{m} g_j(\theta) > 0\} = \Pr\{Z_{\text{ext}} > 0\} \quad (19)$$

因此，系统可靠度

$$R = \Pr\{Z_{\text{ext}} > 0\} = \int_{0}^{+\infty} p_{Z_{\text{ext}}}(z) \mathrm{d}z \quad (20)$$

其中 $p_{Z_{\text{ext}}}(z)$ 为 Z_{ext} 的概率密度函数。

类似地，可以证明其他等价系统。

根据本文第3节的分析，任何结构系统总是可以等效地表示为失效路径与失效机构的组合。因此，上述等价极值事件原理为求解一般结构系统整体可靠度提供了理论基础和新的思路：为了计算结构整体可靠度，只要找到等价极值事件及其概率分布就够了。

对于工程问题，数学上的推理总可以找到物理上的依据，对于上述问题也不例外。事实上，从物理的观点考察，一个串联的弹簧系统，如果考察它的可靠性，应该寻求具有最小强度的那个单元，只要这个单元不失效，整个系统就是安全的。等价极值事件，就是在物理上具有最小强度的那个单元，只是这个单元具有随机强度而已。

从等价极值事件的角度分析系统可靠性，各个多元功能函数之间的相关关系自然已经被包含了进去，因此自然不需要考虑复杂的概率相关问题。实际上，在分析过程中，可以引入如式（2）所示物理准则求取 Z_{ext}。如此，自然不再需要寻求失效机构与失效路径，也自然不再存在传统结构整体可靠度分析的组合爆炸问题。

6. 结构整体可靠度分析：物理综合法

结合概率密度演化理论和等价极值事件原理，即可以解决复杂结构体系整体可靠度分析的问题。在早期，我们是通过关于等价极值构造虚拟随机过程、并进而求解相应的广义概率密度演化方程的方式来解决这一问题的[27,28]。这使得有些同志认为：求解结构整体可靠度纯粹依赖于数学技巧。这与我们研究的初衷是不相符的。

经过进一步的研究、思考，本文作者发现：从纯粹物理原则出发，同样可以构造求解结构整体可靠性的基本方程，并且，这一途径更为高效、更易于理解、也更具有普遍性。

在这一方程中，依据物理准则构建广义概率密度演化方程的概率耗散条件。即：对于给定失效界限 $[Z]$，当满足 $Z_{ext} \leqslant [Z]$ 时，代表点 u_p 的概率密度函数：

$$p_{U_p\Theta}(u_p, \boldsymbol{\theta}, \tau) = 0 \tag{21}$$

这里，代表点 u_p 是指定的结构总体状态观察点，可根据具体分析对象灵活选取，例如：对于高层建筑，可取结构顶点位移。

引入代表点的意义在于：当依据不同的失效准则考虑结构安全与否时，只要有一点满足给定失效准则，则整体结构失效，此时，由 u_p 所携带概率将被耗散。当遍历所有样本之后，系统仍保有的概率即为结构可靠概率，这只要关于 u_p 做全域积分即可，即：

$$R = \int_{-\infty}^{\infty} p_{U_p}(u_p, \tau) du_p \tag{22}$$

引入代表点，还可以避免对所有结构反应量列写、求解概率密度演化方程，而只需要求解关于代表点的广义概率密度演化方程。

在这一基础上，可以将物理方程与概率密度演化方程联合求解。在具体的方程列式上，存在按吸收边界条件加保守系统概率密度方程[29,30]和直接按耗散系统概率密度方程列出基本求解方程组的区别，在本质上，二者是一致的。这里，以耗散系统概率密度演化方程形式[31]表述之。

不妨以具有损伤演化的固体力学方程为背景说明之，当系统输入或材料性质具有随机性时，一般的随机系统求解基本方程为：

$$\left.\begin{aligned}&\nabla \cdot \boldsymbol{\sigma}+\boldsymbol{b}=\rho\ddot{\boldsymbol{u}}+\eta\dot{\boldsymbol{u}}\\&\varepsilon=\frac{1}{2}(\nabla\boldsymbol{u}+\nabla^{\mathrm{T}}\boldsymbol{u})\\&\boldsymbol{\sigma}=(\boldsymbol{I}-\boldsymbol{D}):\boldsymbol{C}_0:(\boldsymbol{\varepsilon}-\boldsymbol{\varepsilon}^{\mathrm{p}})\\&\frac{\partial p_{U_\mathrm{p}\Theta}(u_\mathrm{p},\theta,t)}{\partial t}+\dot{u}_\mathrm{p}\frac{\partial p_{U_\mathrm{p}\Theta}(u_\mathrm{p},\theta,t)}{\partial \theta}=-\mathrm{H}[Z(t)]p_{U_\mathrm{p}\Theta}\end{aligned}\right\} \quad (23)$$

式中，$\boldsymbol{D}$ 为损伤张量矩阵，$\boldsymbol{C}_\mathrm{D}$ 为弹性张量矩阵，$\boldsymbol{\varepsilon}^\mathrm{p}$ 为塑性应变。$\mathrm{H}(\cdot)$ 为筛分算子，且

$$\mathrm{H}[Z(t)]=\begin{cases}0,&Z_\mathrm{ext}>[Z]\\1,&Z_\mathrm{ext}\leqslant[Z]\end{cases} \quad (24)$$

求解上述联立方程组，可以给出联合概率分布 $p_{U_\mathrm{p}\Theta}(u_\mathrm{p},\boldsymbol{\theta},t)$，而

$$p_{U_\mathrm{p}}(u_\mathrm{p},t)=\int_{\Omega_\mathrm{q}}p_{U_\mathrm{p}}(u_\mathrm{p},\boldsymbol{q},t)\mathrm{d}\theta \quad (25)$$

求得概率耗散系统代表点的概率密度演化过程之后，不难由式（22）求出结构的整体可靠度。

显然，上述求解结构整体可靠度的方法具有普遍适用性。即，这一理论可以应用于求解任何复杂结构的整体可靠度，而不仅仅是如传统结构整体可靠度分析那样，只适用于具有理想弹塑性性质的框架结构系统。

注意到概率密度演化方程本质上反映了随机性在物理系统的传播规律。因此，应用物理失效准则，结合物理方程与概率密度演化方程求解结构整体可靠度的方法，可以称之为"物理综合法"。这种综合，涵盖了物理力学机制、物理失效准则、随机性在物理系统中的传播等多方面的含义。

事实上，注意到物理失效准则的多样性，对物理综合法中的"综合"二字还可以具有拓展性的理解。以高层建筑结构抗震可靠性为例，按照结构"小震不坏，中震可修，大震不倒"的设计准则，整体结构抗震可靠度分析中的物理准则即可包括：小震不坏的材料强度准则或变形准则，中震可修的截面屈服准则或构件承载强度准则，大震不倒的结构倒塌准则[32,33]。在不同层次上构造的失效物理准则，构成了结构整体抗震可靠度分析的三阶递阶结构，如图 4 所示。

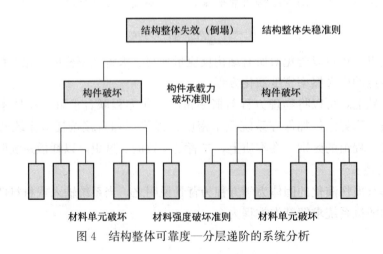

图 4 结构整体可靠度—分层递阶的系统分析

7. 典型工程应用实例

应用上述方法，近年来，本文作者带领学生具体研究了一些典型工程的整体可靠度。这里，仅撷取数例以飨读者。

7.1 上海浦东广电中心抗震可靠性分析

上海浦东广电中心是一栋重要的生命线工程建筑。该建筑共18层，结构总高86.4m，建筑面积2.1万m^2，采用钢筋混凝土框架-剪力墙结构体系。结构底层平面尺寸为38m×36m，标准层尺寸为32m×36m。在结构竖向，按照刚度变化分为3个区，1～7层为一区，8～13层为二区，14～18层为三区。3个区域的剪力墙厚度分别为0.4、0.4、0.3m，所采用的混凝土强度等级分别为C50、C40、C40。

在结构分析中，采用本文作者所建立的随机损伤本构模型，并采用纤维梁单元模拟梁和柱、采用分层壳单元模拟剪力墙和楼板[34]。结构分析模型中共包含53372个单元，252210个自由度。

以本梯队近年来研究发展的随机地震动物理模型为结构随机地震动输入依据[35]，研究结构的抗地震倒塌的可靠性。分析中，采用本梯队近年来发展的能量动力稳定性准则判别结构倒塌[31,32]。

图5给出了在地震过程中结构典型层间位移的时点概率分布，从中可以清晰地看到结构响应概率分布随时间的改变。图6给出了在相当于11度烈度的随机地震动作用下结构整体倒塌概率随时间的变化。由此可见，在非常罕遇的地震动作用下，该结构的倒塌概率高达47.2%。

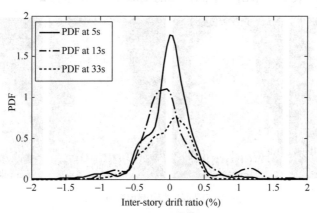

图5 典型时刻结构层间位移角响应的概率密度函数

7.2 火力发电厂大型冷却塔抗震可靠度分析

冷却塔是火力发电厂的重要工程设施。由于电厂装机容量的增大，我国多个地方建设了高度超过200m的大型冷却塔。这些冷却塔大大超过现行设计中总高165m的限制。其抗震可靠性令人担忧。

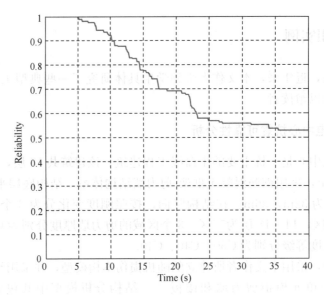

图6 浦东广电中心结构整体抗震可靠度

结合具体工程建设,我们进行了249m高大型冷却塔的抗震整体可靠性研究。该结构采用钢筋混凝土壳体-框桁架结构体系。其中壳体部分高220m,框桁架部分高29m;壳体顶部直径为118m,底部直径为186m,混凝土等级为C45。对壳体采用分层壳单元建模、底部框-桁架柱采用纤维单元建模,整个有限元分析模型共包含24650个单元、190468个自由度。

采用随机地震动输入,均值地震动峰值为0.4g。图7为不同随机地震动样本作用下结构倒塌变形实例,表1则给出了不同样本的倒塌模式与时间。

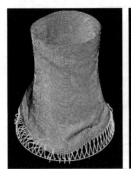

图7 不同的结构倒塌破坏模式

倒塌模式及倒塌时间　　　　　　　　　　表1

样本编号	倒塌模式	倒塌时间（s）
10号	壳体屈曲	24
14号	壳体屈曲	19
16号	交叉斜撑破坏	9
23号	壳体屈曲为主	20

续表

样本编号	倒塌模式	倒塌时间（s）
35号	壳体屈曲	17
38号	壳体屈曲	16
39号	交叉斜撑破坏为主	20
55号	交叉斜撑破坏	18
86号	交叉斜撑破坏为主	14

可见，由于地震动的随机性，结构发生地震倒塌的模式和时间具有显著区别。经概率密度演化分析，该结构发生地震倒塌的概率为9.86%。在9度烈度左右的地震动作用下，这类大型冷却塔有接近百分之十的地震倒塌概率，这不能不引起设计者的高度警惕！

7.3 福建溪尾大桥疲劳可靠度分析

结构整体可靠性分析的物理综合法，不仅可以应用于各类结构的抗灾可靠度分析，而且可以应用于结构全寿命动态可靠度分析。为展示这一可能，以福建溪尾大桥为背景，进行了结构整体疲劳可靠度分析。

溪尾大桥全长308.5m，为预应力混凝土连续梁桥。桥体分为三段：3m×30m+4m×30m+3m×30m。桥面宽19m。混凝土等级为C50，设计寿命为100年。在分析中，采用复合泊松过程模拟随机车辆荷载[36]，以我们近年所发展的混凝土疲劳随机损伤本构模型[37]为基础进行结构疲劳分析。以结构跨中位移为代表点进行概率密度演化分析，并以疲劳损伤限值 $d=0.7$ 规定结构疲劳破坏准则。

图8给出了该桥的整体疲劳寿命随年限的变化，表2则用更为具体的数字给出了更直接的说明。

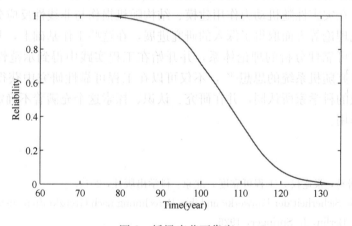

图8 桥梁疲劳可靠度

溪尾大桥疲劳可靠度计算结果　　　　表2

桥梁服役期（年）	可靠度
70	1
75	0.9998

续表

桥梁服役期（年）	可靠度
80	0.9915
85	0.9675
90	0.9327
95	0.8614
100	0.7113
105	0.5433
110	0.3607
115	0.1910
120	0.0795
125	0.0296
130	0.0105

可见：虽然在设计寿命期的前 70 年内，结构整体可靠度可以得到保证，但超过 80 年不久，结构整体可靠度即迅速下降，在达设计预期寿命 100 年时，结构可靠度仅有 71%，换句话说，此时该桥梁疲劳失效概率几近 30%！

8. 结语

保证工程结构可靠性，是结构工程师进行结构设计时的基本目标之一。在长达 50 年的时间里，关于工程结构整体可靠性的研究一直没有获得实质的、可以应用于工程实践的重要进展。本文作者从研究结构的随机损伤及其演化进程开始，历经 20 年努力，带领学生和研究梯队，在灾害性随机动力作用建模、结构随机损伤与非线性反应分析、随机系统的概率密度演化理论等方面取得了深入的研究进展，在这些工作基础上，基本完整地建立了工程结构整体可靠性分析的理论体系，并开始在工程实践中得到示范性应用。作者相信，基于物理研究随机系统的思想[38]，不仅可以在工程可靠性研究中获得成功，而且一定会被相关领域的科学家所认同，并在研究、认识、探索这个充满着不确定性的世界的过程中发挥关键作用。

参考文献

[1] 赵国藩，曹居易，张宽权．工程可靠度．北京：科学出版社，2011．
[2] Mayer M. Die Sicherheit der Bauwerke und ihre Berechnung nach Grenzkräften anstatt nach zulässigen Spannungen. Berlin, J. Springer, 1926.
[3] Freudenthal AM. The safety of structures. ASCE Transactions，112，125-180，1947.
[4] 李继华．可靠性数学．北京：中国建筑工业出版社，1988．
[5] Cornell C A. Structural specification based on second moment reliability, Sym. Int. Assoc. of Bridge and Struc. Engr., London, 1969.
[6] Ditlevsen O. Structural Reliability and the Invariance Problem. Report No. 22, University of Waterloo, Solid Mechanics Division, Waterloo, Canada. 1973.

[7]　Hasofer A M, Lind NC. Exact and invariant second moment code format. Journal of the Engineering Mechanics Division, ASCE, 100: 111-121, 1974.

[8]　Rackwitz R, Fiessler B. Structural reliability under combined random load sequences. Computers & Structures, 9: 489-494, 1978.

[9]　Hohenbichler M, Rackwitz R. Non-normal dependent vectors in structural reliability. Journal of the Engineering Mechanics Division, ASCE, 107: 1127-1238, 1981.

[10]　Lind N C, Consistent Practical Safety Factors, Journal of the Structural Division, ASCE, No. ST6: 1951-1669, 1971.

[11]　中华人民共和国住房和城乡建设部. 工程结构可靠性设计统一标准 GB 50153—2008. 北京: 中国建筑工业出版社, 2009.

[12]　Freudenthal AM, Garrelts JM, Shinozuka M. The analysis of structural safety. Journal of the Structural Division, ASCE, 92(ST1): 267-325, 1966.

[13]　Ang A H-S, Ma H F. On the reliability of structural systems. Proc. 3rd Int. Conf. on Struct. Safety and Reliability, Elsevier, Amsterdam, Netherlands, 295-314. 1981.

[14]　Ditlevsen O. Narrow reliability bounds for structural systems. Journal of Structural Mechanis, 7: 453-472, 1979.

[15]　Thoft-Christensen P, Sorensen, J. D. reliability of structural systems with correlated element, Applied Mathematical Modeling, Vol. 6: 171-178, 1982.

[16]　Murotsu, Y., Okada, H., Taguchi, K., Grimmelt, M., and Yonezawa, M. Automatic generation of stochastically dominant failure modes of frame structures. Struct. Saf. 509 2(1), 17-25, 1984.

[17]　Thoft-Christensen P, Murotsu Y. Application of Structural Systems Reliability Theory. Springer, 1986.

[18]　董聪. 现代结构系统可靠性理论及其应用. 北京: 科学出版社, 2001.

[19]　Lee Y, Song J. Risk analysis of fatigue-induced sequential failures by branch-and-bound method employing system reliability bounds. Journal of Engineering Mechanics, 137(12), 807-821, 2011.

[20]　胡云昌, 郭振邦. 结构系统可靠性分析原理及应用. 天津: 天津大学出版社, 1992.

[21]　Li J, Chen JB. The principle of preservation of probability and the generalized density evolution equation. Structural Safety, 30(1): 65-77, 2008.

[22]　Li J, Chen JB. The number theoretical method in response analysis of nonlinear stochastic structures. Computational Mechanics, 39(6): 693-708, 2007.

[23]　Chen JB, Ghanem R and Li J, Partition of the Probability Space in Probability Density Evolution Analysis of Non-linear Stochastic Structures, Probabilistic Engineering Mechanics, 24(1): 27-42, 2009.

[24]　Xu J, Chen JB and Li J., Probability density evolution analysis of engineering structures via cubature points". Computational Mechanics, 50(1): 135-156, 2012.

[25]　Chen JB, Yang J, Li, J. A GF-discrepancy for point selection in stochastic seismic response analysis of structures with uncertain parameters. Structural Safety, 59: 20-31, 2016.

[26]　Chen JB, Song PY. A generalized L_2-discrepancy for cubature and uncertainty quantification of nonlinear structures. Science China Technological Sciences, 59(6): 941-952, 2016.

[27]　Li J, Chen JB, Fan W. L. The Equivalent Extreme-Value Event and Evaluation of the Structural System Reliability. Structural Safety, 29(2): 112-131, 2007.

[28]　Chen JB, Li J. The extreme value distribution and dynamic reliability analysis of nonlinear structures with uncertain parameters. Structural Safety, 29(2): 77-93, 2007.

[29]　Chen JB, Li J. Dynamic response and reliability analysis of nonlinear stochastic structures. Probabi-

listic Engineering Mechanics, 20(1): 33-44, 2005.

[30] Li Jie, Chen Jian-Bing, Stochastic Dynamics of Structures. Wiley & Sons, 2009.

[31] Li J, Xu J. A new probability dissipative system and its control Equation, Proceedings of IFIP2014, 2014.

[32] Xu J, Li J. An energetic criterion for dynamic instability of structures under arbitrary excitations. International Journal of Structural Stability and Dynamics, 15(2), 1-32, 2015.

[33] Zhou H, Li J, Zhou H, et al. Effective energy criterion for collapse of deteriorating structural systems[J]. Journal of Engineering Mechanics, 143(12), 04017135, 2017.

[34] 李杰，吴建营，陈建兵. 混凝土随机损伤力学. 北京：科学出版社，2014.

[35] Wang D, Li J, Physical random function model of ground motions for engineering purpose, Science China: Technological Sciences, 54(1): 175-182, 2011.

[36] 高若凡，李杰. 基于随机谐和函数的复合泊松过程模拟. 同济大学学报（自然科学版），45(12): 1731-1738, 2017.

[37] Ding Z, Li J. A physically motivated model for fatigue damage of concrete. International Journal of Damage Mechanics, DOI: 10.1177/1056789517726359, 2017.

[38] 李杰. 随机动力系统的物理逼近. 中国科技论文在线，1(9): 95-104, 2006.

16

2 Making Recycled Aggregate Suitable For Structural Concrete

Surendra P. Shah[1], Yuxi Zhao[2], Weilai Zeng[2], Hongru Zhang[2,3]

1. Center for Advanced Cement-Based Materials, Northwestern University, Evanston, IL 60208, USA
2. Institute of Structural Engineering, Zhejiang University, Hangzhou 310058, Zhejiang, PR China
3. College of Civil Engineering, Fozhou University, Fuzhou 350116, Fujian, PR China

Abstract: Previous researches have pointed out that recycled aggregate concrete (RAC) has inferior properties because of the old mortar attached to the recycled aggregate (RA). In order to promote the structural application of RA, the RA's modifications have been investigated, including physical methods, chemical methods and microbial methods. In this paper, two novel modification methods with nano-material and microbial precipitation proposed by authors, to enhance the properties of RA, are introduced. The great improvements in micro and macro properties of RAC were found after the surface treatment with nano-material. Meanwhile the properties enhancements of RA were also revealed with an optimum microbial precipitation modification. Moreover together with natural aggregate concrete, non-modified and nano-material slurry modified RAC were applied in three beams in a real project. In situ strain monitoring in the target beams showed the improved resistance to cracking of RAC after modification. This paper aims to make a contribution to making RAC suitable for structural concrete by modifying RAs.

Keywords: Recycled aggregate concrete; Nano-material modification; Microbial precipitation modification; Field application

Contents

1. Introduction
2. Different approaches of modifying RA
3. Modification with nano-material slurry
   3.1 Strengthening slurries
   3.2 Micro properties test methods
   3.3 Results and discussion
4. Modifications with microbial precipitation
   4.1 Material and bacteria
   4.2 Experimental program
   4.3 Results and discussion
5. Field application of modified RA
   5.1 Materials

    5.2   Information for the applied project
    5.3   Concrete strain monitoring in situ
    5.4   Results and discussion
6. Summary and future work
    6.1   Summary
    6.2   Future work

1. Introduction

Currently in order to satisfy the sustainable aims of development, concrete recycling has raised worldwide interest considering its potential benefits in controlling the over-discharge of construction and demolition (C&D) wastes. According to statistics, in China the discharge of C&D waste is more than 300 million tonnes in 2016, one third of which is discarded concrete. Moreover some natural disasters, like earthquake and tsunami, also results in the quantities of construction waste. Meanwhile with the increasing need of development, natural aggregates become more and more scarce.

To reduce the consumption of natural aggregate and relieve the pressure of land shortage for landfill used for disposing of C&D wastes, concrete waste is recycled as recycled aggregate (RA). At first, a state-of-the-art report on recycled concrete as an aggregate for concrete was given by Nixon, covering the period 1945-1977 [1]. Then RA and recycled aggregate concrete (RAC) were widely researched in America, Europe, Japanese and other developed counties. From 1996 to now, lots of studies on RA and RAC have also been investigated in China.

Aggregate has a huge influence on the properties of concrete. There have been many studies examining the properties of RA and RAC.

Different from NA, old hardened cement mortar attached to RA combined with the crushing process produces higher water absorption and lower crushing value [2]. Besides the larger apparent density and mud content are also found in a large quantity of researches.

Many researchers focused on the influence of old hardened cement mortar attached on microstructure in RAC. C. S. Poon et al. [3] studied the microstructure of ITZs in concrete with NA, recycled normal-strength concrete aggregate, and recycled high performance concrete. Their scanning electron microscopy observation revealed that ITZ in concrete with normal-strength concrete aggregate consisted mainly loose and porous hydrates compared with the other groups. Ryu [4] found that the compressive strength of RC depends on the relative quality of old and new ITZs in concrete. When the water-cement ratio is low, the strength characteristic of the concrete is governed by the effect of old ITZs. When the water-cement ratio is high, the predominant governing factors are the new ITZs. Xiao et al. [5] found that the ratio of the old ITZ's mechanical properties (elastic modulus and strength) to those of the old mortar matrix affects the stress-strain curves and failure processes of modelled recycled aggregate concrete (MRAC). Increasing the ratio resulted in higher strength, but lower ductility. The ratio of new ITZ's mechanical properties of those of the new mortar matix had a negligible effect on the compressive strength of MRAC, while the tensile strength of MRAC increased as this ratio increased.

Considering the inferior micro properties of RAC, the mechanical of RAC was studied. Etxeberia et al. [6] found that when the replacement of RA was 100%, the compressive strength of RAC was 20%~25% lower than that of NAC. Breccolotti et al. [7] believed that the tensile strength of RAC decreased with the increasing RA replacement ratio. Silva et al. [8] held the view that the tensile strength of RAC was 60% lower than that of NAC according to the statistics. They also found that for RAC with 100% RA replacement ratio, the elastic modulus was just 52%~97% of that of NAC [9].

Moreover the durability of RA also caught researchers' attentions. Wang et al. [10] proved that the RAC's ability of resistance to Cl- penetration was 14% lower than NAC. Villagrán-Zaccardi et al. [11, 12] pointed that it was easier for RAC to absorb and permeate chloride ion. Hansen et al. [13] discovered that compared with NAC, the value of air shrinkage was 60% higher for RAC. Sagoe-Crentsil et al. [14] also observed that the shrinkage strain of RAC was 25% higher than that of NAC at the age of 1 year. Based on these researches, Srubar et al. [15] thought the inferior resistance to chloride ion of RAC made it difficult to protect the internal steel bar from corrosion.

The inferiority of RAC in material properties has adversely affected some structural performance of reinforced RAC members or structures. Sato et al. [16] have pointed out that though the ultimate flexural capacity of reinforced RAC beams is comparable to, that of the NAC beams, however, the deflection and the overall crack width of RAC beams under the same moment are significantly larger. Etxeberria et al. [17, 18] have reported that the shear capacity of RAC beams without stirrups is smaller than that of NAC beams. The study carried out by Andrzej et al. [19] has confirmed that though loading capacity of the reinforced RAC beams or columns is comparable to that of the NAC members, however, the effects of RAC's inferiority in material properties on the deformability of reinforced members should be carefully considered, at beam deflection assessment, as well as at the column shortening analysis.

All these research results mentioned above show that RAC possesses substandard mechanical properties, poor durability properties, and inferior structural performance because of the old cement mortar, which is attached to the RA, resulting in the higher water absorption and a greater number of interfacial transition zones within the RAC. Therefore, recycled C & D waste is primarily used as a filler for road bases and backfills [20, 21].

2. Different approaches of modifying RA

Considering the inferiority of RAC in materials properties, and thereby the affected structural performance, especially in the deform ability of reinforced RAC members, as referred above, it is necessary to raise effective raise effective techniques for modifying RAC. Some researchers have attempted to remove the old cement mortar from RA via methods. Other modification methods have been used to strengthen the old cement mortar of the RA.

Shima Hirokazu et al. [22, 23] heated up the demolished concrete in a heater to make cement paste brittle with its dehydration. The heated concrete was then rubbed in two mills to recover the recycled aggregate, while the paste was removed from the surface of aggregate and collected as cement fine powder. While in this method, much energy was consumed to heat and rub concrete.

Kathy Bru et al. [24] used microwave to heat the concrete waste before impact crushing and found that microwave heating always induced an embrittlement of concrete samples which resulted in lower fracture energy, higher fragmentation of samples and higher liberation of aggregates. Akbarnezhad et al. [25] investigated the capability of the microwave-assisted RA beneficiation technique and compared its efficiency with other benefi- ciation methods proposed in available literature. Katz A et al. [26] treated the RA by ultrasonic cleaning with the objective of overcoming the drawbacks and found a 7% improvement in compressive strength after treatment.

Akbarnezhad A et al. [27] proposed an acid treatment testing method to determine the mortar content of RAC and found that most cementitious mortar was able to be removed with a 24-hour treatment. Vivian W. Y. Tam et al. [28] studied three pre-soaking treatment approaches: HCl, H_2SO_4, and H_3PO_4 in reducing the mortar attached to RA. The results revealed that the behavior of RA had improved with reduction in water absorption, without simultaneous exceeding the limits of chloride and sulphate compositions after the treatment. Ha-Seog Kim et al. [29] used sulfuric and hydrochloric acids as the acidic substances. They found that the acid-treated recycled aggregates were superior to the original aggregates and the RAs that were manufactured using natural water in terms of density, absorption ratio and solid volume percentage.

Poon et al. [30] improved the properties of RA by their impregnation with polyvinyl alcohol and found that the water absorption of polyvinyl alcohol impregnated RA was lower when compared to the untreated RA. Their results also showed a decreasing shrinkage and increasing resistance to chloride-ion penetration of the modified RAC.

Bibhuti Bhusan Mukharjee et al. [31] addressed the effect of incorporation of colloidal Nano-Silica on the behavior of concrete containing 100% RA and revealed that compressive strength, tensile strength and Non-Destructive parameters were enhanced due to treatment of Nano-Silica. Wengui Li et al. [32] investigated the effect of nano-particles including nano-silica and nano-limestone on the crack propagation and microstructure properties of RAC. Their results indicated that compared to nano-limestone, nano-silica was more effective in improving the microstructure properties and enhancing the mechanical strength of RAC.

Kou et al. [33] used a CO_2 curing step to improve the properties of concrete prepared with recycled aggregates. They found that not only was there an improvement in the mechanical properties and resistance to chloride ion penetration for the concrete prepared with modified RA. J Zhang et al. [34] attempted to improve the quality of RACs through carbonation of the attached cement paste. The results showed that carbonation increased the density, and decreased the water absorption and crushing values of the RAs. Therefore this treatment increased the flowability and compressive strength and decreased drying shrinkage of the recycled aggregate mortars.

Grabiec et al. [35] used calcium carbonate biodeposition to modify the surface of RA, and found that there was a reduction in the water absorption of the RA. Qiu et al. [36] investigated the factors that influenced MCP on RA, and found that enhanced MCP on RA could be achieved through proper control of culturing and precipitating conditions. Pan et al. [37] modified RA, intended for asphalt mixtures, using MCP, and revealed that the treated RA exhibited superior adhesion behavior to that of the untreated RA. Although it has been shown that MCP can improve the properties of RA, different precipitation methods were used in the aforementioned studies. The use of MCP to improve the properties of RAC should be further investigated to explore whether its application is possible and feasible.

3. Modification with nano-material slurry

An effective and economical modification technique for RAC by applying surface treatment on RA with the use of nanomaterials was proposed by authors [38]. Two types of slurries containing different nanomaterials were prepared and used to treat the surface of RA. The nanoindentation technique was applied to evaluate the mechanical properties of different types of ITZs and mortars in RAC quantitatively on the microscale. The physical properties of RA were tested before and after the surface treatment; the macro properties of the corresponding concrete, using the treated or untreated RA, were also tested and compared.

3.1 Strengthening slurries

Two strengthening slurries with the incorporation of different nanomaterialswere prepared. The first strengthening slurry consisting of nSi and nCa, was labeled as the nSi + nCa slurry. The other strengthening slurry, labeled as the Cement + nSi slurry, reduced the dosage of nano-SiO_2 but added more cement as a substitute, considering the high price of nanomaterials. The proportions of the components contained in the two types of nanoslurries are provided in Table 1.

Proportions of two nanoslurries prepared in this study   Table 1

Parameters	nSi + nCa slurry	Cement + nSi slurry
Water	90	50
Cement	0	95
Nano-SiO$_2$ dispersant	5	5
Nano-CaCO$_3$ slurry	5	0
Superplastizer	0	1.5

Note: All values are in kilograms.

Surface treatment of RA was achieved by soaking the RAs in the corresponding nanoslurries. First, the two types of strengthening slurries with good dispersion were prepared by mixing the corresponding strengthening materials with water for 120 s. After that, the untreated RA was added into the corresponding slurry and soaked for 45 min. The RA was then removed from the bath, and a screen was used to remove the redundant nanoslurry adhering to the RA. The final step of the process was to dry the RA at a temperature of 20 ± 2C and a relative humidity of 64 ± 5% for at least 3days.

The RA samples modified by the nSi + nCa slurry and the Cement + nSi slurry were labeled as R_1 and R_2, respectively. Therefore, four coarse aggregate groups were prepared to include the untreated coarse aggregates, N_0, R_0, R_1 and R_2. As shown in Fig. 1, the untreated and treated RA samples were R_0, R_1, and R_2, from left to right. Fig. 1 demonstrated that, after surface treatment, a coating formed on the surfaces of R_1 and R_2, primarily attributing to the precipitation of the particles contained in the two nanoslurries.

Fig. 1   RA before and after surface treatment with different nanoslurries

3.2 Micro properties test methods

A Hysitron 950 Triboindenter, fitted with a Berkovitch tip (tip radius of 0.6 μm and angle of 142.3°), was used to administer nanoindentation tests on different phases of the RAC at the ages of 3 and 28 days. At the ages of 3 and 28 days, small concrete slices with dimensions of 10×10×5mm were cut from the 100-mm concrete cubes, prepared as the nanoindentation samples. Then these concrete slices were carefully groud and polished before indentation to obtain adequately even and smooth testing surfaces, considering the significant influence of samples' surface roughness on the reliability of nanoindentation tests [39].

The high-resolution probe microscopy installed in the Triboindenter allowed observation and in situ image capturing. Nanoindentation was conducted on or around all the potential weak points in RAC, i.e., the old ITZs, the new ITZs, and the old mortars. The indent areas across old ITZs were 100×100 μm squares. The indent areas across the new ITZs between the old and new cement mortar were enlarged to 150×100 μm to ensure that the new ITZs were completely covered. For both old and new ITZs at each of the testing ages, four indent areas were tested. For old mortars, five indent areas with the dimension of 30×30μm were selected for each group of RACs, labeled as OM-1 to OM-5 according to their distances away from the old mortars' surface, as shown in Fig. 2. Indentations of new mortars were obtained in the same way. Different indent matrices were accordingly designed on indent areas of different dimensions. A 21×11 matrix and a 31×11 matrix were designed for the 100×100μm and 150×100μm indent areas, respectively. A 4×4 matrix was applied for the 30×30μm areas on the old mortar and the new mortar in RAC, as shown in Fig. 3.

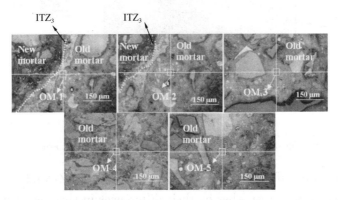

Fig. 2 Indent areas selected on old mortars with different distances away from new ITZs.

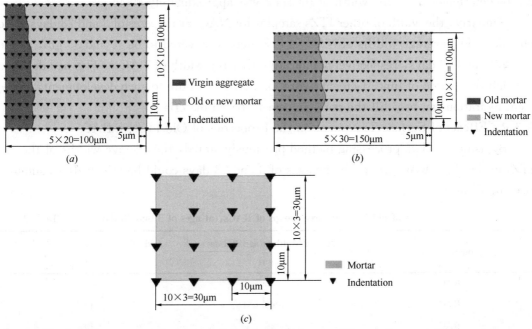

Fig. 3 Indent areas and corresponding indent matrices on different phases: indent matrix
(a) across ITZs in NAC$_0$ and across old ITZs in three groups of RAC;
(b) across new ITZs in RAC; (c) on old and new mortar

3.3 Results and discussion

3.3.1 Properties of ITZ in NAC_0

Fig. 4 shows the modulus distribution within this indent area after data processing with Origin.

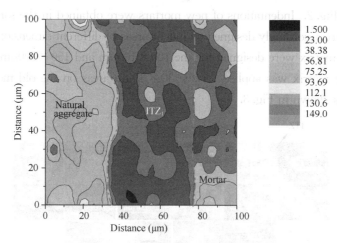

Fig. 4 Modulus distribution map within indent area

In Fig. 4, the three different phases, i.e., aggregate, ITZ, and cement mortar, was easily distinguished from one another by their differences in shading, representing the different modulus ranges. In this way, the boundaries of ITZ contained in this indent area can be determined, and the width of this ITZ was approximately 40-45 μm.

Similarly, the width of other ITZs samples for NAC_0 at the ages of 3 and 28 days were obtained. The average widths of NAC_0's ITZ were discovered to be 40 and 35 μm, at the ages of 3 and 28 days, respectively, indicating that the width of NAC_0's ITZ had declined through this period of time. Cement hydration may contribute to such development, as cement hydration can help fill the pores in cement-based materials.

3.3.2 Effects of Surface Treatment on the Properties of Old ITZs in RACs

By using the data processing method previously stated, the average widths of the old ITZs in the three RAC groups at the ages of 3 and 28 days could be achieved, as summarized in Table 2.

Table 2 Widths of old ITZs in three groups of RACs, at ages of 3 and 28 days

Concrete types	Old ITZ width at age of 3 days (μm)	Old ITZ width at age of 28 days (μm)
RAC_0	45	40
RAC_1	40	40
RAC_2	40	40

According to Table 2, no large differences among the old ITZs could be determined on the basis of width among the three groups of RACs. Meanwhile, the width of the old ITZs contained in all three RAC groups did not develop over time. Such results will be discussed in depth in the following parts of this section.

The probability distributions of the modulus within the ITZs were determined according to statistical analyses. The probability distribution of the modulus within the old ITZs in the three RAC groups at the ages of 3 and 28 days were plotted and compared in Figs. 5. The results obtained from NAC_0's ITZ are also shown in Fig. 8 and served as the control.

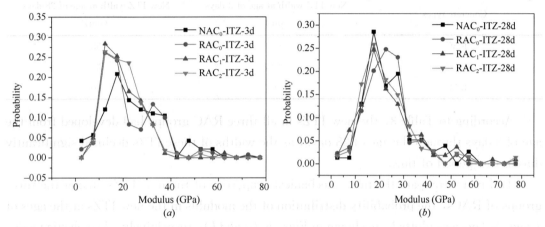

Fig. 5 Probability distribution of modulus of ITZ in NAC and of old ITZ in the three groups of RAC at the age of (a) 3 days; (b) 28 days

Fig. 5 demonstrates that from the age of 3 days through the age of 28 days, the elastic modulus of old ITZs in the three groups of RAC experienced little development because the probability distribution curves of the modulus shown in Fig. 5 (a) were comparable to those shown in Fig. 5 (b) for RAC_0, RAC_1, and RAC_2, respectively. Such results could be explained because cement hydration in these old ITZs was no longer significant because the age of RA was usually several decades and the contained unhydrated cement particles in the adhering cement mortar and the old ITZs had been almost consumed. By contrast, the micromechanical properties of the NAC_0's ITZs had developed as time passed, as seen in not only the decrease in width (40 μm at the age of 3 days; 35 μm at the age of 28 days) but also the gain of elastic modulus [see differences between Figs. 5 (a and b)]. A large amount of unhydrated cement particles existed in NAC_0's ITZs at its relatively early age; the continuous hydration of these cement particles contributed largely to these ITZs' development in mechanical properties.

According to both Table 2 and Fig. 5, at both of the two testing ages, the old ITZs of the three groups of RACs were comparable to one another in either width or probability distribution of the elastic modulus. Such results demonstrate that neither of the two nanoslurries used in this study significantly enhanced the mechanical properties of the old ITZs. Through the surface treatment procedure, neither RA_1 nor RA_2 were soaked in the nanoslurries for long (approximately 45 min). As a result, the nanoparticles and the cement particles were unable to deeply penetrate into the old cement mortar or arrive at the old ITZs inside the RA, so that the old ITZs could not be significantly enhanced.

3.3.3 Effects of Surface Treatment on the Properties of New ITZs in RACs

Similarly, the average widths of the new ITZs contained in the three RAC groups at the ages of 3 and 28 days were achieved, as summarized in Table 3.

Widths of new ITZs in three groups of RACs, at ages of 3 and 28 days   Table 3

Concrete types	New ITZ width at age of 3 days (μm)	New ITZ width at age of 28 days (μm)
RAC_0	65	55
RAC_1	60	50
RAC_2	50	40

According to Table 3, the new ITZs in all three RAC groups had developed from the age of 3 days through the age of 28 days, as the widths of these ITZs declined significantly during this period of time.

To further compare the micromechanical properties of the new ITZs among the three groups of RACs, the probability distribution of the modulus in the new ITZs at the ages of 3 and 28 days was plotted, as shown in Figs. 6 (a and b), respectively. The elastic modulus of the new ITZs in all three groups of RACs had experienced significant development. At the age of 3 days, the peaks of the three probability distribution curves, as shown in Fig. 6 (a), fall in the ranges of 5-10 GPa, 10-15 GPa, and 10-15 GPa for the new ITZs in RAC_0, RAC_1, and RAC_2, respectively. At the age of 28 days, as shown in Fig. 6 (b), the three peaks shift to the right to the higher modulus ranges (10-15 GPa, 15-20 GPa, and 15-20 GPa, respectively). Similarly, continuous cement hydration in these new ITZs may also be the predominant reason for the property development of new ITZs over time, reflected by both the width decrease and the modulus gain.

According to both Table 3 and Figs. 6 (a and b), the properties of the new ITZs in

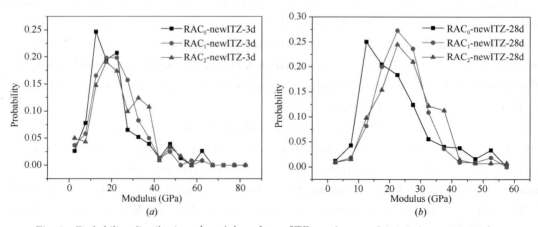

Fig. 6 Probability distribution of modulus of new ITZs at the age of (a) 3 days; (b) 28 days

RAC$_1$ and RAC$_2$ were superior to those of the new ITZs in RAC$_0$. Such results are reflected by the smaller width and the higher elastic modulus level of the new ITZs in RAC$_1$ and RAC$_2$ compared with RAC$_0$. Therefore, the two nanoslurries used in this study have beneficial effects on RAC's new ITZs. Water absorption of RA$_1$ and RA$_2$ were proved to be reduced compared with the untreated RA$_0$. Therefore, in RAC$_1$ and RAC$_2$, the water stored in RA$_1$ and RA$_2$ was reduced and thereby the amount of water penetrating from the old mortar to the surrounding new ITZs was reduced as well. In this way, the water-to-binder ratio in new ITZs was cut down, which may have contributed to the enhancement of the new ITZ. As can be observed in Fig. 1, a thin coating formed around RA$_1$ and RA$_2$ attributable to the precipitation of strengthening particles throughout the 45-min soaking in the nanoslurries.

When the surface-treated RA$_1$ and RA$_2$ were cast into concrete specimens, nanomaterials and cement particles contained in the coating may probably dissolve and penetrate back into the new surrounding ITZs, thereby inducing filling effects and accelerating effects. Hydration of cement particles contained in the Cement + nSi slurry could densify the new ITZs as well new ITZ, Fig. 7 was plotted to illustrate the elastic modulus within the new ITZs' boundaries in relation to the distance away from the surface of old mortars, or in other words, from the new ITZs' boundaries on the old mortar side. Each curve represents the average level of the four randomly selected indent areas. According to Fig. 7, at both the ages of 3 and 28 days, elastic modulus within RAC$_0$'s new ITZ did not illustrate an obvious trend to show whether it increased or decreased away from the old mortar's surface. By contrast, for new ITZs of RAC$_1$ and RAC$_2$, generally, elastic modulus decreased as the distance away from old mortar's surface increased, in spite of fluctuations. Such results further verify that the enhancement of new ITZs of RAC$_1$ and RAC$_2$ may be probably induced by the dissolving and penetration of nanomaterials or cement particles in the thin coating around RA$_1$ and RA$_2$, respectively. However because these parti-

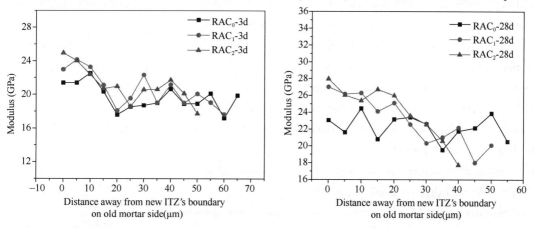

Fig. 7 Modulus development within new ITZs of three group of RACs, in relation to distance away from old mortar's surface, at age of (a) 3 days; (b) 28 days.

cles were not able to penetrate deeply into the new ITZs from the surface of the old mortars (i. e. , in the new ITZs of RAC_1 and RAC_2), the modulus far away from the old mortars' surfaces was no longer significantly enhanced.

Both Figs. 6 and 7 indicate that the enhancing effects of the two nanoslurries on the properties of the new ITZs in the RACs were comparable to one another; the gaps between the RAC_1 and RAC_2 curves are small. Such results prove that by using the Cement + nSi slurry prepared in this study, even though the content of nanosilica is greatly reduced, the beneficial effects on properties of RACs can be ensured compared with the nSi + nCa slurry. Therefore, the designed Cement + nSi slurry may have better prospects in RAC's modification, chiefly considering the smaller dosage of expensive nanomaterials used while guaranteeing the enhancing effects on RAC, primarily on the new ITZ properties. Such results are further discussed subsequently by comparing the modification effects of the two nanoslurries on the macroproperties of RAC.

3.3.4 Effects of Surface Treatment on the Properties of Mortars in RACs

Nanoindentation test results obtained on old and new mortars in the three groups of RACs, at an age of 28 days, are shown and compared in Fig. 8. The old mortars of RAC_1 and RAC_2 seem to be slightly stronger than that of RAC_0. Average modulus of the five indent areas, as previously illustrated in Fig. 2, is shown in Fig. 8 (b). In RAC_1 and RAC_2, the average modulus of the OM-1 area, which was closest to the surface of old mortars, was larger than those farther away; although in RAC_0 all five indent areas did not demonstrate obvious differences in modulus. Such results indicated, that in RAC_1 and RAC_2, old mortars were possibly surface strengthened, thereby explaining why the old mortars in RAC_1 and RAC_2 showed enhanced modulus as indicated by Fig. 8 (a). Throughout the RAs' soaking in nanoslurries, although most strengthening materials precipitated on the RAs' surface, some strengthening materials, primarily the nanoparticles, may penetrate into the adhering old mortar and strengthen the old mortar. However, nanomaterials or cement particles were unable to penetrate deeply into the old mortar, explaining why enhancement of the old mortar was quite limited.

No obvious differences were observed in the probability distributions of the modulus of new mortars among the three concrete groups, as shown in Fig. 8 (c). Such results indicate that the two nanoslurries used did not enhance the properties of the new mortars in RAC. Even though some nanoparticles or cement particles contained in the coatings around R_1 and R_2 could possibly penetrate back into new ITZs, nevertheless, as illustrated by Fig. 7, nanomaterials or cement particles had not crossed new ITZs. Therefore, it was reasonable to conclude that new mortars were not enhanced.

3.3.5 Workability of Fresh Concrete

Slumps of the four groups of fresh concrete are shown in Fig. 9.

Slump test results of the four groups of fresh concrete were consistent with the water absorption results of the coarse aggregate used, as exhibited in Table 4. Therefore, the

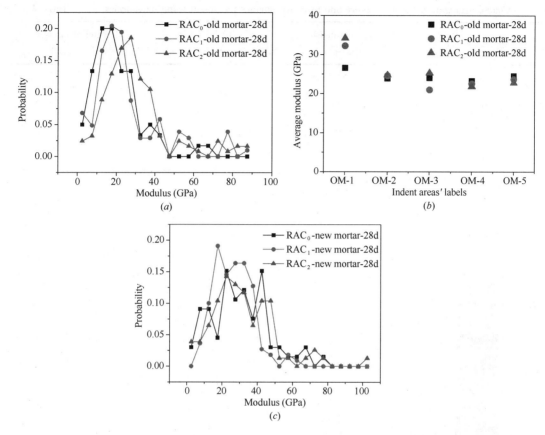

Fig. 8 Nanoindentation test results on old and new mortars in three groups of RACs at age of 28 days: (a) probability distribution of modulus in old mortar; (b) average modulus of indent areas on old mortar with different distances from new ITZs; (c) probability distribution of modulus in new mortar.

differences in water absorption of coarse aggregates directly led to the diversity in workability among the four concrete groups. The beneficial effects of surface treatment, using the two nanoslurries, on the workability of RAC were certified because the slump of RAC_1 and RAC_2 had been significantly improved compared with that of RAC_0. According to Fig. 9, the slump of RAC_1 was comparable to that of NAC_0, indicating that the inferior workability of RAC will no longer limit its application, providing that the proper surface treatments are applied on RA before casting.

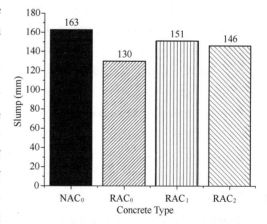

Fig. 9 Slump of four groups of fresh concrete in this study

Coarse aggregate properties before and after the surface treatment using nanslurries   Table 4

Aggregate type	Apparent density (kg/m³)	Crushing value (%)	Water absorption (% by weight)
NA	2763	6.5	1.5
RA_0	2570	10.4	6.4
RA_1	2570	9.4	3.6
RA_2	2581	9.2	4.4

3.3.6 Compressive Strength of Hardened Concrete

The cube compressive strength of the four concrete groups at the age of 28 days is shown in Fig. 10.

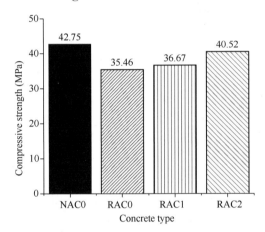

Fig. 10 Cube compressive strength of four concrete groups at age of 28 days

According to Fig. 10, by using the surface-treated RA, the compressive strengths of RAC_1 and RAC_2 were moderately improved. The new ITZs in RAC_1 and RAC_2 were significantly enhanced. This enhancement on the microscale provides the mechanism of the modification effects of the two nanoslurries on RAC's mechanical properties on the macroscale.

However, although the compressive strength of RACs was improved by the surface treatment using nanomaterials, it was still inferior to that of NAC. As stated in the "Introduction," RAC contains more weak points than NAC, i.e., the porous adhering cement mortar, and the more types of ITZs including the old ITZs and the new ITZs. Although the new ITZs were enhanced by the two nanoslurries used, the other two weak points in RAC, i.e., the old ITZs and the old adhering cement mortars, were not strengthened or were merely surface strengthened.

Furthermore, Fig. 10 shows the enhancing effects of the Cement + nSi slurry were better than those of the nSi + nCa slurry, in terms of improving RAC's compressive strength, which is consistent with the more significant enhancement of the new ITZs with the use of the former slurry, as indicated by Fig. 6.

3.3.7 Chloride Ion Diffusion Coefficient of Hardened Concrete

The Cl^- diffusion coefficients of the four concrete groups at an age of 28 days, as obtained from RCM tests, are shown in Fig. 11.

Fig. 11 demonstrates that by employing the surface-treated RA, the resistance to Cl^- diffusion of RAC_1 and RAC_2 was significantly improved. Such results were closely related to the enhancement of the new ITZ in RAC, which helped reduce the penetration paths for Cl^-. The surface-strengthened old mortar may also contribute to preventing Cl^- diffusion. The modification effects of the Cement + nSi slurry on RAC's resistance to Cl^- diffusion were better than those of the nSi + nCa slurry. Such results are consistent with the nanoindentation modulus results obtained from the new ITZs, as indicated by Fig. 6. As illustrated in Fig. 11, the Cl^- diffusion coefficient of RAC_2 decreased to a level that was comparable to that of NAC_0, indicating that the modification effects of the surface treatment using the Cement + nSi slurry were more significant, than those achieved using the nSi + nCa slurry, in terms of improving RAC's resistance to Cl^- diffusion.

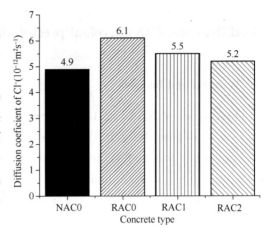

Fig. 11 Diffusion coefficients of Cl^- for four concrete groups at age of 28 days

3.3.8 Relationship between the Microproperties and Macroproperties

Although the old ITZs in RAC were not enhanced, because the nanomaterials or cement particles contained in the two slurries were unable to penetrate far enough into the old mortar to yield the old ITZs, the properties of the new ITZs in RAC were significantly enhanced. The old mortar, however, was likely surface strengthened. The surface-strengthened old mortar contributed to the enhancement of RA's resistance to crushing and water absorption. As a result of the decrease in the water absorption of RAs, the workability of the corresponding concrete, RAC_1 and RAC_2, was improved. The improvement of both new ITZs and old mortar resulted in higher compressive strength and greater resistance to Cl^- of RAC_1 and RAC_2 because the destructive and penetration paths in the two modified RACs were reduced by the nanomaterials.

The enhancing effects of the two nanoslurries on the macroproperties of RAs and RACs were verified to be comparable. In most cases the Cement + nSi slurry, which contained more cement but much less nanomaterials, performed better than the nSi + nCa slurry. These results indicated that the Cement + nSi slurry may be a good alternative, to those strengthening slurries or solutions containing more nanomaterials, in modifying RAC, particularly when considering its advantages in terms of both low cost and the guaranteed modification effects on RAC.

4. Modifications with microbial precipitation

4.1 Material and bacteria

Three types of RA (Fig. 12) are prepared for this study. A series of CCs at size of 30mm were cut from 100×100×100mm concrete specimens, prepared as ideal artificial recycled aggregate for the PM optimization. The other two different sizes of RAs, Large Recycled Aggregate (LRA) ranging from 30 to 40 mm and Small Recycled Aggregate (SRA) from 10 to 20 mm, were purchased from a plant and used for study of modification effect of OPM on RAs.

Fig. 12　CC and RAs: (a) CC, (b) LRA, (c) SRA

S. pasteurii (DSM No. 33) purchased from the German Collection of Microorganism and Cell Cultures (DSMZ), Braunschweig, Germany, was used. Bacteria were activated in culture media in a 30℃ incubator shaker for 2 days. The composition of culture media included 20 g/L tryptone, 5g/L NaCl, and 20 g/L urea recommended by DSMZ. After two-day activation and one-day growth, the bacteria culture media was centrifuged to get highly concentrated bacterial cells.

4.2 Experimental program

4.2.1 Precipitation methods

In order to obtain the better modification effect, adding methods of calcium source, which influenced the formation of MCP, and rolling treatment, which influenced the accumulation of MCP, as significant influence factors on modification effect were experimentally investigated.

Before MCP modification, CCs were dried to constant weight in the oven at the temperature of 40℃ instead of 105 ± 5℃ to avoid change of internal pore structure in CCs at high temperature [40]. The concentrated bacteria cells obtained from centrifuge were diluted to pre-determined cell concentration (108 cell/mL) with the composition of precipitation culture media, which consisted of 3 g/L tryptone, 10 g/L NH4Cl, 20g/L urea and 2.12 g/L Na_2CO_3. Sodium hydroxide (1 N solution) and hydrochloric acid (1 N) were used to adjust the pH of the solution at 9.5.

The whole MCP modification process took 7 days. At the first day, CCs were put into the precipitation culture media containing bacteria, stored in the incubator at the temperature of 30℃ for one day.

4.2.2 Calcium source adding methods

To find the influence of calcium source adding methods, from 2 to 7 day, three different calcium source adding methods were as follows:

(1) PM 1: $CaCl_2$ solution at the pH of 9.5 was poured into the precipitation culture media all once. The calcium ion concentration was 0.55 mol/L in the entire volume of solution.

(2) PM 2: $CaCl_2$ solution at the pH of 9.5 was divided into four parts and was poured into the precipitation culture media at the first four days respectively. $CaCl_2$ solution and culture media were the same as those used in PM 1.

(3) PM 3: CCs were taken out from the precipitation culture media, and put into another precipitation culture media with additional 0.55 mol/L calcium ion at the pH of 9.5.

4.2.3 Rotation treatment

To determine the influence of the rotation treatment on the accumulation of precipitations on the top surface, two concrete cubes were rotated, to ensure that each side of the concrete cube was the top surface for a precipitation period of 1 day. Meanwhile, the other two concrete cubes remained static throughout the entire MCP modification process, as a control experiment.

4.3 Results and discussion

4.3.1 Optimization of the precipitation method

4.3.1.1 Distribution of MCP on the external surfaces

The distributions of MCP on the external surfaces of the concrete cubes treated with different precipitation methods are shown in Fig. 13.

As shown in Fig. 13, white-colored precipitations exist on the surfaces of all the modified concrete cubes; the white-colored precipitations primarily consist of $CaCO_3$. As shown in Fig. 13 (a) and Fig. 13 (c), the layers MCP on the top surfaces of the Precipitation method 1-Control and Precipitation method 2-Control samples are thicker than those of the Precipitation method 1-Rotation and Precipitation method 2-Rotation samples. The thick layer MCP exists solely on the top surfaces of the samples.

In the case of the Precipitation method 3, there is not an apparent difference in the Precipitation method 3-Rotation and Precipitation method 3-Control on the distribution of MCP as shown in Fig. 13 (b) and Fig. 13 (e). Specifically, the rotation treatment has no influence on the Precipitation method 3 group. This also will be discussed in Section 4.3.1.2.

4.3.1.2 Mass increase and water absorption

The mass increases of the samples treated with the different precipitation methods are

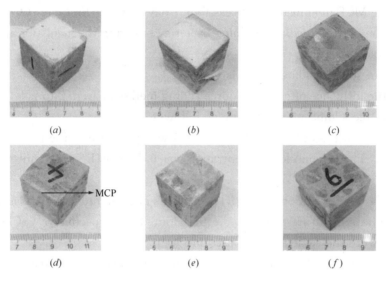

Fig. 13 Distributions of MCP on the external surfaces of the samples treated with different precipitation methods: (a) Precipitation method 1-Control, (b) Precipitation method 2-Control, (c) Precipitation method 3-Control, (d) Precipitation method 1-Rotation, (e) Precipitation method 2-Rotation, (f) Precipitation 3-Rotation.

shown in Fig. 14.

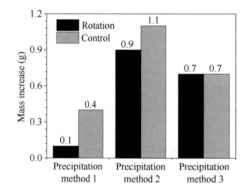

Fig. 14 Mass increases of the samples treated with the different precipitation methods

As shown in Fig. 14, the mass of each concrete cube increases because of the biological treatment. In the cases of Precipitation method 2 and Precipitation method 3, the mass increases are greater than that of Precipitation method 1. In the cases of Precipitation method 1 and Precipitation method 2, the mass increases of the specimens that underwent the rotation treatment are lower than those of the control specimens. While in the case of Precipitation method 3, the rotation treatment appears to have no influence on the mass increase. These results are in accordance with the findings discussed in Section 4.3.1.1.

Fig. 15 shows the water absorption of the specimens, before and after modification.

As shown in Fig. 15, after the modification, the water absorption of each concrete cube decreases, except in the case of the Precipitation method 1-Rotation. The reduction in water absorption is owed to the presence of the microbial carbonate precipitation, which fills the cracks and pores located on the surface of the sample, and covers the old cement mortar. Considering the differences in the initial water absorption of the concrete cubes, the reduction in water absorption is calculated to further compare the modification effects

34

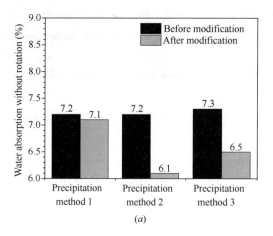 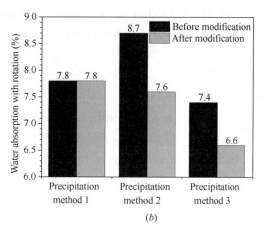

Fig. 15 Water absorption before and after modification: (a) Control, (b) Rotation

of the different precipitation methods, as shown in Fig. 16.

As shown in Fig. 16, the modification effects of Precipitation method 2 and Precipitation method 3 are superior to those of Precipitation method 1. The reason for this will be discussed in 3.3.1.3. In addition, Fig. 16 shows that in the case of all precipitation methods, Δw decreases as a result of the rotation treatment; specifically, the rotation treatment may have a negative effect on the MCP modification.

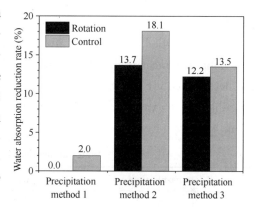

Fig. 16 Water absorption reduction of samples treated with the different precipitation methods

Moreover, in the case of the Precipitation method 1-Rotation, there is an increase in the mass, but there is no reduction in the water absorption. The Precipitation method 3-Control and Precipitation method 3-Rotation exhibit identical increases in mass; however, they exhibit differing reductions in water absorption. This indicates that the results regarding the mass increase cannot be used to accurately reveal the modification effect.

4.3.2 Proposed precipitation method

4.3.2.1 Proposed method

Based on the results discussed in Section 4.3.1, the optimum precipitation method is proposed as follows:

(1) Before the modification process, it is suggested that the RAs are dried to improve their ability to absorb bacteria.

(2) The dried RAs are placed into the precipitation culture media that contains bacteria, and stored within an incubator at an appropriate temperature.

(3) The calcium-source solution is divided equally into several parts.

(4) The RAs that are saturated with bacteria are removed from the precipitation culture media, and directly immersed into another precipitation culture media without bacteria.

(5) The averagely divided calcium-source solution is added into the precipitation culture media in batches.

(6) At the end of modification process, the modified RAs are dried at an appropriate temperature, and are considered ready for test or use.

In this study, the entire optimum precipitation process took place over 7 days. The RAs were dried in an oven at a temperature of 40℃ until they attained a constant weight. They were then placed into a precipitation culture media that contained bacteria, and subsequently stored in an incubator at a temperature of 30℃ for 1 day. The calcium-source solution was divided into five equal portions, and each portion was added to the precipitation culture media over the days that followed, respectively.

4.3.2.2 Verification

To verify the modification effect of the optimum precipitation method, three concrete cubeswere modified. Before and after the modification, the water absorption of each concrete cube was measured. The results are listed in Table 5. The water absorption reduction determined for the samples treated with the different precipitation methods those were not subjected to the rotation treatment, and optimum precipitation method, are shown in Fig. 17.

Table 5　Water absorption of CCs modified with OPM

Before modification	After modification	Δw
7.9 %	5.7 %	41.3 %

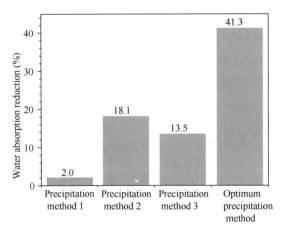

Fig. 17　Water absorption reduction determined for the samples treated with the different precipitation methods, and optimum precipitation method

As shown in Fig. 17, the reduction in the water absorption determined for the concrete cubes modified using the optimum precipitation method is approximately 2 to 3 times those of the concrete cubes modified using Precipitation method 2 and Precipitation method 3; this verifies the effectiveness of the proposed optimum precipitation method.

The concrete cubes are considered artificial RAs. To verify the practical modification effect of optimum precipitation method, manufactured RAs of various sizes were used, as shown in the next section.

4.3.3 Effect of modification on manufactured RAs

The water absorption and water absorption reduction determined for the RAs are shown in Table 6 and Fig. 18.

Water absorption and water absorption reduction of manufactured RAs　　Table 6

	Before modification	After modification	Δw
LRA	4.6 %	3.8 %	20.2 %
SRA	2.0 %	1.8 %	7.5 %

As shown in Fig. 18, following modification, there is a reduction in the water absorption measured for both RAs; this verifies that optimum precipitation method is also effective with regard to the manufactured RAs. The reduction in water absorption determined for the large RA is greater than that of the small RA. Considering the water absorption values determined for the concrete cubes, which are listed in Table 6, it may be concluded that the modification effect improves as the quantity of attached cement mortar increases. This makes sense because a large quantity of cement mortar will absorb a large quantity of bacteria, which would result in the formation of a large quantity of Type 3 precipitations, and enhance the modification effect. This result confirms that the MCP modification can strengthen the weakness of RAs caused by old cement mortar.

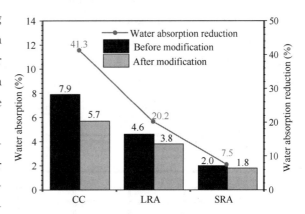

Fig. 18　Water absorption and water absorption reduction of CCs and RAs

5. Field application of modified RA

5.1 Materials

Three concrete groups were prepared, i.e., the commercial natural aggregate concrete (CNAC), the original recycled aggregate concrete (ORAC), and the modified recycled aggregate concrete (MRAC). The percentage replacement of RA was 50% for both ORAC and MRAC.

The coarse aggregate used for preparing CNAC was NA. The RA employed in the ORAC was original recycled aggregate (ORA) without strengthening. The RA used in the MRAC was strengthened recycled aggregate (SRA), which used the designed nano-silica slurry. The ORA was surface treated as mentioned in Section 3.1. All the three groups of coarse aggregates, i.e., NA, ORA, and SRA, were employed in air-dried condition. To make reasonable comparisons among the three concrete groups that were prepared in this study, i.e., CNAC, ORAC, and MRAC, the mixture proportions were kept the same by referring to those of the CNAC mixture.

5.2 Information for the applied project

ORAC and MRAC were applied in two secondary beams in a reinforced concrete building in Hangzhou, China. The CNAC applied in a third beam worked as the control. The three beams were designed to bear similar loads. The applied project was an 8-story frame structure, and the three target beams were located on the roof floor. Apart from the three target beams, other components on this floor, i.e., beams and slabs, were made of CNAC. The component arrangement on this building's roof floor is shown in Fig. 19a. The three target beams on this floor are marked in red and enlarged in detail in Fig. 19b.

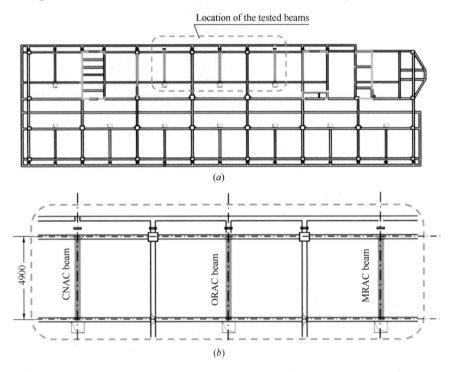

Fig. 19 Locations of the three target beams in the applied project: (a) layout of all columns and beams on the roof floor and (b) detailed locations of the applied CNAC, ORAC, and MRAC beams

The cross-section dimensions of the three target beams were the same: 250 mm × 450 mm. The three beams' reinforcements in the middle of the beam span were also the same, as shown in Fig. 20.

It should be noted that the three target beams, as well as a series 150 mm concrete cubes, using CNAC, ORAC, and MRAC, respectively, were cast on the construction site. After casting, a 28-day on-site moist curing was applied to these concrete cubes and

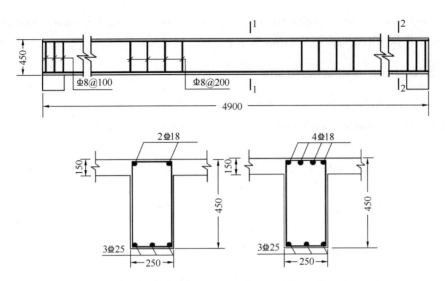

Fig. 20 Sketch of the three target beams (dimensions in mm)

the three target beams; water was poured onto them twice a day and a hessian sheet was used to cover these beams and specimens to slow down water evaporation.

It should also be noted that aninitial arch camber was set for the three target beams' frameworks, prior to casting. The value of the arch camber was set at 1/400 of the horizontal span of the beams, which was 12.25 mm, according to the Code for acceptance of constructional quality of concrete structures (GB 50204—2002). The initial arch camber of large-span beams was set to help slow down the deflection development of beams. The schematic diagram of the arch camber is shown in Fig. 21; l_0 is the initial length of the beam along its axis, while L_0 is the span of the beam, which is Frameworks were removed 14 days after casting, and thereafter the initial arch camber would gradually fall due to loads applied on them; the loads mainly included the self-weight of beams and the slabs supported by the beams, as well as the external loads applied onto the slabs and then transferred to the beams, e.g., the construction loads, and snow loads. As the arch chamber fell down, since the two ends of the arched beams were restrained by columns, therefore, the arched beams would be compressed, which means the length of the arched axes of these beams would decrease from l_0, to approach L_0. After the arch finally lowered down to the horizon, the arch camber would no longer work on controlling the beam de-

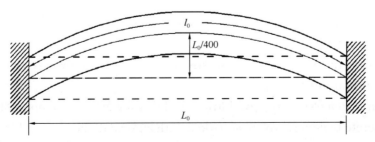

Fig. 21 Schematic diagram of the beams' initial arch cambers

flection. As the arch camber fell, the concrete strain of the target beams would be affected, which will be discussed in Section 5.4.2.

5.3 Concrete strain monitoring in situ

It is common knowledge that concrete cracks when its tensile strain exceeds its ultimate value. The ultimate tensile strain of conventional concrete is considered to be approximately $(1.0-1.5) \times 10^{-4}$.

In this study, the cracking control level for the applied reinforced concrete project was Level III, which means that crack widths below 0.3 mm were acceptable, according to the Code for Design of Concrete Structures (GB 50010—2002). The vertical loads on the three target beams could be treated as uniform loads, including primarily the self-weight of the concrete beams and slabs and other loads that were transferred onto them via their supported slabs. Given such uniform vertical loads, the largest concrete tensile strain was assumed to appear in the middle of the whole spans at the bottom of the beams, which means that this typical position should be set as the control point for concrete tensile strains along the beam span. The largest concrete compressive strain was assumed to appear at the top of those beams, exactly opposite of where the largest tensile strain was assumed to occur. In this study, concrete strains on both the top and bottom surfaces in the middle of the spans of the three beams were monitored and compared with one another.

Concrete strain was monitored by JTM-V5000 vibrating-string strain gauges. For each of the three applied beams, four concrete strain gauges were installed prior to casting, with two of them mounted to the beam framework's bottom longitudinal reinforcement and the other two mounted to the top longitudinal reinforcement. Fig. 22 shows the locations of the four concrete strain gauges inside the three target beams.

A JTM-V10B reading device was employed to read the transferred vibrating frequency of the strings inside these JTM-V5000 concrete in the three beams, the vibrating frequencies of these strain gauges were captured and set as the initial values. The frequencies of these gauges were then monitored in situ and recorded at a series of ages: 1 day, 3 days, 7 days, 14 days, 28 days, 90 days, 180 days, 360 days, and 450 days. A positive ε value represents a concrete tensile strain, while a negative one represents a compressive strain.

5.4 Results and discussion

5.4.1 Mechanical property tests of concrete materials
5.4.1.1 Compressive strength

Fig. 23 shows the cube compressive strength of the three groups of concrete specimens, as obtained at different ages.

The target compressive strength in this study was 30MPa, as required by the applied reinforced concrete building, which is marked with a green dashed line in Fig. 23. This figure shows that the 28-day compressive strength of all three concrete groups failed to reach

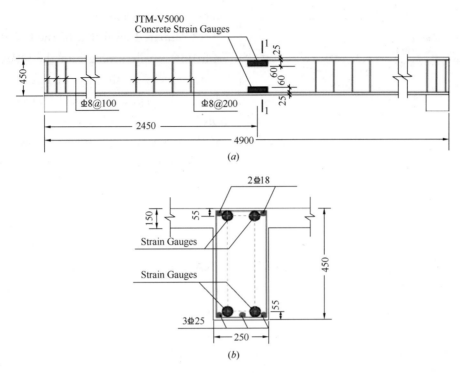

Fig. 22 Locations of the four concrete strain gauges in the three target beams: (a) locations of the four concrete strain gauges along the beam span and (b) locations of the strain gauges in the target beams' cross sections

the target strength value. These concrete cubes were cast and cured on the construction site, along with the three target beams and other components on the same floor. After the casting, three target beams and the concrete cubes had experienced a sudden drop in temperature, which could adversely affect cement hydration in concrete and thereby led to a slow strength gain for all the three concrete groups [41, 42].

Fig. 23 shows the beneficial effects of the employed nano-silica and cement slurry on RAC's compressive strength. The MRAC's compressive strength at early ages (before the age of 28 days) developed faster than that of the ORAC, e.g., at 28 days, the MRAC's compressive strength was 34.22% higher than that of the ORAC.

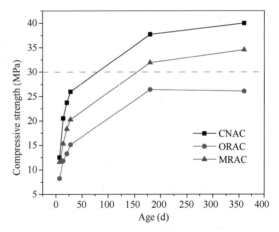

Fig. 23 Compressive strength of the three concrete groups in this study

5.4.1.2 Splitting tensile strength

Fig. 24 shows the splitting tensile strength of the three concrete groups tested at the

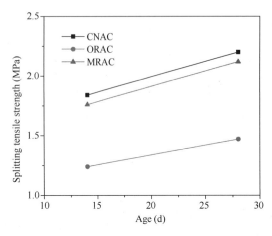

Fig. 24 Splitting tensile strength of the three concrete groups at the age of 14 days and 28 days

ages of 14 days and 28 days.

Compared with that of the CNAC, the splitting tensile strength of the ORAC was much lower. At the ages of 14 days and 28 days, the splitting tensile strengths of ORAC were 67.39% and 66.22% of those of the CNAC, respectively. However, after modification, RAC's splitting tensile strength had been significantly improved. At both testing ages, MRAC's splitting strength was higher than that of the ORAC and was even comparable to that of the CNAC. Concrete's enhanced tensile strength can promote the flexural capacity of reinforced concrete beams and enhance the crack resistance of concrete structures. Therefore, it can be predicted that the structural performance of RAC structures under service may be promoted by applying the nano-silica slurry to RAC's modification, which will be further discussed in Section 5.4.2.

5.4.2 Concrete strain monitoring in applied beams

Fig. 25a and b plot the concrete strains monitored at the bottom and at the top of the three target beams, in the middle of the beam-span, within the ages of 450 days, and 28 days, respectively. The dashed lines represent the concrete strain captured at the bottom, labeled as ε_{bottom}, while the solid lines represent the concrete strain at the top, labeled as ε_{top}.

According to Fig. 25a, from the casting date through the age of 50 days, both ε_{bottom} and ε_{top} of the three beams' middle spans were positive, which represents the tensile strains captured at these positions. As explained in Section 5.3, the strain gauges were mounted at positions where potential maximum compressive and tensile strains of concrete may appear under uniform loads. Therefore, the tensile strains captured at both the top and bottom indicated that there were tensile strains throughout the whole beam, before the age of 50 days. A large amount of free water commonly existed in the freshly cast concrete beams. The free water evaporated gradually as time passed, which resulted in concrete shrinkage. However, concrete shrinkage in these beams were constrained by their adjacent components, i.e., the columns and slabs, which led to tension inside the beams.

The development of tensile strains fluctuated, as the evaporation speed of water varied over time; tensile strains increased at an early age and then declined gradually. As shown more clearly in Fig. 28b, at 7 days or 14 days, peak tensile strains were captured at the bottom surfaces in the three target beams, which were 2.01×10^{-4}, 1.63×10^{-4}, and 1.39×10^{-4} for CNAC, ORAC, and MRAC, respectively. The peak tensile strains of concrete in the CNAC and ORAC beams exceeded the ultimate tensile strain of convention-

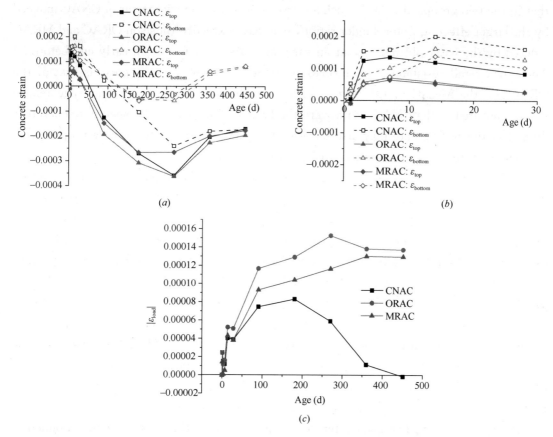

Fig. 25 (a) - (b) Concrete strains at both the top and bottom surfaces of the three target beams in the middle of their spans obtained (a) within 450 days, and (b) within 28 days, respectively; (c) the values of $|\varepsilon_{loads}|$ in the target beams within 450 days

al concrete, which is accepted as $(1.0-1.5) \times 10^{-4}$. Therefore, the concrete at the bottom surfaces of these two beams' middle spans may have cracked. In contrast, the peak tensile strain of concrete in the MRAC beam was still within the safe range. In this study, at an early age (before 28 days), the CNAC beam gained the largest tensile strain, indicating that the shrinkage of CNAC was probably the largest, among the three concrete groups applied in the target beams. Such results unexpectedly conflict with previous studies carried out by other researchers, which have argued that the shrinkage of RAC is generally larger than that of NAC [43-46]. The possible reason can be explained by the adverse effects of CNAC's larger effective water/binder ratio on the early-age mechanical properties of the new mortar. NA usually had a much smaller water absorption than NA [47, 48], therefore, a larger amount of free water existed in the CNAC than in the ORAC and in the MRAC, given the same water and cement amounts among the three concrete groups. Hence, it could be inferred that the effective water/binder ratio in the CNAC was larger than that in the ORAC and the MRAC, which may have brought adverse effects on the early-age strength and elastic modulus of the new mortar in CNAC, compared with that in the two groups of RACs. Such adverse effects on the new mortar of CNAC induced by the larger effective water/binder ration, may have exceeded those on ORAC and MRAC induced by the weak old mortar, at an early age. As a result, at an early age before 28 days, the overall resistance to deformation of the CNAC was not as good as that of the ORAC and the MRAC in the target beams, which explained why the early-age tensile strains captured in the CNAC beam were the largest. Moreover, the sudden temperature drop after the casting date may also have led to CNAC's poor anti-cracking property, as low temperature slowed down concrete's strength gain.

The MRAC's smaller peak tensile strain value captured at the early age indicated that its resistance to shrinkage-induced cracking was better than ORAC's. It was because that the mechanical property of the SRA was better after strengthening by the nano-slurry. Such enhancement may have also improved the elastic modulus of the SRA, or to be more exact, of the old mortar adhering to the SRA, thereby leading to the better deformability of the MRAC, compared with that of the ORAC.

Fig. 28a also illustrates that tensile strains captured at the bottom were always larger than those captured at the top in all target beams. The uniform loads including the three beams' self-weight and the loads transferred from the slabs would lead to tension on the bottom surfaces and compression on the top surfaces of the beams. Therefore, the effects of these loads enhanced the tensile strains caused by constrained concrete shrinkage at the bottom of beams, while those at the top of the beams were reduced.

As can be seen in Fig. 25a, after declining over time, the tensile strains obtained at both the top and bottom of the three beams fell to zero, at 100 days for the CNAC beam and at 125 days for ORAC and MRAC beams, and thereafter transferred to compressive strains. Such results indicate that concrete shrinkage was no longer the predominant influencing factor on concrete strain development for the three target beams. At this time, compression of the three target beams, caused by the falling of the initially arched beams under the applied loads, may be the main reason for the compressive strains. At the age of approximately 300 days, the strains captured at the bottom of the middle spans of the ORAC and MRAC beams began to transfer from compressive strains to tensile ones, which indicated that the axes of the ORAC and MRAC beams may have fell to be horizontal. By contrast, for the CNAC beam, even at the age of 450 days, strains captured at both the top and the bottom were still compressive, implying that the arch camber was still working. Such results implied that under similar loads, the deflection of the CNAC beams developed more slowly.

Based on the analysis listed above, through the 450-days monitoring period, there were three main factors which had worked together to determine the concrete strain development in the three target beams, i. e. , the constrained concrete shrinkage, the falling of the arch camber, and the uniform vertical loads applied onto the beams. The concrete strains induced by the constrained concrete shrinkage and the falling of the arch camber, were positive and negative, respectively, at both the top and the bottom of the target beams, while those strains induced by the vertical loads onto beams were positive at the beam bottom, but negative at the top. Ultimately, the in situ ε_{bottom} and ε_{top}, can be expressed as ($|\varepsilon_{shrinkage}| - |\varepsilon_{arch\,camber}| + |\varepsilon_{loads}|$) and ($|\varepsilon_{shrinkage}| - |\varepsilon_{arch\,camber}| - |\varepsilon_{loads}|$), respectively. $\varepsilon_{shrinkage}$, $\varepsilon_{arch\,camber}$, and ε_{loads} represent the concrete strains in the middle beam span induced by the constrained concrete shrinkage, the compression of the arched beams, and the loads applied onto these beams, respectively. As a result, $|\varepsilon_{loads}|$ can be calculated as follows:

$$|\varepsilon_{loads}| = (u\varepsilon_{bottom} - \varepsilon_{top})/2$$

$|\varepsilon_{loads}|$ can represent the positive strain value induced by the loads applied onto the target beams at the beam bottom in the middle span. According to equation above, the values of $|\varepsilon_{loads}|$, at different ages, in the three target beams, are plotted in Fig. 25c.

Based on Fig. 25c, it can be seen that though fluctuations at the early age, however, after the age of 90 days, the values of $|\varepsilon_{loads}|$ for CNAC were significantly lower than those for ORAC and MRAC, indicating that under the similar loads, the deformation of CNAC at the bottom of the beams was the smallest. Such results demonstrated that the CNAC's elastic modulus against loads was superior to that of ORAC and MRAC, at a relative long age, despite that its early-age resistance to shrinkage-induced deformation was not good, as discussed above. After a relative long age, e. g., 90 days, the new mortar in the three concrete groups may have fully developed, thereby, the new mortar was no longer the weakest point in CNAC. By contrast, at this time, some other weak material phases in RAC, i. e., the old ITZs, the new ITZs, as well as the porous old mortar, may have become the main weak points. There are more weak points in the two groups of RACs than in CNAC, which may have contributed to RAC's inferiority in long-term structural performance. Besides, according to Fig. 10c, the employed nano-silica slurry had induced beneficial effects on long-term deformability of MRAC, as the values of $|\varepsilon_{loads}|$ for the MRAC beam were smaller than those of the ORAC beam.

As an overall summary of the discussion stated above, given the mixture proportions employed in this study, at an early age (before 28 days), the resistance to shrinkage-induced tensile strains of CNAC in the target beam was inferior to that of the ORAC and the MRAC; the larger effective water/binder ratio in CNAC than in the two groups of RACs, as well as the adverse effects of low temperature on deformability of concrete, is the possible reason. However, at a relatively long age (after 90 days), perhaps as a result of the faster gain in mechanical properties, CNAC's resistance to the load-induced deformation in the target beam was the best among the three groups of concrete, leading to the smallest load-induced concrete strains, and the controlled deflection of the target beam. Beneficial effects by using the nano-slurry on the structural performance of MRAC compared with ORAC can be seen from both the better resistance to shrinkage-induced tensile strain at an early age, and to load-induced strains after 90 days. Such in situ experimentation and monitoring of the MRAC will be conducted steadily in the future, and the feasibility of the modified MRAC through the employed modification technique will likely be revealed.

6. Summary and future work

6.1 Summary

Methods to improve the properties of RAs consist of two types. One is to remove the old mortar from RAs, including heating and rubbing, microwave cleaning, acid cleaning, etc. The other is to strengthen the old mortar attached RA, containing polyvinyl alcohol strengthening, carbonation strengthening, nano particle strengthening, etc. This paper introduces two novel modification methods with nano-material slurry and microbial precipitation respectively and proves the feasibility of field application of RAC modified. The main conclusions can be summarized in the following:

After the modification with nano-material slurry, the micro properties of ITZ, i.e., width, elastic modulus, and the macro properties of RAC, i.e., compressive strength, resistance to chloride ion diffusion, slump, were significantly enhanced compared with the unmodified RAC. It was contributed to the enhancement of nano-material slurry on the old mortar attached to RA.

With the comparison and analysis of different precipitation methods, an optimum precipitation method was proposed. The mass increase and the water absorption reduction were found after the optimum microbial precipitation modification. It accounted for microbial precipitation filling the holes and cracks on the surface of RA.

Base on the researches on RA's modification mentioned above, the modified RA was used in the filed application. With the long-term observation on the CNAC, ORAC and MRAC reinforced beams in a real project, it was found that MRAC beams had the better deformability forecasting the better resistance to cracking, which may enhance the durability of reinforced MRAC components and expand the service life, compared with ORAC.

6.2 Future work

The investigation above mainly focus on recycled coarse aggregate, while nowadays with the fast development of construction, the shortage of natural sand is more serious than that of natural coarse aggregate. Therefore lots of researchers are dedicated to finding other replacements of river sand, such as sea sand, glass fine aggregate, recycled concrete fine aggregate and so on.

The properties of recycled fine aggregate (RFA) have been investigated: lower density, greater porosity and higher water absorption. Considering that the inferior properties of RFA will have side effect on the mortar and concrete casted with RFA, it is necessary to improve the properties of RFA.

The authors have tied to modify RFA with microbial precipitation. The decrease of water absorption and the change of morphology were found after the modification. However, this work is on the way and needs to be furthered, including the optimization of modification method, the verification of modification effect on mortar and concrete with modified RFA, the industrial promotion and field application.

Acknowledge

Financial support from the Nation Natural Science Foundation of China (51578489) is gratefully acknowledged.

References

[1] Xiao, J., et al., An overview of study on recycled aggregate concrete in China (1996—2011). Construction and Building Materials, 2012. 31: p. 364-383.
[2] Kisku, N., et al., A critical review and assessment for usage of recycled aggregate as sustainable construction material. Construction and Building Materials, 2017. 131: p. 721-740.
[3] Poon, C. S., Z. H. Shui and L. Lam, Effect of microstructure of ITZ on compressive strength of concrete prepared with recycled aggregates. Construction and Building Materials, 2004. 18(6): p. 461-468.
[4] Ryu J S. Improvement on strength and impermeability of recycled concrete made from crushed concrete coarse aggregate. Journal of Materials Science Letters, 2002, 21(20): 1565-1567.
[5] Xiao, J., et al., Effects of interfacial transition zones on the stress-strain behavior of modeled recycled aggregate concrete. Cement and Concrete Research, 2013. 52: p. 82-99.
[6] Kou S C, Poon C S, Etxeberria M. Influence of recycled aggregates on long term mechanical properties and pore size distribution of concrete. Cement and Concrete Composites, 2011, 33(2): 286-291.
[7] Breccolotti M, Materazzi A L. Structural reliability of bonding between steel rebars and recycled aggregate concrete. Construction and building materials, 2013, 47: 927-934.
[8] Silva R V, de Brito J, Dhir R K. Comparative analysis of existing prediction models on the creep behaviour of recycled aggregate concrete. Engineering Structures, 2015, 100: 31-42.
[9] Silva R V, de Brito J, Dhir R K. Establishing a relationship between modulus of elasticity and compressive strength of recycled aggregate concrete. Journal of Cleaner Production, 2016, 112: 2171-2186.
[10] Wang W, Kou S, Xing F. Deformation properties and direct shear of medium strength concrete prepared with 100% recycled coarse aggregates. Construction and Building Materials, 2013, 48: 187-193.
[11] Villagrán-Zaccardi Y A, Zega C J, Di Maio Á A. Chloride penetration and binding in recycled concrete. Journal of materials in civil engineering, 2008, 20(6): 449-455.
[12] Vázquez E, Barra M, Aponte D, et al. Improvement of the durability of concrete with recycled aggregates in chloride exposed environment. Construction and Building Materials, 2014, 67: 61-67.
[13] Hansen T C. Recycled aggregates and recycled aggregate concrete second state-of-the-art report developments 1945-1985. Materials and structures, 1986, 19(3): 201-246.
[14] Sagoe-Crentsil K K, Brown T, Taylor A H. Performance of concrete made with commercially produced coarse recycled concrete aggregate. Cement and concrete research, 2001, 31(5): 707-712.

[15] Srubar W V. Stochastic service-life modeling of chloride-induced corrosion in recycled-aggregate concrete. Cement and Concrete Composites, 2015, 55: 103-111.

[16] Sato R, Maruyama I, Sogabe T, et al. Flexural behavior of reinforced recycled concrete beams. Journal of Advanced Concrete Technology, 2007, 5(1): 43-61.

[17] Etxeberria M, Mari A R, Vazquez E. Recycled aggregate concrete as structural material[J]. Materials and structures, 2007, 40(5): 529-541.

[18] Etxeberria M, Mari A R, Vazquez E. Recycled aggregate concrete as structural material[J]. Materials and structures, 2007, 40(5): 529-541.

[19] A. B. Ajdukiewicz, A. T. Kliszczewicz, Comparative tests of beams and columns made of recycled aggregate concrete and natural aggregate concrete, J. Adv. Concr. Technol. 5 (2007) 259-273.

[20] Oikonomou N, Recycled concrete aggregates, Cement & Concrete Composites. 27(2) (2005) 315-318.

[21] Chi S P, Chan D, Feasible use of recycled concrete aggregates and crushed clay brick as unbound road sub-base, Construction & Building Materials. 20(8) (2006) 578-585.

[22] Shima H, Matsuhashi R, Yoshida Y, et al. , Life Cycle Analysis of High Quality Recycled Aggregate Produced byHeating and Rubbing Method, Ieej Transactions on Electronics Information & Systems. 123(10) (2003) 1680-1687.

[23] Shima H, Tateyashiki H, Matsuhashi R, et al. , An advanced concrete recycling technology and its applicability assessment through input-output analysis, Journal of Advanced Concrete Technology. 3 (3) (2005) 53-67.

[24] Bru K, TouzÉ S, Bourgeois F, et al. , Assessment of a microwave-assisted recycling process for the recovery of high-quality aggregates from concrete waste, International Journal of Mineral Processing. 126(2) (2014) 90-98.

[25] Akbarnezhad A, Ong K C G, Zhang M H, et al. , Microwave-assisted beneficiation of recycled concrete aggregates, Construction & Building Materials. 25(8) (2011) 3469-3479.

[26] Katz A, Treatments for the Improvement of Recycled Aggregate, Journal of Materials in Civil Engineering. 16(6) (2004) 531-535.

[27] Akbarnezhad, A. , et al. , Acid treatment technique for determining the mortar content of recycled concrete aggregates, Journal of Testing and Evaluation. 41(41) (2013) 441-450.

[28] Tam V W Y, Tam C M, Le K N. Removal of cement mortar remains from recycled aggregate using pre-soaking approaches. Resources, Conservation and Recycling, 2007, 50(1): 82-101.

[29] Kim H S, Kim B, Kim K S, et al. , Quality improvement of recycled aggregates using the acid treatment method and the strength characteristics of the resulting mortar, Journal of Material Cycles & Waste Management. 19(2) (2017) 1-9.

[30] Kou S C, Poon C S, Properties of concrete prepared with PVA-impregnated recycled concrete aggregates, Cement & Concrete Composites. 32(8) (2010) 649-654.

[31] Mukharjee B B, Barai S V. Influence of nano-silica on the properties of recycled aggregate concrete [J]. Construction and Building Materials, 2014, 55: 29-37.

[32] Li W, Huang Z, Cao F, et al. Effects of nano-silica and nano-limestone on flowability and mechanical properties of ultra-high-performance concrete matrix. Construction and Building Materials, 2015, 95: 366-374.

[33] Kou S C, Zhan B J, Poon C S, Use of a CO_2 curing step to improve the properties of concrete prepared with recycled aggregates, Cement & Concrete Composites. 45(1) (2014) 22-28.

[34] Zhang J, Shi C, Li Y, et al. Performance enhancement of recycled concrete aggregates through carbonation. Journal of Materials in Civil Engineering, 2015, 27(11): 04015029.

[35] Grabiec A M, Klama J, Zawal D, et al. , Modification of recycled concrete aggregate by calcium carbonate biodeposition, Construction &. Building Materials. 34(34) (2012) 145-150.

[36] Qiu, Jishen, David Qin Sheng Tng, En-Hua Yang, Surface treatment of recycled concrete aggregates through microbial carbonate precipitation, Construction and Building Materials. 57 (2014) 144-150.

[37] Pan, Zhong-Yao, et al. , Modified recycled concrete aggregates for asphalt mixture using microbial calcite precipitation, RSC Advances. 5(44) (2015) 34854-34863.

[38] Zhang H, Zhao Y, Meng T, et al. Surface treatment on recycled coarse aggregates with nanomaterials. Journal of Materials in Civil Engineering, 2015, 28(2): 04015094.

[39] Miller, M. , Bobko, C. , Vandamme, M. , and Ulm, F. J. (2008). "Surface roughness criteria for cement paste nanoindentation." Cem. Concr. Res. , 38(4), 467-476.

[40] Ma Q, Guo R, Zhao Z, et al. , Mechanical properties of concrete at high temperature—A review, Construction &. Building Materials. 93 (2015) 371-383.

[41] M. Husem, S. Gozutok, The effects of low temperature curing on the compressive strength of ordinary and high performance concrete, Constr. Build. Mater. 19 (2005) 49-53.

[42] G. C. Lee, T. S. Shih, K. C. Chang, Mechanical properties of concrete at low temperature, J. Cold Reg. Eng. 2 (1988) 13-24.

[43] A. Domingo-Cabo, C. Lázaro, F. López-Gayarre, M. A. Serrano-López, P. Serna, J. O. Castaño-Tabares, Creep and shrinkage of recycled aggregate concrete, Constr. Build. Mater. 23 (2009) 2545-2553.

[44] R. S. Ravindrarajah, C. T. Tam, Properties of concrete made with crushed concrete as coarse aggregate, Mag. Concr. Res. 37 (130) (1985) 29-38.

[45] K. Eguchi, K. Teranishi, M. Narikawa, Study on mechanism of drying shrinkage and water loss of recycled aggregate concrete, J. Struct. Constr. Eng. 3 (2003) 1-7.

[46] S. C. Kou, C. S. Poon, D. Chan, Influence of fly ash as cement replacement on the properties of recycled aggregate concrete, J. Mater. Civ. Eng. 19 (2007) 709-717.

[47] C. S. Poon, Z. H. Shui, L. Lam, Effect of microstructure of ITZ on compressive strength of concrete prepared with recycled aggregates, Constr. Build. Mater. 18 (2004) 461-468.

[48] M. S. de Juan, P. A. Gutiérrez, Study on the influence of attached mortar content on the properties of recycled concrete aggregate, Constr. Build. Mater. 23 (2009) 872-877.

3 An overview on the use of 2D nanomaterials in concrete

Ezzatollah Shamsaei[a]*, Felipe Basquiroto de Souza[a], Xupei Yao[a], Shujian Chen[a], Wenhui Duan[a]*

[a] Department of Civil Engineering, Monash University, Clayton, VIC 3800, Australia

* Corresponding author: ezzatollah. shamsaei@monash. edu; wenhui. duan@monash. edu.

Abstract: Two-dimensional (2D) ultrathinnanomaterials have demonstrated great potential as reinforcing materials for high performance composites. Besides the extraordinary mechanical properties, another prominent advantage is the large surface area that improves their interaction with the composite matrix. Extensive research in the past five years demonstrates that little addition of these nanomaterials can present significant influence on the durability and mechanical performance of cement composites. This paperprovides an overview on current research activities on the use of 2D nanomaterials in enhancing the strength and durability of concrete composites. The main challenges and future perspectives in this field are also discussed.

Keywords: Graphene; Boron nitride; Concrete; Cement composites; Strength; Durability.

1. Introduction

It is expected that nanomaterials will be increasingly used in the construction industries [1-6] due to: (1) their promising reinforcing properties and improving or bringing new functionalities to materials at very low addition ratio; (2) the increasing knowledge about their properties and emerging applications; (3) falling nanocomposites pricesdue to the ongoing development of large-scale manufacturing processes. 2D nanosheets, as a new class of nanomaterials, are defined as nanostructures with lateral size larger than 100 nm and thickness less than 5 nm [7-10]. Given their unique mechanical strength, electrical and thermal conductivity, these ultrathin 2D nanomaterials arecurrently one of the hottest research topics in material science and nanotechnology. Extensivestudies on polymer composites incorporated with 2D nanomaterials showed superior mechanical, thermal, gas barrier, and flame retardant properties compared to pure polymers [11].

One of the unique properties of ultrathin 2D nanomaterials that make them different from other types of nanomaterials, such as zero-dimensional (0D) nanoparticles, one-dimensional (1D) nanofibers, and three-dimensional (3D) networks is their large lateral size and ultrathin thickness [12]. Thisendows the 2D nanomaterials with ultrahigh specific surface area, making them highly desirable for surface-active applications. 2D nanomaterials are generally obtained from layered structured materials, such as graphite and hexagonal boron nitride [13], by top-down exfoliation methods [14] (e. g. micromechanical cleavage; chemical exfoliation).

Graphene is composed of a single-layer sheet of carbon atoms closely packed into a 2D (with thickness of 3.35 Å) honeycomb framework [15]. Graphene oxide (GO), the most researched 2D materials in cement composites, also consists of monolayer sheets with a hexagonal carbon network. The main difference to graphene is that GO bears oxygen containing groups on its surface and thus exhibits higher reactivity with cement and superior dispersion in the matrix, making it attractive for structural material composites [16, 17]. Reduced graphene oxide (rGO) has less functional groups in its structure as compared to GO, presenting intermediate properties between GO and graphene. BN is another 2D nanostructure that has recently emerged as potential nanoreinforcement for cement composites. The boron and nitrogen atoms within each layer are covalently bonded and arranged in a hexagonal structure similar to that of graphene. Therefore, BN is often regarded as a graphene analogue, and is commonly known as the 'white graphene' [13]. Figure 1 shows the chemical representation of graphene and BN nanosheets.

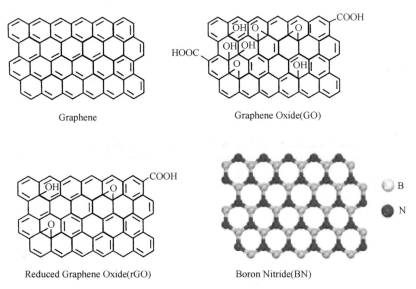

Figure 1 Structural representationof different 2Dnanostructures.

Recent studies have begun to explorethe potential use ofultrathin 2D nanomaterials as reinforcements for cement composites, and substantial progress has been made over the past few years. Graphene and BN have recently demonstrated great potential to enhance crucial properties of construction materials such as mechanical strength and durability. This paper provides an overview on current research activities on properties and applications of 2D nanomaterials (GO, rGO, and BN) to produce cementitious composites with enhanced strength and durability. The main challenges and future perspectives in the use of these ultrathinnanomaterials in the construction field are also addressed.

2. Properties of 2D Nanomaterials

Ultrathin 2D nanomaterials exhibitoutstanding mechanical properties due to the atomic thickness and strongly covalent in-plane bonds. Particularly, graphene and BN display a range of outstanding chemical, mechanical, thermal and electrical properties that can be exploited to reinforce, increase the durability and provide smart functionalities to construction materials. Large lateral size and atomic thickness provide 2D nanomaterials with high specific surface area and ultimate interaction of their surface atoms, making them exceedingly desirable in many practical applications including reinforcements. Also, the large surface area maximizes the interaction between the nanomaterials and the matrix, reducing the addition ratio into the composite. Table 1 summarizes the properties of graphene and BN.

Recent studies have revealed the Young's modulus of graphene to be as high as 1.0 TPa. It also exhibits an extremely high breaking strength of 130 GPa, which is 200 times higher than steel. Such exceptionally high strength suggests that graphene is strongest material ever measured. More importantly, graphene can withstand elastic deformations of more than 20%. For example, wrinkles and folds are usually observed without any rupture. In addition, graphene is chemically inert, presents excellent thermal conductivity and unique electrical and optical properties [18].

Physical and mechanical properties of graphene compared to BN.　　　　Table 1

Properties	Graphene	BN
Modulus of elasticity (GPa)	1000 [19]	865 [20]
Fracture strength (GPa)	130 [19]	70.5 [20]
Elongation at break (%)	20 [19]	12.5 [20]
Specific surface area ($m^2 g^{-1}$)	~2360 [21]	895 [22]
Electron conductivity (Sm^{-1})	10^7-10^8 [23]	Insulator [24]
Thermal conductivity ($W\ m^{-1}\ K^{-1}$)	3000-5300 [25, 26]	~1700-2000 [27]

The tightly-spaced carbon atoms and high electron-density of the graphene atomic structure also makes graphene the thinnest and most impermeable material fabricated to date [23]. Due to this unique impermeability, graphene nanosheets can function as physical barrier for aggressive agents, water and/or air components, and have thus been employed to create composites with enhanced durability, such as anti-corrosion epoxies [28, 29] and concrete with higher durability [30, 31]. Transport properties of graphene nanosheets have also been proven useful to provide smart functionalities to construction materials.

Due to the structural similarity of BN to graphene, it has been predicted that these materials are almost equally strong and much stronger than other 2D nanomaterials. The in-plane stiffness values of a monolayer BN and graphene were reported to be 267 and 335 N m^{-1}, respectively. Therefore, like graphene, BN can be excellent reinforcing fillers for cementitious composites [32]. The calculated thermal conductivity (κ) values for BN range from 300 - 2000 W m^{-1} K^{-1}, which are lower than graphene (1500 - 2500 W m^{-1} K^{-1}) [33]. Although BN shares similar mechanical strength and thermal conductivity properties with graphene, it has a localized electronic states with a large band gap (5 - 6 eV), endowing BN with electrically insulating property, superior temperature stability and acid-base resistance property.

3. Applications of 2D Nanomaterials in Concrete: Strength and Durability Enhancement

2D nanomaterials, with planar structure and high aspect ratio, generate superior potential sites for their interaction and bonding with the cement matrix. GO, in particular, with abundant surface functionalities, shows high dispersibility in water and potential to control and modify the microstructure of the cementitious matrix. Consequently, thesematerials have recently considered as promising alternatives for the nanoscale reinforcement of cement-based composites and structural materials.

Experimental results revealed thatstrength properties of the cementitious materials can be significantlyenhanced byusinga very low weight percentage of the ultrathinnanosheets. For example, Duan et al. [34, 35] reported an improvement of 41%-59% and 15%-33% in flexural and compressive strength, respectively, with only 0.05% (by weight of cement) loading of GO nanosheets in Portland cement (OPC) matrix. The high GO reinforcement was attributed to the strong interfacial adhesion between GO nanosheetsand cement matrix (Figure 2). Graphene materials arrest cracks and prevent them from traver-

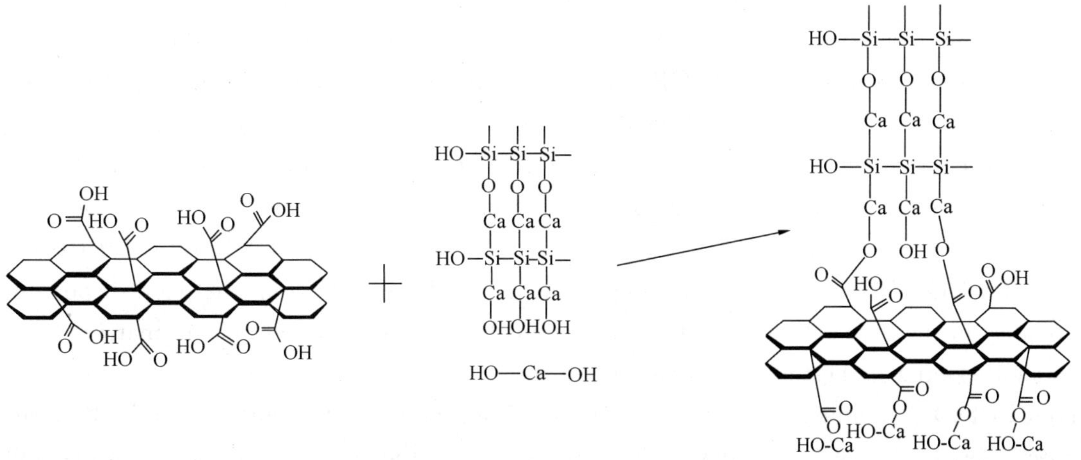

Figure 2 Schematic reaction scheme between carboxylic acid groups and hydration productions (Ca(OH)$_2$ and C-S-H) of cement [34, 37].

sing in a straight-through manner as it normally occurs in aplain matrix (Figure 3a, b). The 2D nanosheets can block and divert the in plane propagation of micro cracks. More recently, an increase of up to 146% in the compressive and 79.5% in the flexural strength was achieved by the addition of water-stabilized graphene dispersions into the cement matrix [36].

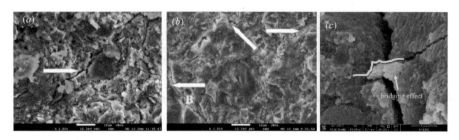

Figure 3 (a) SEM image of plain paste showing a straight-through type crack (arrow), (b) SEM image of GO-cement composite showing a number of fine cracks (arrows) with few branches [34], (c) Bridging effect in BN-cement composite [38].

Compared to other nanomaterials, 2D nanomaterials often display better reinforcement effects. For example, GO showed more uniform dispersion and higher possibility of interfacial adhesion with the hydration products as compared to functionalized carbon nanotubes (FCNTs), which resulted in a better reinforcing action (in terms of compressive strength) [39]. With respect to control specimens, the highest enhancement of 77% in compressive strength was obtained for GO composites at 1 wt% loading compared to that of 58% for FCNT composites at 0.5 wt% loading. In terms of tensile strength, FCNT composites showed slightly higher enhancement (40%) compared to GO (37.5%) composites with the same loadings (0.25 wt%).

The investigation on the usage of BN in cement matrix is at relatively new stage [9, 10, 38, 40], but it has been suggested as one of the most efficient reinforcing materials for cementitious composites. With only about 0.003% addition of BN, the compressive strength and tensile strength of OPC were improved by 13% and 8%, respectively. Besides the nucleation effect as indicated by hydration heat, pore structure refinement and chemical bonding were also found as the main reinforcing mechanisms of BN in cement matrix. In contrast, in order to obtain a similar strength improvement, the required dosages are 1-6 wt% for nanosilica, 0.3-0.5 wt% for carbon nanotubes (CNTs), and 0.01-0.3 wt% for GO. The superior reinforcing efficiency of BN results from the good dispersion stability in the pore solution and the interactions between BN and the cement matrix [9]. In another study, the flexural strength and compressive strength of reactive powder concrete (RPC) containing 0.5wt% of BN achieved an increase of about 16% and 13%, respectively, at the age of 28 days [38]. BN, due to the two-dimensional structure and high strength, was observed to play a bridging role in the RPC and hinder the development of cracks (Figure 3c).

Recent literature, shows that addition of 2D nanomaterials also have significanteffects on the durability-related properties of cementitious materials, a major concern in construction industry. Concrete is susceptible to various forms of deterioration, which can significantly reduce its service life. 2D nanomaterials have been employed to improve the resistance of cement composites to aggressive elements (such as CO_2, and Cl^-), calcium leaching, freezing and thawing cycles, thermal cracking and to create self-sensing smart structures. However, the number of studies are still very limitedand further in-depth investigations on themechanisms of the influence of these materials on concrete long-term properties are needed.

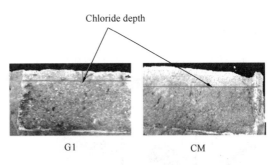

Figure 4 Chloride front of the control mortar (CM) compared to mortar containing 0.01 % GO (G1). Chloride penetration was approximately 26mm for CM, whereas only 5mm for G1 [30].

Concrete durability to aggressive agents is mainly governed by its pore structure and permeability. Recent research shows that incorporating a small content of graphene materials can increase the durability of concrete by optimizing the pore structure and altering the transport properties of the cement matrix [30, 31, 41-45]. Mohammed et al. [30] reported that mortars with 0.01%wt of GO presented a five-fold reduction in chloride penetration compared to plain mortars (Figure 4). Furthermore, addition of GO in mortars decreased their initial water sorptivity ratio. It was suggested that the layered structures of GO may also trap and reduce the chloride ingression depth. More recently, aremarkable decrease in water permeability by nearly 400% compared to reference concrete was achieved by the addition of water-stabilized graphene dispersions [36]. An improvement of 100% has been reported in the chloride penetration resistance of RPC containing appropriate amount of BN [38].

The freeze-thaw resistance of concrete can also be increased by incorporation of 2D nanomaterials. Mohammed et al. [46] observed that mortars containing 0.06% GO presented a weight loss of about 0.25% after 540 freeze-thaw cycles compared to about 0.8% for the reference sample. Cement mixes with GO also displayed less scaling damages on the specimen's surfaces. In addition to the higher compressive strength, the enhancement on freeze-thaw resistance of mixes with GO was associated to the increase of nanopores over mesopores, as lower temperatures are required to freeze water in smaller pores [47]. The generated nano-pores also created rooms for osmotic pressure release, hindering the propagation of microcrackand limiting the frost damage.

4. Enhancement Mechanisms

The reinforcement mechanisms of the 2D nanomaterials, especially GO, on cementitious materials have been proposed in several studies. A hypothesis, proposed by Duan et al. [34], for such efficient reinforcement of GO is that the nanosheets have significant seeding effect for cement hydration kinetics, providing abundant additional sites for nucleation and growth of hydration products. Other researchers [48, 49] proposed that the enhanced toughness of cement composites reinforced with GO is due to a possible template effect of the nanosheets. A space interlocking mechanism was also proposed by Lu et al. [50] for the GO/CNTs/cement paste system.

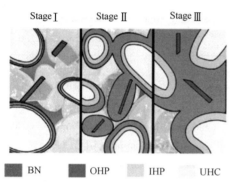

Figure 5 Schematic drawing showing the potential effect of BN nucleation on the microstructure of cement paste. BN: boron nitride nanosheet; OHP: outer hydration products; IHP: inner hydration product; UHC: un hydrated cement [9].

For BN nanomaterials, Zhang et al. [38] attributed their strengthening mechanisms to nucleation, filling, bridging, lubrication and layered blocking effects. Through nucleation effect, BN not only accelerate the early cement hydration and promote more C-S-H formation, but also fill pores and prevent the growth of CH crystals, thus resulting in a denser microstructure. Furthermore, the superior mechanical and thermal properties of BN endow cement composites with high strength, toughness and durability [38]. A schematics for the nucleation effect of BN nanosheets in OPC materials is shown in Figure 5.

Although the preceding studies provide great insight of the reinforcement mechanism of GO on the micro to macroscopic scale properties of GO-cement composites, the understanding of how GO interact and modify the cement matrix at nanoscale (1 to 100 nm), in particular its influence on the C-S-H nanostructure, is still limited. Studies at nanometre levelare often hampered by the complexity associated to studying cement at this scale. Due to the insufficient evidential experimental results, the most recent efforts to resolve the interfacial molecular structure and bonding characteristics between 2D nanomaterials and C-S-H are based on computer modelling [51]. Chen et al. [52] used MD to simulate the pull-out behaviour of graphene in a C-S-H matrix. It was reported that graphene's pull-out

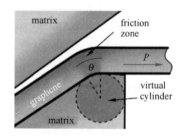

Figure 6 Illustration of snubbing effect during pulling out graphene from matrix at a specific angle [55].

from the matrix, differently from that generally observed for microfibres [53, 54], is not governed by friction only. Due to its 2D geometry, graphene presents an additional constant resistance force risen from the unbalanced adhesion forces near the crack surface. Chen et al. [55] also investigated the snubbing friction between graphene and the contacted matrix (Figure 5). It was found that the snubbing friction magnitude increases with the snubbing angle, friction coefficient, and tension. This particular friction is thus another important behaviour to be considered for designing 2D nanosheets-cement composites.

5. Challenges

Some challenges are associated with the synthesis of 2D nanomaterials. The fabrication of 2D nanomaterials is still a recent technology and their production scale is thus small and relatively expensive. Although the production of GO have been established [56] and these products are a commercial reality, there have been only few bulk quantity methods proposed for production of BN nanosheets [57]. BN can be exfoliated from h-BN by either chemical or physical approaches. In the chemical approach, functional groups are covalently attached to the surface of individual layers to overcome van der Waals attraction. The extensive chemical treatment usually introduces a significant number of surface defects that can degrade the mechanical and chemical stability of the nanostructures. On the other hand, physical approaches utilize direct peeling, ball milling, or ultrasonication to exfoliate nanosheets from their bulk sources. However, direct peeling is not suitable to produce the nanosheets in large quantities. Ball-milling techniques exfoliate the layers by applying shear forces and it also introduces a significant number of surface defects. Ultrasonication can fabricate large scale nanosheets, but it often needs harsher operating conditions and the use of organic solvents [9].

Ensuring the good dispersion of 2D nanomaterials in the cement matrix isanother challenge due to the high surface energy and strong Van der Waals interactions between nanosheets (Figure 6). Any agglomeration of nanomaterials in the pore solution decreases the reinforcing efficiency due to the lower specific surface area [58]. Furthermore, stress concentration and introduction of pores from agglomerated nanomaterials also compromises the material's mechanical strength. Some researchers have recently attempted, with relative success, to improve the dispersion of nanomaterials in cement matrix by either mechanical (e.g. silica fume) or steric separation (e.g. superplasticizers)

Figure 7 (a) GO in aqueous solution; (b) cement pore solution and; (c) GO aggregates in cement pore solution [62].

of the nanosheets [59-62].

The negative impact on the workability (rheological behaviour) is also a key challenge of incorporating 2D nanomaterials to cement composites [63]. The flocculation between GO and cement grains by electrostatic interactions can entrap large amount of free water and thus decrease the fluidity of the mixture. Utilizing fly ash or silica fume was found to be useful to offset the reduction on the workability of GO cement pastes [64, 65].

6. Conclusions

2D nanomaterials including graphene materials and BN applied in construction composites is currently an active area of research. These nanomaterials have demonstrated the capability to transform conventional construction materials into smarter, stronger and more durable advanced composites. The mechanical reinforcement mechanisms are often associated to the nanomaterials' nucleation effects on cement hydration and capability to interlock different hydration products. On the other hand, the improvements on durability, in particular on permeability to aggressive agents and freeze thaw resistance, are attributed to the nanomaterials' ability in tuning the matrix porosity, reducing the number of larger pores while increasing that of nanopores.

However, the use of 2D nanomaterials have not become widespread in the construction field yet. Before this can happen, challenges that still hinder their implementation in the construction materials need to be resolved. One of the main challenges is associated to difficulties regarding the synthesis of high quality and low cost 2D nanomaterials at industrial scale. Other challenges are difficulties of uniform dispersion of nanomaterials in the cement matrix and the negative effect of the nanomaterials on the workability of fresh cement mixes. Further understanding on the 2D nanomaterials nanoscale interaction with cement matrix are required to provide answers for the current problems.

Acknowledgements

The authors are grateful for the financial support of the Australian Research Council for conducting this study.

References

1. Sanchez, F. and K. Sobolev, *Nanotechnology in concrete-a review*. Construction and building materials, 2010. **24**(11): p. 2060-2071.
2. Lee, J., S. Mahendra, and P. J. Alvarez, *Nanomaterials in the construction industry: a review of their applications and environmental health and safety considerations*. ACS nano, 2010. **4**(7): p. 3580-3590.
3. Hanus, M. J. and A. T. Harris, *Nanotechnology innovations for the construction industry*. Progress in materials science, 2013. **58**(7): p. 1056-1102.
4. Lu, S.-N., et al., *Applications of nanostructured carbon materials in constructions: The state of the*

art. Journal of Nanomaterials, 2015. **2015**: p. 6.

5. Han, B., et al., *Review of nanocarbon-engineered multifunctional cementitious composites*. Composites Part A: Applied Science and Manufacturing, 2015. **70**: p. 69-81.

6. Shamsaei, E., et al., *Graphene-based nanosheets for stronger and more durable concrete: A review*. Construction and Building Materials, 2018. **183**: p. 642-660.

7. Li, D. and R. B. Kaner, *Graphene-based materials*. Nat Nanotechnol, 2008. **3**: p. 101.

8. Stankovich, S., et al., *Graphene-based composite materials*. nature, 2006. **442**(7100): p. 282-286.

9. Wang, W., et al., *Exfoliation and dispersion of boron nitride nanosheets to enhance ordinary Portland cement paste*. Nanoscale, 2018.

10. Rafiee, M. A., et al., *Hexagonal boron nitride and graphite oxide reinforced multifunctional porous cement composites*. Advanced Functional Materials, 2013. **23**(45): p. 5624-5630.

11. Ribeiro, H., et al., *Hybrid 2D nanostructures for mechanical reinforcement and thermal conductivity enhancement in polymer composites*. Composites Science and Technology, 2018. **159**: p. 103-110.

12. Zhang, H., *Ultrathin Two-Dimensional Nanomaterials*. ACS Nano, 2015. **9**(10): p. 9451-9469.

13. Tan, C., et al., *Recent advances in ultrathin two-dimensional nanomaterials*. Chemical reviews, 2017. **117**(9): p. 6225-6331.

14. Tan, C., et al., *Recent Advances in Ultrathin Two-Dimensional Nanomaterials*. Chemical reviews, 2017.

15. Novoselov, K. S., et al., *Electric field effect in atomically thin carbon films*. science, 2004. **306**(5696): p. 666-669.

16. Compton, O. C. and S. T. Nguyen, *Graphene oxide, highly reduced graphene oxide, and graphene: versatile building blocks for carbon - based materials*. small, 2010. **6**(6): p. 711-723.

17. McAllister, M. J., et al., *Single sheet functionalized graphene by oxidation and thermal expansion of graphite*. Chemistry of materials, 2007. **19**(18): p. 4396-4404.

18. Giménez-Pérez, A., et al., *Synthesis of N-doped and non-doped partially oxidised graphene membranes supported over ceramic materials*. Journal of materials science, 2016. **51**(18): p. 8346-8360.

19. Lee, C., et al., *Measurement of the elastic properties and intrinsic strength of monolayer graphene*. science, 2008. **321**(5887): p. 385-388.

20. Falin, A., et al., *Mechanical properties of atomically thin boron nitride and the role of interlayer interactions*. Nature Communications, 2017. **8**: p. 15815.

21. Tiwari, A., *Graphene materials: fundamentals and emerging applications*. 2015: John Wiley & Sons.

22. Wu, X.-f., et al., *Boron Nitride Nanoparticles with High Specific Surface Area: Preparation by a Calcination Method and Application in Epoxy Resin*. Journal of Inorganic and Organometallic Polymers and Materials, 2017. **27**(5): p. 1142-1147.

23. Marinho, B., et al., *Electrical conductivity of compacts of graphene, multi-wall carbon nanotubes, carbon black, and graphite powder*. Powder technology, 2012. **221**: p. 351-358.

24. Lin, Y., T. V. Williams, and J. W. Connell, *Soluble, exfoliated hexagonal boron nitride nanosheets*. The Journal of Physical Chemistry Letters, 2009. 1(1): p. 277-283.

25. Balandin, A. A., et al., *Superior thermal conductivity of single-layer graphene*. Nano letters, 2008. **8**(3): p. 902-907.

26. Chen, S., et al., *Thermal conductivity of isotopically modified graphene*. Nature materials, 2012. **11**(3): p. 203.

27. Ouyang, T., et al., *Thermal transport in hexagonal boron nitride nanoribbons*. Nanotechnology, 2010. **21**(24): p. 245701.
28. Yu, Z., et al., *Fabrication of graphene oxide-alumina hybrids to reinforce the anti-corrosion performance of composite epoxy coatings*. Applied Surface Science, 2015. **351**: p. 986-996.
29. Ramezanzadeh, B., et al., *Enhancement of barrier and corrosion protection performance of an epoxy coating through wet transfer of amino functionalized graphene oxide*. Corrosion Science, 2016. **103**: p. 283-304.
30. Mohammed, A., et al., *Incorporating graphene oxide in cement composites: A study of transport properties*. Construction and Building Materials, 2015. **84**: p. 341-347.
31. Du, H., H. J. Gao, and S. Dai Pang, *Improvement in concrete resistance against water and chloride ingress by adding graphene nanoplatelet*. Cement and Concrete Research, 2016. **83**: p. 114-123.
32. Ribeiro, H., et al., *Hybrid 2D nanostructures for mechanical reinforcement and thermal conductivity enhancement in polymer composites*. Composites Science and Technology, 2018. **159**: p. 103-110.
33. Lin, Y. and J. W. Connell, *Advances in 2D boron nitride nanostructures: nanosheets, nanoribbons, nanomeshes, and hybrids with graphene*. Nanoscale, 2012. **4**(22): p. 6908-6939.
34. Pan, Z., et al., *Mechanical properties and microstructure of a graphene oxide-cement composite*. Cement and Concrete Composites, 2015. **58**: p. 140-147.
35. Pan, Z., *Graphene oxide reinforced cement and concrete*. 2012, WO Patent App. PCT/AU2012/001, 582.
36. Dimov, D., et al., *Ultrahigh Performance Nanoengineered Graphene-Concrete Composites for Multifunctional Applications*. Advanced Functional Materials, 2018: p. 1705183.
37. Li, G. Y., P. M. Wang, and X. Zhao, *Mechanical behavior and microstructure of cement composites incorporating surface-treated multi-walled carbon nanotubes*. Carbon, 2005. **43**(6): p. 1239-1245.
38. Zhang, W., et al., *Nano boron nitride modified reactive powder concrete*. Construction and Building Materials, 2018. **179**: p. 186-197.
39. Sharma, S., N. Kothiyal, and M. Chitkara, *Enhanced mechanical performance of cement nanocomposite reinforced with graphene oxide synthesized from mechanically milled graphite and its comparison with carbon nanotubes reinforced nanocomposite*. RSC Advances, 2016. **6**(106): p. 103993-104009.
40. He, Q. and X. Yu. *Effect of Boron Nitride on the Mechanical and Electrochemical Properties of Cement Composite*. in *ASME 2015 International Mechanical Engineering Congress and Exposition*. 2015. American Society of Mechanical Engineers.
41. del Carmen Camacho, M., et al., *Mechanical properties and durability of CNT cement composites*. Materials, 2014. **7**(3): p. 1640-1651.
42. Safiuddin, M., et al., *State-of-the-art report on use of nano-materials in concrete*. International Journal of Pavement Engineering, 2014. **15**(10): p. 940-949.
43. Shekari, A. and M. Razzaghi, *Influence of nano particles on durability and mechanical properties of high performance concrete*. Procedia Engineering, 2011. **14**: p. 3036-3041.
44. Beigi, M. H., et al., *An experimental survey on combined effects of fibers and nanosilica on the mechanical, rheological, and durability properties of self-compacting concrete*. Materials & Design, 2013. **50**: p. 1019-1029.
45. Wang, Q., et al., *Influence of graphene oxide additions on the microstructure and mechanical strength of cement*. New Carbon Materials, 2015. **30**(4): p. 349-356.
46. Mohammed, A., et al., *Graphene Oxide Impact on Hardened Cement Expressed in Enhanced Freeze-

Thaw Resistance. Journal of Materials in Civil Engineering, 2016. **28**(9): p. 04016072.

47. Youssef, M., R. J.-M. Pellenq, and B. Yildiz, *Glassy nature of water in an ultraconfining disordered material: the case of calcium-silicate-hydrate*. Journal of the American Chemical Society, 2011. **133**(8): p. 2499-2510.

48. Lv, S., et al., *Effect of graphene oxide nanosheets of microstructure and mechanical properties of cement composites*. Construction and building materials, 2013. **49**: p. 121-127.

49. Cao, M.-l., H.-x. Zhang, and C. Zhang, *Effect of graphene on mechanical properties of cement mortars*. Journal of Central South University, 2016. **23**: p. 919-925.

50. Lu, Z., et al., *Mechanism of cement paste reinforced by graphene oxide/carbon nanotubes composites with enhanced mechanical properties*. RSC Advances, 2015. **5**(122): p. 100598-100605.

51. Sanchez, F. and L. Zhang, *Molecular dynamics modeling of the interface between surface functionalized graphitic structures and calcium-silicate-hydrate: interaction energies, structure, and dynamics*. Journal of colloid and interface science, 2008. **323**(2): p. 349-358.

52. Chen, S. J., et al., *Reinforcing mechanism of graphene at atomic level: Friction, crack surface adhesion and 2D geometry*. Carbon, 2017. **114**: p. 557-565.

53. Chen, S. J., et al., *Predicting the influence of ultrasonication energy on the reinforcing efficiency of carbon nanotubes*. Carbon, 2014. **77**: p. 1-10.

54. C. Li, V., *Postcrack scaling relations for fiber reinforced cementitious composites*. Journal of Materials in Civil Engineering, 1992. **4**(1): p. 41-57.

55. Chen, S. J., et al., *Snubbing effect in atomic scale friction of graphene*. Composites Part B: Engineering, 2018. **136**: p. 119-125.

56. Dimiev, A. M. and J. M. Tour, *Mechanism of graphene oxide formation*. ACS nano, 2014. **8**(3): p. 3060-3068.

57. Li, L. H., et al., *Large-scale mechanical peeling of boron nitride nanosheets by low-energy ball milling*. Journal of materials chemistry, 2011. **21**(32): p. 11862-11866.

58. Korayem, A., et al., *A review of dispersion of nanoparticles in cementitious matrices: Nanoparticle geometry perspective*. Construction and Building Materials, 2017. **153**: p. 346-357.

59. Li, X., et al., *Incorporation of graphene oxide and silica fume into cement paste: A study of dispersion and compressive strength*. Construction and Building Materials, 2016. **123**: p. 327-335.

60. Bai, S., et al., *Enhancement of mechanical and electrical properties of graphene/cement composite due to improved dispersion of graphene by addition of silica fume*. Construction and Building Materials, 2018. **164**: p. 433-441.

61. Chuah, S., et al., *Investigation on dispersion of graphene oxide in cement composite using different surfactant treatments*. Construction and Building Materials, 2018. **161**: p. 519-527.

62. Zhao, L., et al., *Investigation of dispersion behavior of GO modified by different water reducing agents in cement pore solution*. Carbon, 2018. **127**: p. 255-269.

63. Lv, S., et al., *Fabrication of polycarboxylate/graphene oxide nanosheet composites by copolymerization for reinforcing and toughening cement composites*. Cement and Concrete Composites, 2016. **66**: p. 1-9.

64. Shang, Y., et al., *Effect of graphene oxide on the rheological properties of cement pastes*. Construction and Building Materials, 2015. **96**: p. 20-28.

65. Wang, Q., et al., *Effect of fly ash on rheological properties of graphene oxide cement paste*. Construction and Building Materials, 2017. **138**: p. 35-44.

4 Sustainable Concrete For The Next Century: Multi-recycled Aggregate Concrete

Jorge de Brito[1], Luís Evangelista[2,3], Rui V. Silva[4]

1. CERIS-ICIST, Instituto Superior Técnico, Universidade de Lisboa, Av. Rovisco Pais, 1049-001 Lisboa, Portugal; e-mail: jb@civil.ist.utl.pt
2. CERIS-ICIST, Instituto Superior Técnico, Universidade de Lisboa, Av. Rovisco Pais, 1049-001 Lisboa, Portugal
3. Assistant Professor, ISEL, Instituto Superior de Engenharia de Lisboa, R. Conselheiro Emídio Navarro, 1950-062 Lisboa, Portugal; e-mail: evangelista@dec.isel.pt
4. CERIS-ICIST, Instituto Superior Técnico, Universidade de Lisboa, Av. Rovisco Pais, 1049-001 Lisboa, Portugal; e-mail: rui.v.silva@tecnico.ulisboa.pt

Abstract: This study presents the results of a laboratory investigation on the influence of incorporating coarse recycled concrete aggregates obtained from concrete subjected to up to three cycles of recycling on the mechanical and durability properties of concrete. The coarse natural aggregate fraction was partially (25% by volume) and completely replaced (100% by volume) with coarse recycled concrete aggregates. Aggregate characterization tests include particle size distribution analysis, water absorption, particle density, shape index, and Los Angeles abrasion. Fresh and hardened concrete tests include fresh density, slump, compressive strength, splitting tensile strength, modulus of elasticity, resistance to abrasion, shrinkage, water absorption by immersion and capillary action, carbonation and chloride ion penetration. The results showed that the increasing amount of mortar adhered to recycled concrete aggregates, as the number of recycling cycles increased, resulted in a decline in the mechanical and durability properties of concrete. Nevertheless, since this performance loss stabilized with the aggregate's mortar content and physical properties, there was an asymptotic-like trend in most of the concrete's properties, thereby presenting a predictable decline in performance that can facilitate the design and application of recycled concrete aggregates in structural concrete manufacture.

Keywords: recycled aggregates; concrete; multiple recycling; mechanical; durability

1. Introduction

As people are becoming more aware of the increasing consumption of the World's natural resources, increasing concerns have been raised to maintain them for future generations. The construction industry is one of the economic sectors with the highest accountability in the consumption of natural resources and generation of construction and demolition waste (CDW), representing a large quantity of the total amount of waste generated in the European Union (Eurostat, 2017).

Within CDW, the most common type of waste is concrete (Mália et al., 2013) and thus greater attention has been focused towards its recycling. One of the practical and cost-effective ways of using waste concrete is by processing it into aggregates and use them in new concrete, which significantly reduces the amount of consumed natural resources. Many investigations about the properties of coarse recycled concrete aggregates (RCA) and the behaviour of recycled aggregate concrete (RAC) have been carried out (de Brito and Silva, 2016; Silva et al., 2017). However, the literature shows that there is still a gap concerning the application of RCA sourced from a concrete that has been subjected to several cycles of recycling. Therefore, this study presents the results of an experimental campaign on the physical properties of coarse RCA from different cycles and their influence on the mechanical and durability performance of the multi-recycled RAC.

2. Experimental Campaign

2.1 Materials

Siliceous sand, with size up to 4 mm, was used for the fine fraction. Coarse limestone natural aggregates (NA) were used. These were comprised of fine gravel (4-8 mm), medium gravel (5.6-11.2 mm) and coarse gravel (11.2-22.4 mm). Coarse RCA were sourced from crushed concrete elements produced at the laboratory, with the same mix design as the test mixes. Class 42.5 R CEM I cement and tap water were used.

2.2 Mix design

A total of seven mixes were formulated, with the same mix design (Table 1). The specimens were designed and produced according to EN-206 (2013) and LNEC-E464 (2007), using the Faury's method. All coarse aggregates were sieved and separated in their size fractions to eliminate any variation in particle size distribution and to fit the Faury's reference curve. The slump was kept constant in all mixes (125 ± 15 mm). This was made possible by the application of the water compensation method, which increased the total w/c ratio while maintaining the effective w/c ratio (Ferreira et al., 2011). The control concrete (CC), made entirely with NA, was the first mix to be produced and led to the first RCA (RCA1). Subsequently, 1RAC25 and 1RAC100 were produced with RCA1 at replacement ratios of 25% and 100%, respectively. In the second cycle, crushing 1RAC100 specimens produced RCA2, which was used to produce 2RAC25 and 2RAC100. The same occurred in the third generation of concrete.

	Concrete composition						**Table 1**
Components	CC	1RAC25	1RAC100	2RAC25	2RAC100	3RAC25	3RAC100
Coarse NA	975.4	731.7	—	731.7	—	731.7	—
Coarse RCA	—	217.6	871.6	203.7	816.2	185.1	797.4

(continued)

Components	CC	1RAC25	1RAC100	2RAC25	2RAC100	3RAC25	3RAC100
Fine sand				250.7			
Coarse sand				472.4			
Cement				350.0			
Water				193.6			
Compensation water	—	40.3	10.1	48.6	12.1	54.9	13.7
w/c_{eff}				0.55			
Total w/c	0.55	0.58	0.67	0.59	0.69	0.59	0.71

2.3 Test Methods

The particle size distribution of fine NA was tested in accordance with EN-933-1 (2012), whereas coarse aggregates were evaluated according to their water absorption (EN-1097-5, 2008), bulk density (EN-1097-3, 1998), Los Angeles abrasion (EN-1097-1, 2011) and shape index (EN-933-4, 2008). Slump and density of fresh concrete were determined in accordance with EN-12350-2 (2009) and EN-12350-6 (2009), respectively. Hardened concrete specimens were evaluated according to their compressive strength (EN-12390-3, 2009), splitting tensile strength (EN-12390-6, 2009), modulus of elasticity (EN-12390-13, 2013), resistance to abrasion (DIN-52108, 2010), shrinkage (LNEC-E398, 1993), water absorption by immersion and by capillary action (LNEC-E394 (1993) and LNEC-E393 (1993), respectively), carbonation (LNEC-E391, 1993) and chloride ion penetration (NT-Build-492, 1999).

3. Results and Discussion

This section contains the description and analysis to the results obtained throughout the experimental campaign. The changes in the aggregates' properties when subjected to a multiple recycling process and the effect of their incorporation are studied in the mechanical and durability related performance of concrete. For a better understanding of the results, this section is divided in four sub-sections. Firstly, the results concerning aggregates' characterization tests are presented. Secondly, the properties of concrete in the fresh state, whereas the third and fourth subsections are related to the study of the mechanical and durability properties of concrete in their hardened state, respectively.

3.1 Aggregate Properties

The density and water absorption of all aggregates were determined in accordance with EN-1097-6 (2013). The results are presented in Table 2. For NA, the density and water absorption values were close to those disclosed by the supplier, between 2640 kg/m³ and 2770 kg/m³, and between 0.7 e 1.9%, respectively. The LNEC-E471 (2009) specification states that RA must have a dry density of more than 2200 kg/m³ and a water absorption of less than 7% in order to be used in concrete production. Only RCA1 met the requirements of that specification.

Density and water absorption of aggregates   Table 2

Aggregate type	Apparent density-ρ_a (kg/m³)	Rodded dry density-ρ_{rd} (kg/m³)	Saturated and surface dry density- ρ_{ssd} (kg/m³)	24-h water absorption- WA_{24} (%)
NA	2668	2593	2621	1.09
RCA1	2668	2319	2450	5.64
RCA2	2630	2175	2348	7.95
RCA3	2672	2125	2330	9.64

The results in Table 2 show that all RCA present rodded dry (ρ_{rd}) and saturated and surface-dry (ρ_{ssd}) density values lower than those of NA. These values decrease with increasing number of recycling cycles. This is due to the increasing amount of old mortar adhered to the RCA since it is less dense than NA. This can be clearly seen in Figures 1 to 4, wherein the amount of mortar adhered increases as the number of recycling cycles increases.

Figure 1 NA   Figure 2 RCA1

Figure 3 RCA2   Figure 4 RCA3

Concerning the aggregates' water absorption, all RCA presented higher values than those of NA, the extent of which increased with the number of recycling cycles. This property is also well correlated with the material's higher density, due to the old mortar' s greater porosity. The increase in water uptake by RCA was noticeable after the first cycle (4.55% increase), decreasing in the following cycles (up by 2.31% after the second cycle and 1.69% after the third cycle). In other words, as the number of recycling cycles increases, RCA absorb more water, but at an asymptotic rate (Figure 5). The same occurs in the RCA's rodded dry density and saturated and surface dry density; after each cycle, their values decrease asymptotically (Figures 6 and Figure 7). Based on the data obtained in the 24-hour water absorption tests, the rodded dry density and the saturated and surface-dry density, nonlinear regressions can be obtained using an exponential asymptotic model with very strong coefficients of correlation (R^2 values over 0.99).

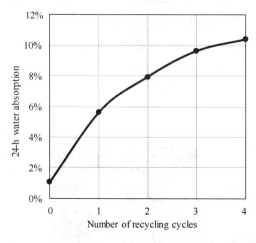

Figure 5 24-hour water absorption of aggregates with increasing number of recycling cycles

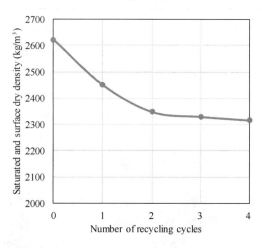

Figure 6 Saturated and surface dry density of aggregates with increasing number of recycling cycles

Because of the RCA's considerably higher water absorption, this property needs to be considered in the concrete's mix design, so that the workability of the resulting mixes is the same as that of the control mixes. For this reason, the RCA's water absorption over time was evaluated, the results of which can be seen in Figure 8. The results of the water absorption for RCA1 are in accordance with those typically seen in the literature for coarse RCA, wherein most of the total water absorption capacity occurs after about five minutes from the beginning of the test. Comparing with the results of the two other aggregates, during the first minutes, RCA1 absorbed water much faster than RCA2 and RCA3. However, after about 30 minutes, all RCA behaved similar-

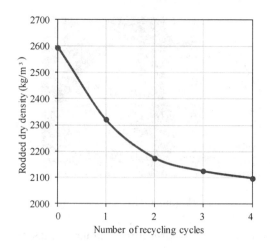

Figure 7 Rodded dry density of aggregates with increasing number of recycling cycles

ly, absorbing water slowly and gradually. An analysis to the 10-min water absorption of coarse RCA showed that RCA1 absorbed around 80% of its 24-hour absorption capacity, whereas, RCA2 and RCA3 did around 65% and 60%, respectively. This behaviour can be observed in fine RCA, which also present higher water absorption values. Again, this can be explained by the progressively higher content of porous adhered mortar with increasing number of recycling cycles.

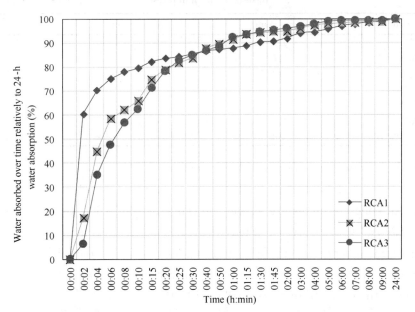

Figure 8　Water absorption of aggregates over time

Concerning the RCA's resistance to fragmentation, these always present greater mass loss than NA (Table 3). A fast stabilization of this property can be observed after the second cycle. This may indicate that the increment of adhered mortar for RCA3 is so small that it becomes insignificant for this test in comparison with the two previous RCA with lower adhered mortar content. By applying the asymptotic exponential nonlinear regression model, the values presented in the second (41.2%) and third (40.9%) cycles are within the estimated range for the final value (41.23±0.41%), thereby suggesting a stabilization of this property after the second recycling cycle.

Resistance to fragmentation and shape index of aggregates　　　　Table 3

Property	NA	RCA1	RCA2	RCA3
Los Angeles wear (%)	27.9	38.8	41.2	40.9
Shape index (%)	19	18	19	18

3.2　Fresh concrete properties

The slump test was performed according to EN-12350-2 (2009). The workability of all concrete mixes had been established as a fixed parameter in the range of 125 ± 5 mm belonging to class S3 (100 mm-150 mm) of EN-206 (2013). To prevent stiff unworkable mixes due to high water absorption of RCA, compensating water was added during the mixing process, the amount of which was higher as the number of recycling cycles increased. Naturally, this led to an increase of the mixes' total w/c ratio, whilst the effective w/c ratio was maintained as estimated by the constant slump values (Table 4).

Results of the slump test   Table 4

Concrete	Slump (mm)	Effective w/c ratio	Total w/c ratio
CC	125	0.55	0.55
1RAC100	126	0.55	0.67
1RAC25	126	0.55	0.58
2RAC100	121	0.55	0.69
2RAC25	120	0.55	0.59
3RAC100	129	0.55	0.71
3RAC25	120	0.55	0.59

The fresh density of concrete was determined in accordance with EN-12350-6 (2009). This test was carried out immediately after the mixing process and the results can be seen in Table 5. All concrete mixes are within the range of 2000-2600 kg/m^3 for normal concrete as per EN-206 (2013). As expected, the density of fresh concrete decreased with increasing replacement ratios and more so with higher number of recycling cycles. This occurs as a result of the RCA's lower density when compared to that of their natural counterparts, due to the cementitious paste adhered to the former.

Results of the density test   Table 5

Concrete	Density (kg/m^3)	Relative difference to CC (%)
CC	2414	—
1RAC100	2319	−3.91
1RAC25	2385	−1.18
2RAC100	2280	−5.55
2RAC25	2384	−1.24
3RAC100	2265	−6.17
3RAC25	2376	−1.57

3.3 Mechanical performance

This section presents the test results of the mechanical properties of all hardened concrete mixes. The following properties were evaluated: compressive strength, splitting tensile strength, modulus of elasticity and resistance to abrasion. A comparative analysis was carried out between the mechanical properties of multi-recycled concrete mixes and those of the control concrete mixes.

3.3.1 Compressive strength

The compressive strength test, which was made in accordance with EN-12390-3 (2009), was carried out at 7, 28 and 56 days. In this study, exclusively for the 28-day compressive strength, the 3RAC100 specimens were crushed to obtain RCA4, which were subsequently used to produce a 4th generation of RAC (4RAC100). This mix was produced with the same composition as that of the previous generations. The average values of the compressive strength for different curing ages ($f_{cm\,7}$, $f_{cm\,28}$, $f_{cm\,56}$), and the variation thereof in relation to CC (V_{CC}) are given in Table 6. As expected, the compressive strength decreased linearly with increasing replacement ratios of NA with RCA. This is clearer after the second and third cycles of recycling (where the coefficients of correlations are closer to 1.0). From Table 6, a clear reduction in compressive strength can also be observed with increasing number of recycling cycles. By applying an exponential asymptotic model, high coefficients of correlation were obtained and it was possible to establish that the compressive strength will probably have a maximum loss of about 13% (Table 7). The incorporation of 25% RCA, in mixes 1RAC25, 2RAC25 and 3RAC25, caused only slight variation when compared with the control concrete, suggesting little influence of the recycling process on the material's performance. This is probably due to the small quantity of recycled materials in the overall concrete volume. Nevertheless, a small decline in performance was observed in mixes containing RCA subjected to a greater number of recycling cycles (Figure 9).

Results of the compressive strength test　　　　　　　　　　　　　　　　Table 6

Concrete	$f_{cm\,7}$ (MPa)	V_{CC} (%)	$f_{cm\,28}$ (MPa)	V_{CC} (%)	$f_{cm\,56}$ (MPa)	V_{CC} (%)
CC	46.2	—	55.9	—	63.8	—
1RAC100	44.0	−4.8	54.1	−3.2	59.0	−7.5
1RAC25	47.6	3.0	59.7	6.8	65.0	1.9
2RAC100	43.3	−6.3	53.3	−4.7	57.6	−9.7
2RAC25	47.0	1.7	55.9	0.0	60.7	−4.9
3RAC100	40.3	−12.8	48.6	−13.1	56.2	−11.9
3RAC25	45.2	−2.1	55.9	0.0	62.7	−1.7
4RAC100	—	—	47.8	−14.5	—	—

Estimation of the final compressive strength for each of the concrete mixes　　　Table 7

Variable	CC	1RAC100	1RAC25	2RAC100	2RAC25	3RAC100	3RAC25
f_{cm7}	46.2	44.0	47.6	43.3	47.0	40.3	45.2
f_{cm28}	55.9	54.1	59.7	53.3	55.9	48.6	55.9
f_{cm56}	63.8	59.0	65.0	57.6	60.7	56.2	62.7
f_{cm} final calculated	74.3	61.3	67.1	59.3	63.4	70.4	67.9
R^2	1	0.99995	0.99993	0.99992	0.99996	1	0.99998

3.3.2 Splitting tensile strength

The splitting tensile strength was determined in accordance with EN-12390-6 (2009). The values of this property are presented in Table 8. For concrete specimens with 100% coarse RCA, there was a noticeable decrease in splitting tensile strength from CC to 1RAC100 (from 4.18 MPa to 3.80 MPa), and less significant from 1RAC100 to 2RAC100 and 3RAC100 (3.70 MPa and 3.55 MPa, respectively). Applying the asymptotic expo-

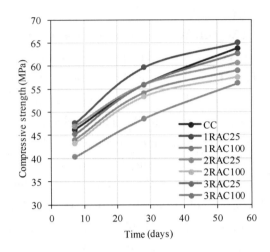

Figure 9 Evolution of compressive strength over time

nential method to this property, the existence of a horizontal asymptote can be observed with a high coefficient of correlation ($R^2 = 0.977$). This behaviour can be justified by the fact that the quality of the RCA decreases with increasing number of cycles of recycling. This results in a higher number of interfacial transition zones between the new mortar and the old adhered mortar from previous cycles. Apart from this causing more fragile sections wherein the rupture may occur, the progressively higher porosity of the RCA also leads to a reduction of the materials' resistance to loading. By applying the exponential asymptotic model, it was established that this property stabilized after the third cycle of recycling. Considering a much higher number of cycles of recycling, a maximum loss of 19.4% is expected.

Results of the splitting tensile strength test Table 8

Concrete	Splitting tensile strength (MPa)	V_{CC} (%)
CC	4.18	—
1RAC100	3.80	−9.1
1RAC25	4.00	−4.2
2RAC100	3.70	−11.4
2RAC25	3.70	−11.4
3RAC100	3.55	−15.1
3RAC25	3.40	−18.6

Concerning the RAC specimens containing 25% coarse RCA, contrary to what was expected, it was perceived that there was a linear decrease in splitting tensile strength with increasing number of recycling cycles, with a coefficient of correlation of 0.9879. This behaviour suggests that the decreasing quality of RCA with increasing number of recycling cycles has a greater relative impact on the splitting tensile strength in lower (25%) rather than in higher (100%) replacement ratios. Except for the first recycling cycle, the splitting tensile strength of concrete specimens with replacement ratio of RCA of 25% is equal to or less than that of specimens with 100% coarse RCA.

In spite of the splitting tensile strength losses after three cycles of recycling (the splitting tensile strength was 10%, 12% and 14% lower for mixes with 100% RCA after the three consecutive cycles), Figure 10 presents the relationship between the mean tensile strength (f_{ctm}) and characteristic compressive strength (f_{ck}). The results suggest that all mixes follow more or less the same relationship as that proposed by Eurocode 2 (EN-1992-

1-1, 2008) for conventional concrete. These results are in line with those presented in the study of Silva et al. (2015a).

3.3.3 Modulus of elasticity

The determination of the concrete specimens' modulus of elasticity was performed in accordance with EN-12390-13 (2013), after 28 days of curing. Table 9 presents the results of this test and its variation in relation to CC (V_{CC}). As expected, the modulus of elasticity was lower in all RAC specimens and proportional to the replacement ratio of NA with RCA. A maxi-

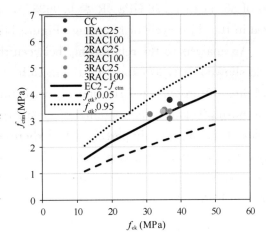

Figure 10 Relationship between tensile strength and compressive strength of concrete

mum reduction of 32.3% was recorded in this property, corresponding to the complete replacement of NA with RCA3. The aggregates ability to withstand deformations is controlled by their stiffness, which in turn is influenced by density. Since the latter increases with increasing number of recycling cycles, then RCA with lower stiffness are expected.

Results of the modulus of elasticity test   Table 9

Concrete	Modulus of elasticity (GPa)	V_{CC} (%)
CC	45.4	—
1RAC100	34.5	−24.0
1RAC25	41.2	−9.2
2RAC100	30.6	−32.7
2RAC25	41.0	−9.7
3RAC100	30.7	−32.3
3RAC25	39.5	−13.0

In concrete specimens with 100% RCA, there was a notable decrease in elastic modulus from CC to 1RAC100 (24.0%), lower from 1RAC100 to 2RAC100 (11.3%) and there was a marginal difference from 2RAC100 to 3RAC100, suggesting stabilization of the property from the second recycling cycle on. Applying the asymptotic exponential model, it was perceived that the asymptote was in the range of 29.17 GPa to 30.83 GPa and presented high coefficients of correlation ($R^2 = 0.99636$), thus confirming the stabilization of the property after the second recycling cycle.

For concrete specimens with 25% RCA, a marked decrease of the 1RAC25 mix's modulus of elasticity was observed when compared with the CC (9.2%), whereas negligible differences were observed for specimens made with RCA2 and RCA3. Using the asymptotic exponential model, the existence of a horizontal asymptote was observed for values of 38.44 to 40.76 GPa ($R^2 = 0.96337$), which indicates that the property only stabilized in the third cycle of recycling (Figure 11).

An analysis to the relationship between the modulus of elasticity and the replacement ratio showed that, as expected, the moduli of elasticity were lower for RAC, progressing in a directly proportional relationship with the replacement ratio. Figure 12 presents the relationship between the modulus of elasticity and compressive strength of the evaluated mixes. It is immediately perceived that there is a considerable drop in the modulus of elasticity for the same compressive strength. This is due to the greater deformability of RCA in comparison with NA. Nevertheless, the values for RAC specimens are within those proposed in Eurocode 2 for structural concrete. These findings are in line with those presented by Silva et al. (2016).

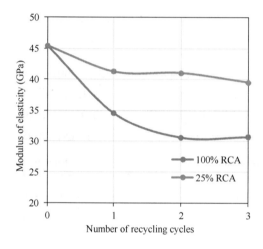

Figure 11 Development of the modulus of elasticity of concrete with increasing number of recycling cycles

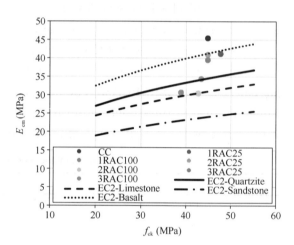

Figure 12 Relationship between the modulus of elasticity and compressive strength

3.3.4 Abrasion resistance

The resistance to abrasion test was carried out in accordance with DIN-52108 (2010). The mass losses due to abrasion and the variation in relation to the CC (V_{CC}) are presented in Table 10. The results show that, with the exception of the specimens obtained using the RCA from the third recycling cycle (where 3RAC25 showed a higher wear than 3RAC100), the mass loss by abrasion increased with increasing replacement ratios and more so if RCA were obtained from concrete subjected to several cycles of recycling. In concrete specimens with 100% RCA, a significant increase in abrasion wear from CC to 1RAC100 was observed (26.16%) but appeared to stabilize after the first cycle of recycling. The asymptotic exponential model shows an excellent correlation with a R^2 close to 1, suggesting that there is stabilization of the property at the value of 3.00 mm. For specimens with replacement ratio of 25%, there was a linear decrease ($R^2 = 0.8996$) in the abrasion resistance with the increasing number of recycling cycles. The resistance to abrasion generally increases with the aggregates' hardness and tenacity. It also depends on the wear resistance of the hardened cement mortar adhered to the RCA. Even though the bond between the new cementitious matrix and the RCA is more resistant than that observed in the conventional concrete, RCA present a lower wear resistance than NA. As previously shown, the RCA's resistance to fragmentation decreases as it is progressively recycled. However, after each cycle, their porosity increases, which may improve the bond strength between the aggregates and the new cement mortar. The combination of these two factors may justify the stabilization of this property after the first recycling cycle.

Results of the resistance to abrasion test   Table 10

Concrete	Mass loss (mm)	V_{CC} (%)
CC	2.37	—
1RAC100	2.99	26.16
1RAC25	2.70	13.92
2RAC100	3.00	26.58
2RAC25	2.70	13.92
3RAC100	3.01	27.00
3RAC25	3.05	28.69

3.4 Durability Performance

This section presents the results of the durability tests carried out on hardened concrete specimens. The following properties were evaluated: water absorption by immersion, water absorption by capillary action, shrinkage, resistance to carbonation and resistance to chloride ion penetration. A comparison of the results was made between the various types of concrete produced in this study.

3.4.1 Shrinkage

The evaluation of the dimensional variation of concrete due to shrinkage was carried out in accordance with LNEC-E398 (1993). The specimens' measurements started just after the concrete's demoulding and were performed over a duration of 91 days. The quality of the concrete's cementitious matrix has a considerable influence on this property; as its stiffness increases, so does its capacity to restrain the deformations due to shrinkage. The use of RCA in the composition of concrete implies an overall decrease in stiffness due to the adhered mortar, which is much more deformable than the NA. Because of this, it is expected that this property presents higher values with increasing replacement ratios of NA with RCA and more so if the latter were subjected to more cycles of recycling.

In Figure 13, this property shows a non-linear evolution as a function of time. As expected, RCA-containing concrete presented higher shrinkage values. It was observed that 1RAC100, 2RAC100 and 3RAC100 presented shrinkage values of around 34%, 48% and

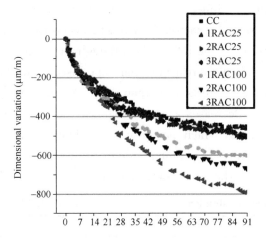

Figure 13 Shrinkage of concrete over time

75% higher, respectively, than those of the CC. At the same time, 1RAC25, 2RAC25 and 3RAC25 presented an increase in shrinkage of around 3%, 11% and 12%, respectively, when compared to the CC. It should be noted that the shrinkage values obtained in the first 7 days were essentially the same regardless of the type of concrete. This may be due to the fact that the RCA absorbed and retained some water in their pores, releasing it gradually in the first days of curing thereby contributing to an internal curing. After having applied the asymptotic model, good coefficients of correlations were observed. However, it was also observed that stabilization of this property was not reached after three cycles of recycling. Since a decrease in density and an increase in water absorption were observed in RCA3 due to the still increasing content of adhered mortar with increasing number of cycles, future generations of RAC are likely to present greater deformability.

3.4.2 Water absorption by immersion and by capillary action

The evaluation of the absorption capacity of concrete by immersion was made in accordance with LNEC-E394 (1993). This property is mainly influenced by the material's open porosity. Due to the RCA's higher porosity, the absorption of concrete is expected to increase with increasing replacement ratios and more so with the RCA's number of recycling cycles. The average values for water absorption by immersion for each type of concrete, as well as their comparison with the control concrete's values, are presented in Table 11. As expected, the water absorption by immersion increased with increasing incorporation level and number of recycling cycles; however, the increment of the water absorption diminishes progressively with increasing number of cycles. The 2RAC100 concrete specimens showed an increase in water absorption of 3.07% in relation to that of the 1RAC100 concrete, whereas the 3RAC100 specimens showed a 0.05% increase when compared to that of the 2RAC100. A similar behaviour was observed in specimens with incorporation levels of 25% coarse RCA, wherein there was an increase of 0.43% from 1RAC25 to 2RAC25 specimens and 0.37% from 2RAC25 to 3RAC25 specimens. These values indicate stabilization of this property as a function of the number of recycling cycles. This is also reinforced by the high coefficients of correlation after having applied the exponential asymptotic model. Furthermore, since the RCA's water absorption also showed an asymptotic behaviour, this allowed an indirect estimation of the RCA's open porosity as there is a strong interdependence between the water absorption by immersion and the RCA's water absorption.

Water absorption by immersion   Table 11

Concrete	Water absorption by immersion	
	%	V_{CC}
CC	13.5	—
1RAC25	15.3	1.13
2RAC25	15.7	1.17
3RAC25	16.1	1.19
1RAC100	18.6	1.38
2RAC100	21.7	1.61
3RAC100	21.7	1.61

    The water absorption by capillary action was determined in accordance with LNEC-E393 (1993). Table 12 presents the results of the test after 3 h, 6 h, 24 h, 48 h and 72 h. The relationship between the water absorbed by the RAC and that of the control concrete is also presented (VCC). The results show that RAC exhibited a higher water absorption than that of the control concrete, the magnitude of which increased with increasing replacement ratio and number of recycling cycles. Specimens with 100% coarse RCA showed increases of 120%, 169% and 212% after the first, second and third cycles of recycling, respectively. After each consecutive recycling cycle, RCA are covered with a greater amount of porous adhered mortar, which contribute to an increasingly permeable microstructure. Similar to that observed in the water absorption by immersion test, very good coefficients of correlation were observed by applying the asymptotic exponential model, suggesting stabilization of the property. However, the results suggest that this was not achieved within the three cycles of recycling.

Water absorption by capillary action   Table 12

Concrete	Water absorption by capillary action ($\times 10^{-3}$ g/mm^2)				V_{CC}
CC	1.00	1.35	2.36	3.43	—
1RAC25	1.84	2.40	4.09	5.84	1.70
2RAC25	1.86	2.43	4.15	6.25	1.82
3RAC25	1.95	2.54	4.33	6.57	1.92
1RAC100	2.55	3.40	5.32	7.53	2.20
2RAC100	3.06	4.11	6.55	9.21	2.69
3RAC100	3.23	4.30	7.10	10.70	3.12

3.4.3 Carbonation and chloride ion penetration

    Figure 14a presents the results of the resistance to carbonation test according to LNEC-E391 (1993). Concrete with increasing incorporation of RCA exhibited higher carbonation depths than those of the control mixes and even greater after each consecutive cycle of recycling; specimens with 100% RCA1 and RCA3 showed increases of about 30% and 80%, respectively. Since the concrete's resistance to carbonation essentially depends on its diffusion capacity and because RCA have greater porosity than NA, it is expected that mixes with higher replacement ratios show greater penetration of CO_2. Figure 14b shows a similar trend for chloride ion penetration; 1RAC100, 2RAC100 and 3RAC100 presented increases in chloride ion penetration of about 26%, 40% and 47%, respectively, relative to the control concrete. This performance loss is a direct result of the progressively higher adhered mortar content in RCA, which contributes to an increasingly porous microstructure. These findings are in agreement with the authors' previous studies (Silva et al., 2015b; Silva et al., 2015c; Silva et al., 2015d).

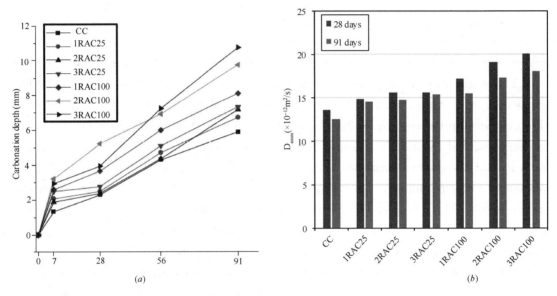

Figure 14   Carbonation (a) and chloride ion penetration (b)

4. Conclusions

The main objective of this study was to evaluate the influence of incorporating coarse RCA obtained from concrete subjected to multiple recycling cyclesand to ascertain the recyclability of the material over time. As expected from the findings in the literature, there was an overall decline in performance of concrete with the incorporation of RCA. The variation of this performance loss was significantly influenced by the number of recycling cycles that the RCA had been subjected to. With increasing number of cycles, the RCA presented progressively higher water absorption, lower density and lower resistance to fragmentation, due to the gradually higher content of mortar adhered to RCA. Because of this, the mechanical and durability properties of concrete were affected. However, since there is a stabilization of the quantity of mortar adhered to RCA and thus of their physical properties, asymptotic-like trends were observed for most of the properties of concrete, suggesting stabilization after a certain number of recycling cycles. From a practical point of view, this study's finding allows defining a "minimum performance aggregate", for which specific losses in performance can be expected in each of the main properties of concrete. By knowing the number of recycling cycles that concrete was subjected to as well as their properties, future designers may be able to estimate the material's maximum loss in performance and design it to several applications with a high safety level.

5. Acknowledgments

The authors are grateful for the support of the CERIS-ICIST Research Institute, IST, University of Lisbon and FCT (Foundation for Science and Technology).

6. References

[1] De Brito J & Silva RV (2016) Current status on the use of recycled aggregates in concrete: Where do we go from here? RILEM Technical Letters, 1: 1-5.
[2] DIN-52108 (2010) Testing of inorganic non-metallic materials-Wear test using the grinding wheel according to Böhme-Grinding wheel method. Deutsches Institut für Normung (German Institute for Standardization), Germany, 10 p.
[3] EN-206 (2013) Concrete-Specification, performance, production and conformity. Comité Européen de Normalisation (CEN), Brussels, Belgium, 98 p.
[4] EN-933-1 (2012) Tests for geometrical properties of aggregates-Part 1: Determination of particle size distribution. Sieving method. Comité Européen de Normalisation (CEN), Brussels, Belgium, 22 p.
[5] EN-933-4 (2008) Tests for geometrical properties of aggregates-Part 4: Determination of particle shape. Shape index. Comité Européen de Normalisation (CEN), Brussels, Belgium, 14 p.
[6] EN-1097-1 (2011) Tests for mechanical and physical properties of aggregates-Part 1: Determination of the resistance to wear (micro-Deval). Comité Européen de Normalisation (CEN), Brussels, Belgium, 18 p.
[7] EN-1097-3 (1998) Tests for mechanical and physical properties of aggregates-Part 3: Determination of loose bulk density and voids. Comité Européen de Normalisation (CEN), Brussels, Belgium, 10 p.
[8] EN-1097-5 (2008) Tests for mechanical and physical properties of aggregates-Part 5: Determination of the water content by drying in a ventilated oven. Comité Européen de Normalisation (CEN), Brussels, Belgium, 54 p.
[9] EN-1097-6 (2013) Tests for mechanical and physical properties of aggregates-Part 6: Determination of particle density and water absorption. Comité Européen de Normalisation (CEN), Brussels, Belgium, 54 p.
[10] EN-1992-1-1 (2008) Eurocode 2-Design of concrete structures: Part 1-1: General rules and rules for buildings. Comité Européen de Normalisation (CEN), Brussels, Belgium, 259 p.
[11] EN-12350-2 (2009) Testing fresh concrete-Part 2: Slump-test. Comité Européen de Normalisation (CEN), Brussels, Belgium, 12 p.
[12] EN-12350-6 (2009) Testing fresh concrete-Part 6: Density. Comité Européen de Normalisation (CEN), Brussels, Belgium, 14 p.
[13] EN-12390-3 (2009) Testing hardened concrete-Part 3: Compressive strength of test specimens. Comité Européen de Normalisation (CEN), Brussels, Belgium, 22 p.
[14] EN-12390-6 (2009) Testing hardened concrete-Part 6: Tensile splitting strength of test specimens. Comité Européen de Normalisation (CEN), Brussels, Belgium, 14 p.

[15] EN-12390-13 (2013) Testing hardened concrete. Determination of secant modulus of elasticity in compression. Comité Européen de Normalisation (CEN), Brussels, Belgium, 18 p.

[16] Eurostat (2017) Waste statistics in Europe [Online]. Available: epp. eurostat. ec. europa. eu [Accessed 13/06/2017].

[17] Ferreira L, de Brito J & Barra M (2011) Influence of the pre-saturation of recycled coarse concrete aggregates on concrete properties. Magazine of Concrete Research, 63(8): 617-627.

[18] LNEC-E391 (1993) Concrete: determination of carbonation resistance (in Portuguese). National Laboratory in Civil Engineering (LNEC-Laboratório Nacional de Engenharia Civil) Lisbon, Portugal, 2 p.

[19] LNEC-E393 (1993) Concrete: determination of water absorption by capillarity (in Portuguese). National Laboratory in Civil Engineering (LNEC-Laboratório Nacional de Engenharia Civil) Lisbon, Portugal, 2 p.

[20] LNEC-E394 (1993) Concrete: determination of water absorption by immersion-testing at atmospheric pressure (in Portuguese). National Laboratory in Civil Engineering (LNEC-Laboratório Nacional de Engenharia Civil) Lisbon, Portugal, 2 p.

[21] LNEC-E398 (1993) Concrete: determination of drying shrinkage and expansion (in Portuguese). National Laboratory in Civil Engineering (LNEC-Laboratório Nacional de Engenharia Civil) Lisbon, Portugal, 2 p.

[22] LNEC-E464 (2007) Prescriptive methodology for a design working life of 50 and of 100 years under the environmental exposure. National Laboratory of Civil Engineering (Laboratório Nacional de Engenharia Civil-LNEC), Lisboa, Portugal, 16 p.

[23] LNEC-E471 (2009) Guide for the use of coarse recycled aggregates in concrete (in Portuguese). National Laboratory of Civil Engineering (Laboratório Nacional de Engenharia Civil-LNEC), Portugal, 6 p.

[24] Mália M, de Brito J, Pinheiro MD & Bravo M (2013) Construction and demolition waste indicators. Waste Management & Research, 31(3): 241-255.

[25] NT-Build-492 (1999) Concrete, mortar and cement-based repair materials: Chloride migration coefficient from non-steady-state migration experiments. Nordtest, Espoo, Finland, 8 p.

[26] Silva RV, de Brito J & Dhir RK (2015a) Tensile strength behaviour of recycled aggregate concrete. Construction and Building Materials, 83: 108-118.

[27] Silva RV, de Brito J & Dhir RK (2016) Establishing a relationship between modulus of elasticity and compressive strength of recycled aggregate concrete. Journal of Cleaner Production, 112: 2171-2186.

[28] Silva RV, de Brito J & Dhir RK (2017) Availability and processing of recycled aggregates within the construction and demolition supply chain: A review. Journal of Cleaner Production, 143: 598-614.

[29] Silva RV, de Brito J, Neves R & Dhir R (2015b) Prediction of chloride ion penetration of recycled aggregate concrete. Materials Research, 18(2): 427-440.

[30] Silva RV, Neves R, de Brito J & Dhir RK (2015c) Carbonation behaviour of recycled aggregate concrete. Cement and Concrete Composites, 62: 22-32.

[31] Silva RV, Silva A, de Brito J & Neves R (2015d) Statistical modelling of carbonation in concrete incorporating recycled aggregates. Journal of Materials in Civil Engineering, 28(1).

5 Properties of Concrete With Recycled Construction and Demolition Waste: A Research Experience in Belgium

Zengfeng Zhao[1], Luc Courard[1], Frédéric Michel[1], Simon Delvoie[1],
Mohamed ElKarim Bouarroudj[1], Charlotte Colman[1], Jianzhuang Xiao[2]

1. Urban and Environmental Engineering, GeMME Building Materials
University of Liège, Liège, Belgium
e-mail: Zengfeng.Zhao@uliege.be, Luc.Courard@uliege.be

2. Department of Structural Engineering, Tongji University, Shanghai, China
e-mail: jzx@tongji.edu.cn

Abstract: Construction waste management is a quite important economic and environmental deal for our societies. More than 2 million tons demolition and construction waste are annually produced only in Wallonia, Southern Region of Belgium; recycling has clearly to be promoted. Waste concrete blocks were crushed in the laboratory by a jaw crusher and the different fractions of laboratory produced recycled concrete aggregate (RCA) were characterized by measuring the hardened cement paste content, the density, the porosity and the water absorption. Results clearly show that, the recycled sands possessed significantly higher cement paste content and higher water absorption than coarse RCA. Then, concrete blocks with different substitutions (0%, 30%, 100%) of natural aggregate by the same volume fraction of RCA were manufactured. The fresh properties (slump, density, air content), and mechanical properties (compressive strength) were studied. The compressive strength of concrete decreased as the substitution of RCA increased. Results show that the compressive strength of concrete made with 100% RCA could reach 8 MPa after 28 days. Therefore, the use of RCA obtained from old blocks in the production of new blocks can be envisaged depending on their class of exposure and the grade requirement. Moreover, the influence of the fine recycled concrete aggregates (FRCA) on the mechanical and durability properties of concrete was studied. The industrial FRCA produced from recycling center was used to cast concrete. The concretes with different substitutions (0%, 30%, 100%) of natural sand by the FRCA were manufactured. Mechanical properties (compressive strength) and durability properties (capillary absorption, carbonation depth, and freeze/thaw resistance) were investigated. The results confirm that the compressive strength of concrete decreased as the substitution of FRCA increased. Durability of concrete could be strongly influenced by the high porosity and water absorption of fine recycled concrete aggregates.

Keywords: Concrete; Recycling; Coarse Aggregates; Fine particles; Durability

1. Introduction

Large quantities of construction and demolition waste (CDW) are produced each year (Colman et al., 2018; Delvoie et al., 2018). So far, only a small fraction of these waste concrete are reused as aggregate for concrete production (Topcu and Sengel, 2004). Recycled concrete aggregates are composed of a mix of natural aggregates and hardened adherent cement paste. The latter is usually much more porous than natural aggregates (Zhao et al., 2013) and leads to a large water demand which makes RCA harder to recycle into concrete (Courard et al., 2010; Evangelista and De Brito, 2014; Zhao et al., 2017a). Properties of RCA such as water absorption, porosity can deeply influence the properties of fresh concrete as well as mechanical properties and durability of concrete made with RCA (Hansen and Narud, 1983; Poon et al., 2004; Khatib, 2005; He et al., 2012; Xiao et al., 2015; Zhao et al., 2016; Zhao et al., 2018). The hardened cement paste content and its properties have a decisive influence on the properties of RCA, such as density, porosity, and water absorption (Zhao et al., 2015b). The determination of adherent cement paste and mortar is however difficult to carry out experimentally.

The CONREPAD research project focuses on the relationship between the properties of original industrial produced concrete to the different physical properties of RCA. In this study, RCA from industrial produced blocks (RCA _ Blocks) and slabs (RCA _ Slabs) were crushed by laboratory jaw crusher and then separated into four granular fractions (0/2, 2/6.3, 6.3/14, 14/20 mm). Each granular fraction of RCA was characterized in order to study the influence of granular fraction and the origin of recycled concrete aggregates on their properties (Zhao et al., 2017b). Real RCA from recycling plant were also used to compare with these two laboratory produced RCA from industrial concretes (Zhao et al., 2017b). The feasibility of concrete blocks made with different substitutions of natural aggregate by the same volume fraction of RCA was investigated. Moreover, the influence of the fine recycled concrete aggregates (FRCA) on the mechanical and durability properties of concrete was investigated.

2 Use of coarse recycled aggregates for construction blocks production

2.1 Materials and testing

Waste concrete blocks (C8/10) were collected from Prefer Company (Belgium) and then crushed in a laboratory jaw crusher retaining the same jaw opening for all products. After crushing, RCA _ Blocks were separated into four granular fractions (0/2, 2/6.3, 6.3/14, 14/20 mm). RCA were characterized by measuring the hardened cement paste content, the density, the porosity and the water absorption. Only the fraction 2/6.3 mm was used for the manufacture of new concrete blocks.

New concrete blocks with different substitution rates (0%, 30%, 100%) of natural aggregate by the same fraction of RCA (only fraction 2/6.3 mm) were manufactured. Table 1 shows the composition of new concrete blocks. CEM III/A 42.5 and water to cement ratio of 0.7 were used for the new concrete blocks. Natural calcareous aggregate (noted as

NA 2/7) and natural river sand (noted as NS 0/2) were used for the manufacture of concretes. The water absorption of RCA 2/6.3 was 5.0% and its apparent particle density was 2.52 g/cm^3 according to the standard EN 1097-6 (while it was 0.68% and 2.7 g/cm^3 for natural aggregate). Natural aggregate and recycled aggregate were used in air dried condition. The absorbed water was adjusted according to the water content of the aggregates and their water absorption. A half of the total water was added to pre-saturate the aggregate in the mixer for 5 minutes before the addition of cement. The other half of the water was added after introduction of the cement.

Cement paste content of RCA was measured by the salicylic acid dissolution (Zhao et al., 2013). Salicylic acid allows the dissolution of most phases contained in OPC cement paste (C_2S, C_3S, ettringite, portlandite and C-S-H for example) but not of the main phases contained in natural aggregates and especially limestone. The water absorption coefficient of three coarse fractions of RCA was determined according to EN 1097-6. The water absorption coefficient of the fraction 0/2 mm of RCA was determined on the basis of the relationship between water absorption and cement paste content (Zhao et al., 2017b). The specimens were cast with the vibration table and stored in laboratory conditions. After 24h, they were demoulded and stored in curing room (20±2°C and a relative humidity 95±5%). The compressive strength of concrete was measured according to EN 12390-3 on cubic samples (150mm x 150mm x 150mm), performed after 1 day, 7 days and 28 days of curing in curing room.

Table 1 Compositions of concrete blocks (1m^3)

	B_RCA0	B_RCA30	B_RCA100
NA 2/7 (kg)	1080.0	754.0	0.0
RCA 2/6.3 (kg)	0.0	302.0	1008.0
NS 0/2 (kg)	825.0	825.0	825.0
Cement (kg)	150.0	150.0	150.0
Efficient water (kg)	105.0	105.0	105.0
Absorbed water (kg)	13.1	26.0	56.2
W_{eff}/C	0.7	0.7	0.7

2.2 Concrete properties

Figure 1 shows the cement paste content (CPC) and water absorption of RCA as a function of granular fraction. As can be seen, CPC of fraction 0/2 mm was larger compared with the three coarse fractions of RCA, while the values obtained for the three coarser fractions were similar. The fraction 0/2 mm of RCA revealed a larger value of water absorption in comparison with the three coarse fractions of RCA, while the values obtained were similar for the three coarser fractions. Recycled sands thus possessed higher cement paste contents than the coarse recycled aggregates, which may heavily penalize their use properties (such as water absorption, porosity) comparing with coarse recycled aggregates.

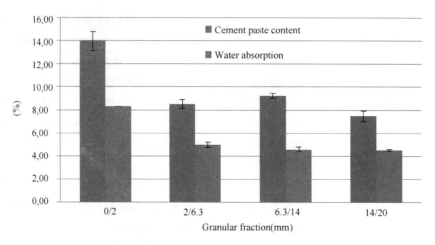

Figure 1 Cement paste content and water absorption of RCA as a function of granular fraction

The workability of three concretes is low (zero-slump, means that dry consistency showing no measurable slump after removal of the slump cone), which is conventionally observed in the industrial environment for the manufacture of blocks (using mechanical vibration for putting on a caisson). Figure 2 shows the compressive strength of the various mixes at different ages. The compressive strengths of concretes with RCA were lower than those of concrete with natural aggregates. The compressive strength of concrete made with 100% RCA at 28 days deceased 14.4% comparing with the reference concrete, while the concrete made with 30% RCA at 28 days decreased 7.2%. These lower mechanical strengths are certainly caused by the poorer properties of RCA in comparison to natural aggregates used; the presence of adherent cement paste leading to higher porosity comparing with the natural aggregate. The compressive strength of concrete made with 100% RCA could reach 8 MPa after 28 days.

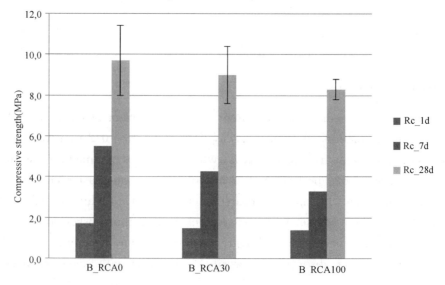

Figure 2 Compressive strength (Rc) of new concrete blocks at different ages

3. Durability properties of concrete made with fine recycled concrete aggregates

3.1 Introduction

The fine fraction of RCA, essentially composed of mortar and hardened cement paste, possesses a large water demand which makes it harder to recycle into concrete compared to coarser RCA. The influence of the fine recycled concrete aggregates (FRCA) on the mechanical and durability properties of concrete was studied (Zhao et al., 2017a). The industrial FRCA produced from recycling center was used into concrete. The concretes with different substitutions (0%, 30%, 100%) of natural sand by the FRCA were manufactured, and fresh properties (slump), mechanical properties (compressive strength), and durability properties (capillary absorption, carbonation depth, and freeze/thaw resistance) of these concretes were studied. The results showed that the compressive strength of concrete decreased as the substitution of FRCA increased. Durability of concrete could be strongly influenced by the high porosity and water absorption of fine recycled concrete aggregates.

3.2 Materials and testing

The cement used in concrete was an Ordinary Portland Cement (CEM I 52.5 N). Two calcareous natural aggregates (noted as NA2/7, NA6/14) were used for the manufacture of concretes. The water absorption of these two aggregates was 0.68% and 0.32% respectively for NA2/7 and NA6/14 and their apparent particle density was 2.70 g/cm^3 according to the standard EN 1097-6. A natural river sand (noted as NS0/2) was used for the manufacture of concretes with a water absorption of 0.70%.

Recycled concrete aggregates (RCA) (0/31.5mm) were provided by crushing waste concrete in the recycling center and only the fine fraction 0/2mm (noted as FRCA0/2). The water absorption of FRCA0/2 was 8.8% and its apparent particle density was 2.47 g/cm^3. The sieve analysis showed that the grain size distribution of NS and used FRCA was comparable.

The concretes with different substitutions (0%, 30%, 100%) of natural sand by the same volume of FRCA were manufactured (Table 2). Natural aggregates and recycled aggregates were used in air dried condition. The absorbed water was adjusted according to the water content of the aggregates and their water absorption. A half of the total water was added to pre-saturate the aggregate in the mixer for 5 minutes before the addition of cement. The other half of the water was added after introduction of the cement.

After the mixing, the slump of fresh concrete was measured with the Abrams cone according to EN 12350-2. The air content of fresh concrete was measured according to EN 12350-7. The specimens were cast with the vibration table and stored in laboratory conditions and covered with plastic film in order to avoid evaporation of water. After 24h, they were demoulded and stored in curing room (20±2°C and a relative humidity 95±5%).

Compositions of concretes (1m³)　　　　Table 2

	B_FRCA0	B_FRCA30	B_FRCA100
NA 6/14 (kg)	550.0	550.0	550.0
NA 2/7 (kg)	775.0	775.0	775.0
NS 0/2 (kg)	600.0	420.0	0.0
FRCA 0/2 (kg)	0.0	168.0	559.2
Cement (kg)	320.0	320.0	320.0
Efficient water (kg)	160.0	160.0	160.0
Absorbed water (kg)	11.2	24.7	56.2
Superplasticizer (kg)	3.4	3.4	3.4
W_{eff}/C	0.5	0.5	0.5

The compressive strength of concrete was measured according to EN 12390-3 on cubic samples (150mm x 150mm x 150mm), performed after 7 days and 28 days of curing incuring room. The capillary water absorption of concrete was measured on cubic samples (100mm x 100mm x 100mm) according to NBN 15-217. The lower side of the specimen was placed into water and removed and weighed at various time intervals. The porosity of concrete was evaluated by the total immersion test in water.

The carbonation of concrete was evaluated on prismatic specimens (100mm × 100mm × 400mm). After curing for 28 days in the curing room, the specimen was stored in a room at 20±2°C and a relative humidity of 60±5% until constant mass. Then it was stored in the carbonation room with a CO_2 concentration of 3% by volume for 4 weeks. After each week, specimens were taken out the carbonation chamber and split. The fresh split surface was sprayed with a phenolphthalein pH indicator. In the carbonated part of the specimens, where the alkalinity was reduced, no coloration occurred. Thus, the average depth of the colorless phenolphthalein region was measured from three points in each side.

In order to evaluate the resistance to freezing, the specimens (100mm × 100mm × 100mm) were subjected to 14 freeze-thaw cycles (24h cycle from −15°C to +10°C) according to NBN B 05-203. The freezing was carried out at −15°C in air and thawing was undertaken in the water at 10°C. The mass of all the specimens were recorded to show the weight loss trend.

3.3 Fresh properties of concretes

Figure 3 shows the variation of slump for the three concretes after 0 and 30 minutes. It can be seen that the initial slump slightly decreased for concretes with recycled sand. After 30 minutes, the slump of all types of concrete decreased whatever the different substitutions. It is also shown that the rate of slump loss was larger as the substitution increased, which could be due to the higher water quantity absorbed by the higher percentage of recycled sand after the mix.

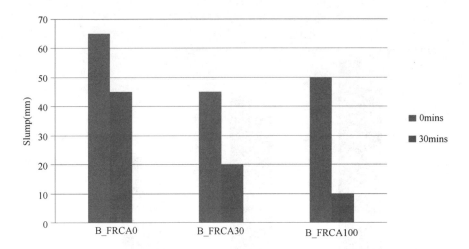

Figure 3 Change of slump as a function of three concretes

The air content of concrete increased (2.2% for concrete B_FRCA0, 3.3% for concrete B_FRCA30, and 5.5% for concrete B_FRCA100) when the substitution of recycled sand increased. Increased air content is also known to occur in lightweight aggregate concrete, which shows some similarities with concrete made with recycled aggregate (Amnon, 2003).

3.4 Mechanical and durability performances

Figure 4 shows the compressive strength of the various mixes at different ages. The compressive strengths of concretes with FRCA were lower than those of concrete with natural sand. The compressive strength of concrete made with 100% FRCA at 28 days deceased 48.2% comparing with the reference concrete, while the concrete made with 30% FRCA at 28 days decreased 15.9% comparing with the reference concrete. These lower mechanical strengths are certainly caused by the poorer properties of FRCA in comparison to natural sand used; the presence of adherent cement paste leading to higher porosity comparing with the natural sand (Zhao et al., 2015a).

Figure 5 presents the rate of capillary absorption as a function of types of concretes. As can be seen in this figure, the rates of absorption of the recycled concrete were much larger than the reference concrete. The coefficient of capillary absorption of concrete B_FRCA100 was 0.38 kg/m^2/h$^{0.5}$, while it was 0.11 and 0.14 g/m^2/h$^{0.5}$ for the reference concrete and B_FRCA30 respectively. The higher capillary absorptions of recycled concrete were certainly caused by the incorporation of FRCA, which had higher porosity comparing with the natural sand, leading to the higher porosity of concrete. It appeared that the capillary absorption of concrete was little affected by the presence of FRCA up to 30%. This was confirmed by the results of porosity of concrete measured by immersion (the total porosity estimated by water absorption for the concrete B_FRCA100 was

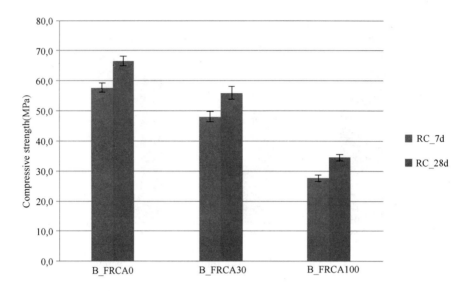

Figure 4　Compressive strength (Rc) of concretes at different ages

9.5%, while it was estimated as 4.2% and 5.3% for the reference concrete and B_FRCA30 respectively). The rate of absorption, rather than the total absorption is mainly affected by the structure and size distribution of the pores in the concrete.

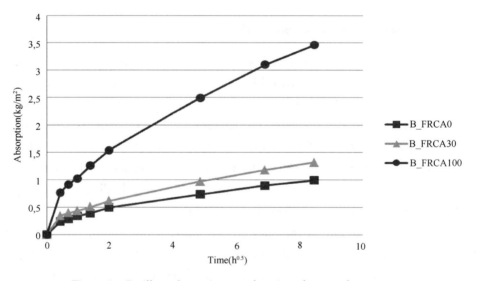

Figure 5　Capillary absorption as a function of types of concretes

Figure 6 shows the depth of carbonation as a function of time in the carbonation room for all types of concretes. For the first 14 days, the carbonation depths of B_FRCA0 and B_FRCA30 were zero, while it was 5 mm for the concrete made with 100% recycled sand. After 28 days, the carbonation depth of B_FRCA100 was 9 mm, while it was lower than 2 mm for the B_FRCA30, and zero for the reference concrete. The higher depth of carbonation of the recycled concrete could be due to the higher porosity in concrete, inducing a faster diffusion of CO_2 into the concrete.

After 14 cycles of freeze-thaw, the visual specimen's examination did not allow detecting any significant deterioration for all the concretes. The mass loss is presented in Figure 7. The recycled concretes had lower freeze-thaw resistance comparing with the reference concrete, which was due to higher porosity in the recycled concrete.

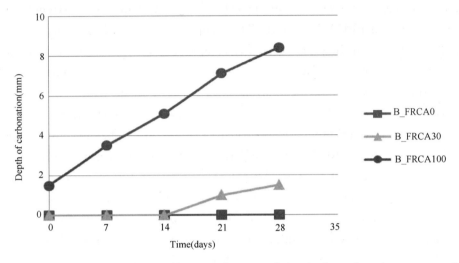

Figure 6   Depth of carbonation as a function of time in the carbonation room

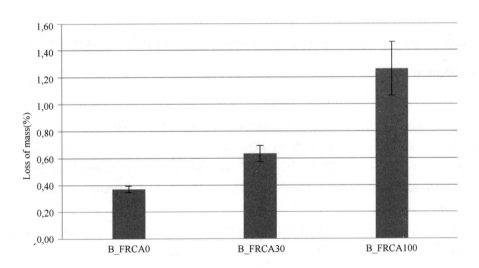

Figure 7   Mass loss due to the freeze-thaw cycles

4. Conclusions

The feasibility of using RCA obtained from old waste concrete blocks in the production of new concrete blocks is studied. Results clearly show that, the recycled sands possessed significantly higher cement paste content and higher water absorption than coarse

RCA. The compressive strength of concrete blocks decreased as the substitution of RCA increased. The compressive strength of concrete made with 100% RCA could reach 8 MPa after 28 days without increasing the cement content of the concrete mix. Therefore, the use of RCA obtained from old waste blocks in the production of new blocks can be envisaged depending on their class of exposure and the grade requirement.

Concrete design with recycled materials shows that the compressive strength decreases as the substitution of FRCA increased. The compressive strength of concrete made with 100% FRCA deceased in the range of 48.2% comparing with the reference concrete, while the concrete made with 30% FRCA decreased up to 15.9% comparing with the reference concrete. However, the compressive strength of concrete made with 100% FRCA could reach 35MPa after 28 days. Durability of concrete could be strongly influenced by the high porosity and water absorption of recycled concrete aggregates. The durability properties of concrete made with 30% FRCA were comparable to the reference concrete, especially for capillary absorption and carbonation. Therefore, the use of FRCA in concrete structures can be envisaged depending on their class of exposure and the concrete grade requirement (for example the concrete C25/30 with no risk of corrosion or attack). Substitution rate of natural sand up to 30% is acceptable, while for the substitution rate higher than 30%, mechanical properties of concrete should be checked while the effects on durability should be also monitored for specific applications.

Investigations performed here above clearly show that an adapted design opens opportunities for recycling CDW as aggregate and sand in the formulation of concrete.

Acknowledgment

The authors are grateful to INTERREG FWVL, which provided financial support to this study as part of VALDEM research project entitled "*Solutions intégrées de valorisation des flux « matériaux » issus de la démolition. Approche transfrontalière vers une économie circulaire*" and INTERREG NWE through SeRaMCo research project entitled "*Secondary Raw Materials for Concrete Precast Products (introducing new products, applying the circular economy)*".

Authors warmly thank PREFER Company for providing materials. The authors would like also to thank the Government of Wallonia (DGO6) for financial support through the project Beware Academia CONRePaD *Concrete design with recycled concrete aggregates by means of Packing Density Method* (2014-2016).

References

[1] Amnon, K. (2003). Properties of concrete made with recycled aggregate from partially hydrated old concrete, Cem Concr Res. 33 (2003) 703-711.

[2] Colman, C., Zhao, Z., Michel, F., Remond, S., Bulteel, D., Courard, L. (2018) Gypsum residues in recycled materials: characterization of fine recycled aggregates. 4th International Conference

on Service Life Design for Infrastructures SLD4, Delft, 26-29 August.

[3] Courard L., Michel F., Delhez P. (2010). Use of Concrete Road Recycled Aggregates for Roller Compacted Concrete. Construction and Building Materials, 24, 390-395.

[4] Delvoie, S., Zhao Z., Courard L., Michel F., De Cort G., Lauret S. Report WPT1 Market analysis and formal regulations in NWE, Progessive report Interreg SeRaMCo Project, January 2018, 15p.

[5] Evangelista L., De Brito J. (2014). Concrete with fine recycled aggregates: a review. European Journal of Environmental and Civil Engineering, 18, 129-172.

[6] Hansen T., Narud H. (1983). Strength of recycled concrete made from crushed concrete coarse aggregate. Concrete International, 5, 79-83.

[7] He H., Le N. F., Stroeven P. (2012). Particulate structure and microstructure evolution of concrete investigated by DEM Part 1: Aggregate and binder packing. HERON, 57, 119-132.

[8] Khatib J. M. (2005). Properties of concrete incorporating fine recycled aggregate. Cement and Concrete Research, 35, 763-769.

[9] Poon C. S., Shui Z. H., Lam L., Fok H., Kou S. C. (2004). Influence of moisture states of natural and recycled aggregates on the slump and compressive strength of concrete. Cement and Concrete Research, 34, 31-36.

[10] Topcu I. B., Sengel S. (2004). Properties of concretes produced with waste concrete aggregate. Cement and Concrete Research, 34, 1307-1312.

[11] Zhao Z., Remond S., Damidot D., Xu W. (2013). Influence of hardened cement paste content on the water absorption of fine recycled concrete aggregates. Journal of Sustainable Cement-Based Materials, 2, 186-203.

[12] Zhao Z., Remond S., Damidot D., Xu W. (2014). Teneur en pâte de ciment et coefficient d'absorption d'eau des sables recyclés. Quinzième édition des Journées Scientifiques du Regroupement Francophone pour la Recherche et la Formation sur le Béton (RF)^{2}B, Douai, France, 03-04 Juillet, 101-111.

[13] Zhao Z., Remond S., Damidot D., Xu W. (2015a). Influence of fine recycled concrete aggregates on the properties of mortars. Construction and Building Materials, 81, pp 179-186.

[14] Zhao Z., Damidot D., Remond S., Courard L. (2015b). Toward the quantification of the cement paste content of fine recycled concrete aggregates by salicylic acid dissolution corrected by a theoretical approach. The 14th International Congress on the Chemistry of Cement Beijing, China, 13-16 October 2015.

[15] Zhao Z., Courard L., Michel F., Remond S., Damidot D. (2016). Mechanical and durability properties of concrete made with fine recycled concrete aggregates. 3rd International Conferences on Microstructure Related Durability of Cementitious Composites, Nanjing, China, 24-26 October 2016.

[16] Zhao Z., Courard L., Michel F., Remond S., Damidot D. (2017a). Properties of concrete blocks made with recycled concrete aggregates: from block wastes to new blocks. HISER International Conference on Advances in Recycling and Management of Construction and Demolition Waste, Delft, Netherland, 21-23 June 2017.

[17] Zhao Z., Courard L., Michel F., Remond S., Damidot D. (2017b). Influence of granular fraction and origin of recycled concrete aggregates on their properties. European Journal of Environmental and Civil Engineering, 1-11.

[18] Zhao Z., Courard L., Michel F., Remond S., Damidot D., Delvoie S. (2018) Effects of limestone

filler on the behavior of recycled concrete aggregates. 4th International Conference on Service Life Design for Infrastructures SLD4, Delft, 26-29 August.

[19] Xiao J. Z., Fan Y. H., Tam V. W. Y. (2015). On creep characteristics of cement paste, mortar and recycled aggregate concrete. European Journal of Environmental and Civil Engineering, 19, 1234-1252.

6 地聚合物混凝土收缩控制剂研究

李柱国,冈田朋友,桥爪进

摘 要:地聚合物混凝土的缩重合反应伴随着水的排出,而且目前还没有降低碱激发剂水溶液用量的添加剂。因此,常温养护的地聚合物混凝土的干燥收缩很大,容易造成收缩裂缝。为促进地聚合物混凝土的实用化,作者在开发了地聚合物缓凝剂之后又开发了适用于地聚合物的收缩控制剂。本文将详细介绍地聚合物收缩控制剂的添加对粉煤灰和高炉矿渣复合使用型地聚合物混凝土的流动性,可使时间和强度的影响以及常温养护时体积变化的减少效果。

关键词:地聚合物混凝土;干燥收缩;收缩控制剂;缓凝剂;体积变化

Chemical Admixtures for Controlling Shrinkage of Geopolymer Concrete

Zhuguo Li, Tomohisa Okada, Susumu Hashizume

Abstract: The polycondensation reaction of geopolymer concrete is accompanied by water discharge, and there is currently no admixture for reducing the amount of alkali activator aqueous solution used in concreter of unit volume. Hence, the dry shrinkage of geopolymer concrete cured at room temperature is large so that shrinkage cracks may be caused. To promote the practical use of geopolymer concrete, the authors have developed shrinkage reducing agents suitable for geopolymer besides the geopolymer retarder. This paper introduces in detail the effects of the addition of the shrinkage control agents on the fluidity, working time, and strength development of geopolymer concrete using fly ash and blast furnace slag (FA-BFS based GP-C), and the reducing effect of volume change of FA-BFS based GP-C during curing at room temperature.

Keywords: Geopolymer concrete; Shrinkage; Shrinkage reducing agent; Retarder; Volume change

1. Introduction

In the manufacture of Portland cement that is one of main raw materials of concrete, CO_2 is emitted in large quantities due to thermal decomposition of limestone and consumption of energy. Out of 0.725t of CO_2 emission intensity of cement, 58% is due to decomposition of limestone. Coal ash, sewage sludge incineration ash, and municipal solid waste incineration ash substitute the whole amount of clay, but limestone has been still essential and accounts for 70% of cement raw materials. Therefore, even if excellent energy-saving technology is introduced, it can reduce CO_2 emissions by up to 40%. Therefore, the study and application of low carbon binder of not needing limestone are extremely important tasks for sustainability of construction industry.

Geopolymer (GP) is an inorganic material hardening due to a polycondensation reaction of SiO_4 and AlO_4 tetrahedras that are eluted from aluminosilicate powder under stimulation with an alkaline solution, e. g. water glass, NaOH, KOH, Na_2CO_3, etc.[1]. Initially, metakaolin powder is mainly used as aluminosilicate material, but from the viewpoints of saving natural resources and landfill, and recycling of wastes, waste or by-product-based geopolymers has recently been drawing attention, their research are actively carried out in the world[2].

The GP using fly ash (FA) discharged from a coal-fired power plant has been reported much since the 90's of last century. For other waste utilization, Yamaguchi et al. developed GP using sewage sludge incineration ash melted slag[3]. The authors has been successfully prepares GPs using municipal refuse incineration ash slag, fluidized bed coal ash, and paper sludge incineration ash[4-7]. Although it is not easy to prepare high-strength GP in normal temperature environment using only these waste powders due to foaming, inactivity, porousness of these wastes, it is possible to produce high-strength GP by mixing blast furnace slag (BFS).

In recent years, with the increase in electricity demand, operation of a coal-fired power plant with excellent fuel procurement stability and economic efficiency has been promoted in Japan, and as a result coal ash has been increasingly discharged, and it is expected to increase in the near future[8]. Although it is possible to manufacture FA-GP at a high curing temperature of above 60℃., the setting of GP using FA with low Cao is slow under a normal temperature, and it is difficult to obtain a practical strength above 20MPa. However, if BFS is mixed, high strength GP can be produced even in a normal temperature environment. From the viewpoints of mass utilization of FA and curing in the ambient air, it is preferable to prepare GP (abbreviated as FA-BFS based GP) using FA and BFS in combination. The CO_2 emission intensity of concrete (called FA-BFS based GP-C here), which uses FA-BFS based GP as a binder, is 50%~60% of ordinary Portland cement concrete (OPC-C) with equivalent compressive strength and slump, or is 80 to 90% of concrete using slag cement[9].

Since the polycondensation reaction is accompanied by water discharge, drying shrinkage of GP-C is large due to dissipation of water in the case of room temperature curing[10]. The control of drying shrinkage is one of main problems waiting to be solved for the practical use of GP-C.

The authors have already developed dry shrinkage reducing agents (SRA) for geopolymer[11,12], which are mainly composed of the polyester, and polyether derivatives, respectively. These derivatives have smaller molecular mass than those used in general SRA for OPC-C. In particular, the former SRA for GP using the polyester derivative has a considerably small shrinkage reduction effect for OPC-C. In this study, we investigated the effects of the SRA on the flow ability, setting time, compressive strength, and volume change of FA-BFS based GP-C in detail.

2. Experimental Program

2.1 Raw materials used

Table 1 shows the raw materials used, and Table 2 shows the chemical composition analysis results of active fillers (AF) by X-ray fluorescent analysis. Compared to blast furnace slag (BFS), fly ash (FA) had a large content of SiO_2 of 58.76%, but CaO was small. The CaO content of BFS was as much as 43.20% compared with FA.

Raw materials used Table 1

Sort	Symbol		Name and main composition	Property
Active filler	AF	FA	Fly ash	Density: 2.24, Specific surface area: $3550 cm^2/g$
		BFS	Ground blast furnace slag	Density: 2.88, Specific surface area: $4290 cm^2/g$
Alkaline solution	GPW		Aqueous solution of JIS No. 1watergals and NaOH	Density: 1.315
Fine aggregate	S1		Toyoura siliceous sand	Density: 2.64
	S2		Reviver sand	Density in saturated surface-dry condition: 2.60, Water absorption: 1.46%
Coarse aggregate	G		Crushed limestone	Density in saturated surface-dry condition: 2.70, Water absorption: 0.40%
Shrinkage recurring agent	SRA-1		Polyester derivative	Density: 1.15
	SRA-2		Polyether derivative	Density: 0.93
	SRA-3		Polyester derivative with small molecule mass	Density: 1.10

As shrinkage reducing agents (SRA), two kinds of organic compound containing polyester derivative as main component (Hereafter referred to as SRA 1 and SRA 3, respectively), and one kind (hereafter referred to as SRA2) of organic compound containing polyether derivative as a main component were used in the FA-BFS based GP-C. The retarder and the SRA were additionally added to GP mortar (GP-M) or GP-C. The dosages of both the retarder and the SRA were 5% of the mass of AF.

Chemical compositions of active fillers (AF)   Table 2

AF \ Oxide	SiO_2	TiO_2	Al_2O_3	Fe_2O_3	MnO	CaO	MgO	K_2O	P_2O_5	SO_3	Others
Fly ash	58.76	1.27	26.13	6.23	0.05	2.65	1.44	0.63	0.15	0.24	2.45
Blast furnace slag	33.40	2.10	15.50	0.52	0.38	43.10	5.20	0.66	0.01	0.04	—

Alkaline solution (GPW, density: 1.315g/cm³) was mixtures of water glass aqueous solution (WG) and sodium hydroxide solution (NH) according to the volume ratio of 3 : 1. The WG was prepared by diluting JIS (Japanese Industrial Standards) No.1 grade water glass with distilled water by a volume ratio of 1 : 1. The concentration of NH was 10 moles. Mix proportions of FA-BFS based GP-C and GP-M are shown in Table 3.

Mix proportions of FA-BFS based GP materials   Table 3

Type	GPW/AF (wt%)	WG/NaOH (vol.)	BFS/AF (wt%)	Unit mass (kg/m³)				
				GPW	FA	BFS	S (Type)	G
Mortar	52.1	2.50	30	328	404	223	1350 (S1)	—
Concrete	50.0	2.50	30	200	280	120	756 (S2)	1000

2.2 Mixing, and property measurement of in fresh state

GP mortar was mixed by aHobart type mixer of 5 L capacity. A gravity type concrete mixer with a capacity of 100 L was used for mixing the FA-BFS based GP-C. The active fillers and fine aggregate were firstly put into the concrete mixer and mixed for 1 minute. Then, the alkali activator (GPW) and admixture were added and mixed for 2 minutes to get GP mortar. In case of concrete, finally coarse aggregate was added to the mortar and further mixed for 2 minutes to get GP concrete. Retarder (LST), SRA 1, SRA 2, and SRA 3 were dissolved in GPW in advance.

After the mixing of the GP-M, the flow value of GP-M was measured according to JIS R 5201 (Physical Testing Methods for Cement). And after the GP-C was mixed, the slump test in conformity with JIS A 1101 (Method of Test for Slump of Concrete) and the air content test in accordance with JIS A 1118 (Method of Test for Air Content of Fresh Concrete by Volumetric Method) were conducted. Also, the temperature of GP-C was measured in accordance with JIS A 1156 (Method of Measurement for Temperature of Fresh Concrete). Since the method for measuring the working time of GP-C has not yet been established, in this study, as shown in Fig. 3, the smoothed sample surface of concrete or mortar was penetrated with a rod, generally used in the slump test, by its own weight at temperature of $20 \pm 2°C$. The elapsed time till trace was clearly remained was measured and it was taken as working time. The working time (Te) was measured starting from the time when the mixing of the active fillers and the alkaline solution was started.

2.3 Preparation, curing, mechanical test of GP-C specimens

Immediately aftermixing, GP-C was filled in plastic cylindrical moulds of diameter 10 cm × height 20 cm, and compacted with a vibration rod to prepare specimens for compressive strength test. Then, the top surfaces of the specimens were sealed with a food wrap film. After cured at room temperature of 20℃ for 24 hours, the specimens were demoulded and continually cured to required ages in the air of 20℃ and R. H. 60%. Once the required age reached, the end surfaces of the specimens were polished with a polisher and the compressive strengths were measured according to JIS A 1108. The compressive strength was an average value of three specimens. The tests were carried at 3, 5, 7, and 28 days of age.

2.4 Measurement of shrinkage strain

GP-M was mixed and then filled in diameter 5cm × height 10cm plastic cylindrical molds to prepare specimens. The embedded type strain gauges with a length of 50mm were embedded into the center of the specimens. The top surfaces of the specimens were sealed with a food wrap film. After sealing the specimens, curing was carried out at 20℃ for 24 hours, and the specimens were demolded. Then, room temperature curing was lasted for 2 weeks under the conditions of 20 ± 2℃, R. H. 60 ± 5%. Using a data logger connected to the embedded strain gauges, the change in the lengths of the specimens was measured immediately after the preparation of the specimens.

The cylindrical specimens of GP-C were prepared by the method described in Section 2.3, and cured in room temperature of 20 ± 2℃, R. H. 60 ± 5%, then demoulded after 6 hours curing. Then, strain gauges having a length of 60 mm (PL 60) and a length of 90 mm (PL 90) were adhered adjacently on the surface of the center of the specimen in the vertical direction. Next, the specimens were sealed all over with curing tape and the curing was continued in a room of 20 ± 2℃, R. H. 60 ± 5%. The curing tape was removed at 1 day (condition A), 3 days (condition B), 5 days (condition C), 7 days (condition D) age. Measurement of shrinkage strain was started after the gauges were adhered. In order to examine the influence of the measurement starting time of shrinkage strain, some specimens were demoulded at 3 days age and the strain gauges were stuck, accordingly the measurement was started from 3 days age (condition E). The shrinkage strains became almost stable after 6 weeks and the measurement was completed.

3. Test Results and Discussion

3.1 Properties of GP-M and GP-C in fresh state

Fig. 1 shows the flow values of Series B-M with no any of admixture added, Series R-M with 5% added LST, and Series R-SRA1~3-M with 5% added LST in combination of

5% SRA1, SRA2, and SRA 3, respectively. The flow values immediately after mixing were almost the same as that of Series B-M, even when only the LST was added singly or the LST and the SRA were used together. It was found that the addition of any of the retarder of GP or the shrinkage reducing agent of GP did not affect the fluidity of GP-M.

Fig. 2 shows the slump, air content and concrete temperature of Series R-C with only LST added to GP – C by 5%, and Series R-SRA2-C with LST and SRA2 added by 5%, respectively. As shown in Fig. 5, the slump just after mixing tended to increase with the addition of SRA2. However, even when SRA2 was added, the air content of GP-C did not show a large change, and it was 1.2 to 1.3%. Also, comparing the temperature of GP-C with the room temperature, it is found that the reaction heat of GP-C is small since the two temperatures were very near. From these results, it was confirmed that the addition of the shrinkage reducing agent SR2 to FA-BFS based GP-C does not adversely affect the properties of freshly mixed GP-C.

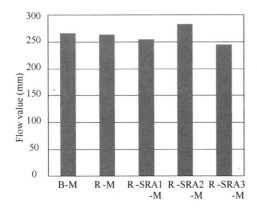

Fig. 1 Effects of admixtures on the fluidity of GP-M

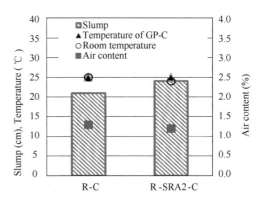

Fig. 2 Effects of SRA2 on the properties of fresh GP-C

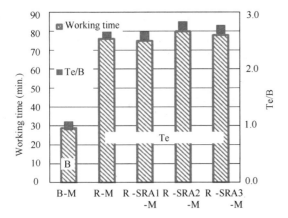

Fig. 3 Working time of GP-M with the addition of admixtures

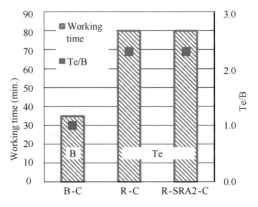

Fig. 4 Effect of SRA2 on the working time of GP-C

3.2 Working times of GP-M and GP-C

Fig. 3 shows the working times of GP mortars. For Series R-M, the LST was added by 5% of mass ratio of AF. Series R-SRA1-M, R-SRA2-M, and R-SRA3-M used 5% of the LST and 5% of the SRA1, SRA2, and SRA3, respectively. Fig. 4 shows the working times of Series B-C with no admixture added, Series R-C with 5% LST added, and Series R-SRA2-C with 5% respective addition of LST and SRA2. As shown in Figs. 3 and 4, the working time ratio (Te/B) of the samples using the retarder was 2.3~2.8 times of the samples without addition of the reatrder (LST). Moreover, the working time of GP-C with only the retarder addition was almost the same as in the case of using the retarder and the SRA in combination. Hence, it was found that the use of the SRA in addition to the retarder does not harm the retardation effect of the retarder.

3.3 Compressive strength of FA-BFS based GP-C using SRA2

Fig. 6 shows the test results of compressive strength at 3, 5, 7 and 28 days ages for Series R-C with only 5% LST added, and Series R-SRA2-C with 5% respective addition of LST and SRA2. As shown in the figure, the compressive strength of the series R-C was higher by 1.1 to 1.2 times than that of the series R-SRA2-C specimens until the age of 3 to 7 days. However, at the age of 28 days, the compressive strength when the SRA2 was added was the same as the case where no shrinkage reducing agent was added. Therefore, it can be concluded that the compressive strength of FA-BFS based GP-C does not decrease even if the SRA2 was further added together with the retarder (LST).

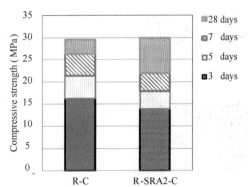

Fig. 5 Effect of addition of SRA2 on the compressive strength of GP-C

3.4 Shrinkage of FA-BFS based GP-M and GP-C

Fig. 6 shows the shrinkage strains of the GP mortar specimens cured at room temperature. For Series R-M, the LST was added by 5% of mass ratio of AF. Series R-SRA1-M, R-SRA2-M, and R-SRA3-M used 5% of the LST and 5% of the SRA1, SRA2, and SRA3, respectively. The measurement of shrinkage strain started after demolding at 1day age. As shown in Fig. 6, when the shrinkage reducing agent of GP was not added, the shrinkage largely occurred at young age, and approached to a stable value around the age of 6 days. The shrinkage strain was not less than 3500×10^{-6} for 2 weeks when no shrinkage reducing agent was added, but the shrinkage strain of the specimen with addition of the SRA2 containing polyether derivative as its main component was about 1400×10^{-6} at 2 weeks age, about 60% of shrinkage strain was reduced. Moreover, in cases of adding the SRA1 or the SRA3 containing different polyester derivative as their main component, the shrinkage strain of the GP-M specimen was greatly reduced too. When the SRA3 was added, the shrinkage strain was about 500×10^{-6} at 2 weeks age, but when the SRA1 was used, the expansion strain of about 300×10^{-6} occurred at 3 weeks age. The reason of expansion has been under investigation.

Fig. 7~Fig. 11 show the shrinkage strains of the specimens of Series R-C with the retarder added by 5% of AF, and Series R-SRA2-C with the respective addition of LST and SRA2 by 5% of AF. The specimens were cured at room temperature of 20℃, R. H. 60%, but demoulded at different ages (1, 3, 5, 7days), and the measurement of shrinkage strain was started at 6 hours age (Figs. 7~10) and 3 days age (Fig. 11), respectively. As shown in these figures, in the case where no shrinkage reducing agent of GP was added (Series R-C PL-60, Series R-C PL-90), shrinkage largely occurred at young age. Also, it was found that the shrinkage occurred even during the sealing curing by the curing tape. Regarding the length of the strain gauge, the distribution range of the measured values was wider and widely fluctuated when the short gauge (PL 60) was used. This is because the measurement value of the short gauge is more likely to be affected by the sticking position. For example, if the gauge were near coarse aggregates, the measurement value would be small.

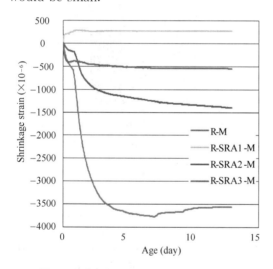

Fig. 6 Shrinkage strain of GP-M using three kinds of SRA

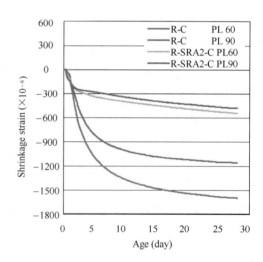

Fig. 7 Shrinkage strain of GP-C using the SRA2 (1 day sealing curing)

Sealing curing period with the curing tape was 1, 3, 5, 7 days in conditions A, B, C and D, respectively. In the case of the condition E, as described above, it was demolished at 3days age, and after the strain gauges were stuck, the shrinkage strain was measured from 3days age. According to Fig. 10~Fig. 11, regardless of the length of the gauge, the shrinkage strains tended to be smaller as the sealing curing period was longer, but there was no great difference if the sealing curing period was longer than 3 days. As shown in Fig. 11, the shrinkage strain increased sharply after the specimens were demoulded at 3 days, but the final mean shrinkage strains, which were measured from 3 days age, were almost the same as those of the specimens, which were measured from 6 hours age (see Fig. 8), no matter for Series R-C or Series R-SRA2-C.

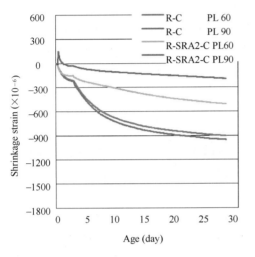

Fig. 8 Shrinkage strain of GP-C using the SRA2 (3 days sealing curing)

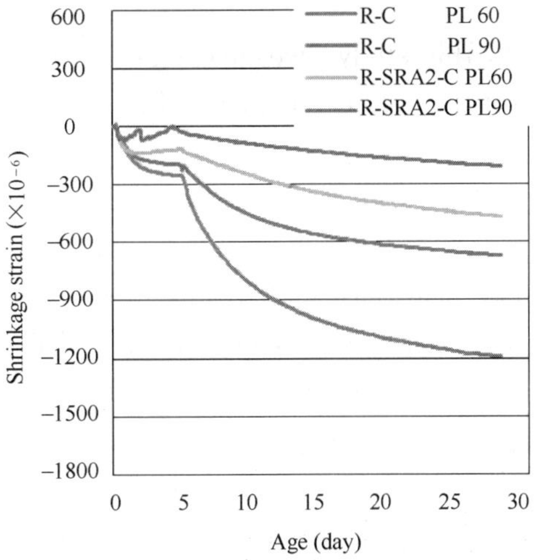

Fig. 9 Shrinkage strain of GP-C using the SRA2 (5 days sealing curing)

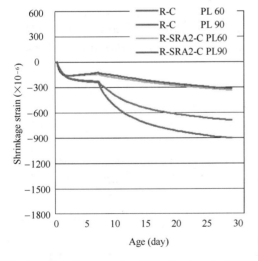
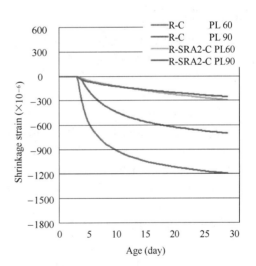

Fig. 10 Shrinkage strain of GP-C using the SRA2 (7 days sealing curing)

Fig. 11 Shrinkage strain of GP-C using the SRA2 (demoulded and measured at 3 days age)

The measured shrinkage strains varied with the sealing curing period and the demolding time, but the mean shrinkage strains for 4 weeks were almost larger than 800×10^{-6} when the shrinkage reducing agent for GP was not added. The shrinkage strain larger than 800×10^{-6} is not permitted in Japan since large shrinkage would result in concrete cracks. On the contrary, the shrinkage strains of the GP-C specimens with SRA2 added were almost smaller than 300×10^{-6} in 4 weeks, regardless of the sealing curing period and the demolding time, about 75% shrinkage strain can be reduced.

4. Conclusions

In this study, we investigated the applicability of the shrinkage reducing agents to control the volume change of FA-BFS based GP. The obtained results are summarized as follows.

(1) In addition to the retarder of GP, further addition of any of the shrinkage reducing agents of GP did not adversely affect the setting property, fluidity, air content, and reaction heat release of freshly mixed FA-BFS based GP mortar or concrete.

(2) The Compressive strength of FA-BFS based GP concrete with addition of any of the shrinkage reducing agents of GP did not decrease, compared with the normal concrete without addition of the shrinkage reducing agent.

(3) The addition of the shrinkage reducing agents containing polyester derivative or polyether derivative can greatly reduce the shrinkage of GP concrete. When the shrinkage reducing agent containing a polyester derivative having a special component is added, the GP specimen may slightly expand.

References

[1] Li Z.: Geopolymer and its practical application in construction materials, Part 1, Hardening mechanism and reaction products, *JTCCM (Japan Testing Center for Construction Materials) Journal*, Vol. 52, pp. 2-7, 2016. 9.

[2] Li Z.: Geopolymer and its practical application in construction materials, Part 2, Raw materials and utilization situation, *JTCCM Journal*, Vol. 52, pp. 2-7, 2016. 12.

[3] Yamaguchi N., Kisou K., and Ikeda K.: Geopolymer using sewage sludge molten slag as active filler, Japanese Patent No. 5435255, 2013. 12. 20.

[4] Li Z. Ikeda K., and Zhang Y.: Development of geopolymer concrete using ground molten slag of municipal waste incineration residue, *Proc. of Annual Research Meeting Chugoku Chapter*, AIJ, Vol. 36, pp. 57-60, 2013. 3.

[5] Li Z., and Ikeda K. et al: Method for solidification of low calcium fluidized bed coal ash, Japanese Patent Application No. 2015-254394, 2015. 12. 25.

[6] Li Z., and Ikeda K.: Geopolymer using papermaking sludge incineration ash and its application to treatment of radioactively contaminated water, *Proc. of the Japan Concrete Institute*, Vol. 37, No. 1, pp. 2337-1342, 2016. 7.

[7] Li Z., Ohnuki T., and Ikeda K.: Development of paper sludge ash-based geopolymer and application to treatment of hazardous water contaminated with radioisotopes, *Materials*, Vol. 9, No. 8, paper No. 633, pp. 1-17, 2016. 7.

[8] Coal Energy Center: http://www.jcoal.or.jp/coaldb/tech/coalash.

[9] Quantitative assessment of environmental impact of geopolymer concrete, *Proc. of Symposium of*

Current status and Issues of Geopolymer Technology in Construction Field, pp. 43-50, 2016. 6.

[10] Nagai T., Li Z., Takagaito N., and Okada T.: Fundamental study on drying shrinkage characteristics of geopolymer concrete, *Proc. of the Japan Concrete Institute*, Vol. 38, No. 1, pp. 2301-2306, 2016. 7.

[11] Li Z., Okada T. Kitazato S., Hashitsume S., and Suga A.: Shrinkage reducing agent for geopolymer, Japanese Patent Application No. 2016-96908, 2016. 5. 13.

[12] Li Z., Okada T. Kitazato S., Hashitsume S., and Suga A.: High performance shrinkage reducing agent for geopolymer, Japanese Patent Application No. 20176-047669, 2017. 3. 13.

7 基于 IMU 的混凝土结构隐蔽工程质量实时监控系统

李 恒[1]，杨新聪[1,2]

（1. 香港理工大学建筑及房地产系；2. 哈尔滨工业大学）

摘 要：混凝土结构隐蔽工程的施工过程往往难以人工观察测量，因此而导致隐蔽工程的施工质量难以通过常规工程的过程控制理论和方法来保障。针对隐蔽工程施工质量过程控制系统和方法缺乏的挑战，本文提出了利用惯性测量单元对隐蔽工程的施工全过程进行实时监控与记录的理论和方法。该系统包括数据搜集、处理和可视化三个模块，可对建筑构件、材料、施工设备及工具、施工人员等的位置和姿态进行定位和追踪。同时该系还整合了环境可测变量，例如温度、湿度等，为施工人员和管理人员提供隐蔽工程关键控制要素的可视化信息，实现对隐蔽工程的工程质量进行自动化、智能化的控制与管理。最后本文在混凝土振捣实验中应用并实践了该系统，结果表明该系统稳定且效率较高，具有实用价值。

关键词：隐蔽工程；惯性测量单元；自动化；智能化

IMU-based Real-time Monitoring System for Construction Quality of Concealed Concrete Projects

Li Heng, Yang Xincong

Abstract: The process of concealed projects in concrete structures is always out of monitoring and management on construction sites, therefore the methodology and techniques for general quality management cannot be implemented immediately and directly. To address the challenge of quality management of concealed projects, this paper proposed an IMU-based method and monitoring system for concealed projects. This system consists of micro accelerators, gyros, temperature, and magnetic sensors, aiming to track and locate the posture and position of construction components, materials, equipment and workforce. Meanwhile, it is integrated with environmental sensors of the temperature and humidity. To validate the proposed system, the experiment based on the concrete vibration was conducted. The results show the effectiveness and potential applications in future.

Keywords: Concealed projects; inertial measurement unit; automation; artificial intelligence

随着科学技术的发展和人民生活水平的提升，我国的城市化进程不断加快，智慧城市成为未来城市规划的发展目标。然而作为建设新城市建设核心引擎的建筑业，其劳动生产力水平、科技成果的转化效率等还停留在较低层次，面临着巨大的转型压力和挑战。近些年来，以"互联网+"为代表的新兴技术彻底改变了许多传统产业的生产和发展的模式，有效地提高了生产效率和管理水平，同时降低了生产成本和生产事故率。作为劳动密集型产业的建筑业应借助先进的科学技术与理论方法，将云计算、大数据、物联网等应用到建筑产品的生产加工过程中，以信息技术为手段以工作协同与信息共享为载体，实现自动化、智能化的建筑过程与全天候、全方位的全员参与，进而促进建筑业劳动生产效率和管理效率的提高。

1. 引言

在混凝土建筑结构工程中，根据建造过程可否直接观测分成可见工程和隐蔽工程，其中可见工程是指直接暴露在可操作环境中的工程，例如房屋的地上结构，港口桥梁的水上结构等，而隐蔽工程是指不可直接观察和测量的工程，例如房屋的地下结构，港口桥梁的水下结构等。对于可见工程，管理人员可以采用人工观察和测量的方法对施工过程进行质量、安全的监控和管理，例如施工监理人员观察建筑构件表面的平滑程度来检验相应的抹灰质量等。然而对于隐蔽工程，管理人员往往采用破坏性的抽样原位检验或者预留试样检验等方法对隐蔽工程的施工结果进行监控与管理，例如施工人员对混凝土浇筑的预留试样进行强度检验来推测真实浇筑混凝土的力学性能等。

由于建造过程难以或不能直接观察和测量的本质属性，混凝土建筑结构中的隐蔽工程相比可见工程具有施工难操作、过程难控制、结果难预料和事故难处理的"四难"特点。这四个特点导致现有的常规工程质量、安全管理的理论、技术和方法难以高效应用于隐蔽工程，其不足主要体现在以下几个方面：（1）常规工程的过程质量控制技术手段和方法难以适用于隐蔽工程，表观检测与直接性能测试都无法在隐蔽工程中直接使用；（2）常规工程的过程质量控制管理理论和方法难以应用于隐蔽工程，基于统计抽样理论的质量管理和控制无法在隐蔽工程中直接使用；（3）常规工程的过程质量控制难以匹配新工艺、新技术和新材料的发展，传统施工工艺的过程质量控制方法基于人工判断与历史经验，对着新的施工工艺无法应用[1]；（4）常规工程的过程质量控制难以为建筑建造自动化、智能化奠定基础，传统对照设计和施工方案的质量控制检验结果依赖于隐蔽工程验收记录，纸质版文件无法实现建造过程的信息化管理。

综上，近年以来，在全国各地工程项目中的隐蔽工程因为建造过程中的质量控制管理存在漏洞或缺陷，埋下了质量隐患，其中的个别项目在工程建设完成后，出现了结构整体性差、局部结构薄弱、渗漏、水电管线阻塞，一方面导致建筑结构的耐久度和舒适度下降，使用功能受限，另一方面也导致不得不进行二次施工、结构鉴定与修复加固、防渗处理等，给工程效益带来损失。我国住房和城乡建设部因而多次发文，在全国范围内开展工程质量提升行动和建筑施工安全专项治理活动，落实推进建筑业质量变革同时推动建筑业安全发展，完善建筑行业风向防控机制，保障工程质量。

由于既有的质量管理理论和方法难以保障隐蔽工程又好又快的建造实施，因此如何对混凝土建筑结构中的隐蔽工程进行过程质量控制，从而实现对于质量事故的主动预防和及时处理是目前亟待解决的科学研究问题。

科学的理论方法和先进的技术手段可以为隐蔽工程的建造质量进行全天候、全方位的精细化质量监控与管理提供有力支撑与保障。因此本文旨在通过对混凝土建筑结构隐蔽工程的施工过程进行数据的采集，处理和决策，探索并研发隐蔽工程各个建造环节的可视化过程控制与管理系统，实现对隐蔽工程施工过程的全程可控，当出现建造偏差或质量缺陷时，自动纠正或者发出警报，保证建造偏差始终处于容许范围之内，同时存储相关质量缺陷的形成过程，减少后续相似类型的质量问题及隐患。

2. 研究背景

建筑工程质量一直是建筑业监控与管理的核心之一，尽管社会和经济不断发展，但是由于工程质量问题引起的安全事故和经济损失仍旧不在少数。利用先进信息技术来提高施工过程的监控与管理，一直是建筑业学界与业界的研究热点和发展重点之一[2-4]。一般说来，对施工过程进行实时监控主要有技术方案，按照建筑构件或者施工人员是否需要附加信息发射器或信息载体可分成两种，介入式和轻介入式。

最早的建筑业施工过程的信息化载体是条形码，早在20世纪90年代初期，Bell和McCullouch[5]就将二维码引入建筑业，用于快速准确地将施工过程信息录入计算机系统。Rasdorf和Herbert[6]将二维码在施工过程中的应用细化，形成了施工信息管理系统（Construction information management system），在该系统中整合了施工规划、工程量、造价和文档管理等内容。然而，尽管条形码成本低、便于携带和生产，但是其在复杂的建筑施工环境中难以提供稳定的数据信息，因此在工程项目领域的应用并未推广。近些年，预制化混凝土建筑结构的出现与大量应用，具有构件完整、湿作业少、环境整洁的优点，条形码又再次被研究人员发掘出其应用的潜力。Cheng和Chen[7]将二维码和全球定位系统升级整合，研究并开发了自动化装配进度更新系统ArcSched，实现了低成本地对装配化建筑构件的运输与吊装的全过程进行可视化控制与管理。

由于条形码在潮湿、污浊的环境中难以准确识别，更加稳定可靠的信息化载体 RFID（Radio frequency identification）被引入建筑业。Jaselskis等[8]率先在混凝土结构工程中应用RFID，在施工现场建立扫描出入口，在混凝土运输车顶等安放RFID标签，实现对于混凝土运输、进场、浇筑的过程定位追踪与监控。Song的研究团队[9]在此基础上，进一步将RFID的识别精度提高与GPS整合，形成基于RFID的离散区域施工定位系统，通过在建筑材料和建筑设备上附加标签，实现对于建筑材料的全过程定位追踪。Goodrum等[10]利用RFID来监测施工用具以实现对施工用具的定期维护与保养，由于RFID定位系统的精度不高，难以实现连续位置的追踪与定位，在21世纪初，Teizer[11, 12]等应用了UWB(Ultrawideband)来提高施工实时对施工人员、设备和材料的追踪与定位，根据现场空间的分布等，主要同于对施工人员进行危险区域预警，提高建筑业安全管理水平。Li[13]的研究团队等着力于减少标签能耗，减少可重复使用标签的更换次数，选取了低功耗蓝牙技术（Blue tooth energy）来传递数据信号，并建立了施工主控式质量安全管理系统，并在此基础上应用于装配式建筑结构施工，同时探索了智能化的区域划分与辨识原理。刘文平[14]将GPS、WUB定位技术与BIM结合，进一步提出了施工安全事故预防与预警的原型系统。此外、ZigBee、WiFi、普通蓝牙等也被研究人员相继引入建筑行业，适合于不同尺度、不同监控目标和要求的施工过程控制与管理体系[15-17]。

以上介入式建筑结构施工过程质量实时监控系统的最大缺陷是对施工人员的正常工作动

作增加了额外工作负重,且对建筑构件需要考虑标签是否需要重复使用等问题。与介入式不同,非介入式的建筑结构施工过程质量实时监控系统不需要在建筑构件或者施工人员身上附加标签,其安装使用效率更高,常用的技术手段包括施工高精度相机探测、激光探测与测量(Laser detection and ranging)等。其中使用高精度摄像机是较为常用的施工过程检测与管理手段,具有高精度、低成本和简单有效的特点[18],Teizer和Vela[19]讨论了在工程项目现场使用静态相机和动态相机的必要性,Yang[20]等利用监控相机对塔吊的循环吊装过程进行监控,实现自动化地对于混凝土灌注等施工过程的辨识与效率统计,Brilakis[21]等将计算机视觉的识别范围拓宽到施工过程中几乎所有相关的构件与设备,通过已知不同摄像机的角度与位置,将二维坐标转换为三维坐标实现高效率的施工过程监控。

然而,无论介入式还是非介入式的建筑结构施工过程质量实时监控系统,针对混凝土建筑结构中隐蔽工程的施工过程却难以高效稳定的工作。其原因是:(1)介入式建筑结构施工过程质量实时监控系统需要信号发射端实时发射信号,同时信号接收端实时接收信号并处理实现对建筑构件、施工人员等的追踪与定位,其必要条件是信号的通讯范围内有足够多的已知参考点或锚点。然而混凝土建筑结构中隐蔽工程往往隐藏在钢筋混凝土之内,具有较强的信号屏蔽作用,基于信号强度等测距手段会受到极大干扰而导致最终的定位不准或者系统失效;(2)非介入式建筑结构施工过程质量实时监控系统则是基于信号反射原理实现对于目标建筑构件或者材料的测距,其必要条件是目标对象无遮挡、无重叠。然而由于混凝土建筑结构中隐蔽工程的不可见、不可测的本质属性,导致追踪和定位基本无效。

3. 隐蔽工程施工过程质量实时监控的原理与方法

在空难事故发生后,在救援过程中人们往往需要寻找飞行数据记录仪(也被称为黑匣子)。飞行数据记录仪其实是一种事故分析研究仪器,通过记录飞机的高度、速度、航向、耗油量、发动机及飞行系统的工作状态和参数等,通过其提供的证据人们可以了解事故的发生原因与过程。受到黑匣子工作原理的启发,在混凝土建筑结构隐蔽工程中类似也可采用,即通过实时监测并记录隐蔽工程在施工过程中的各项参数指标,例如施工人员动作行为、施工现场环境温湿度、施工工具或者器械的振动与位姿等,一方面可以实时对施工过程进行控制与管理,另一方面存储相应数据变化,当出现隐蔽工程质量安全隐患时,可供施工人员和管理人员确定问题产生的原因与过程。

为确定隐蔽工程施工过程的必要监控项目与参数,本文从监控目标出发,分析影响因素确定监控项目,根据可控可测技术手段,决定监控要素。

3.1 隐蔽工程施工质量监控目标

根据我国《建筑工程施工质量验收统一标准》GB 50300—2013[22],工程质量验收应通过施工单位的自行检查、由工程质量验收责任方组织,工程建设相关单位参与的抽样检验,其技术文件应进行审核,并根据设计文件和相关标准以书面形式对工程质量是否达到合作做出确认。隐蔽工程在隐蔽前应由施工单位通知监理单位进行验收(也称为中间验收),并应形成验收文件,验收合格后方可继续施工。具体到混凝土建筑结构工程,根据《混凝土结构工程施工质量验收规范》[23],其中隐蔽工程质量验收应包括下列主要内容:

在钢筋分项工程中，纵横向钢筋、箍筋的牌号、规格、数量、位置均应符合设计要求，且钢筋的连接方式等应符合锚固、连接要求；在现浇结构分项工程中，应主要进行位置、尺寸和表观质量缺陷的检验，避免蜂窝、孔洞、夹渣、疏松等质量隐患；在装配式结构分项工程中，应主要进行粗糙面检验、钢筋和预埋件的检验；当建筑结构一体化施工时，还应考虑到与结构构件相互连接的其他材料，例如保温、节能材料等，其品种、规格、厚度、基层连接等应满足设计要求，不应出现脱层、空鼓和开裂。综上，隐蔽工程的施工质量监控目标可以大致的分成几个种类，如图1所示。

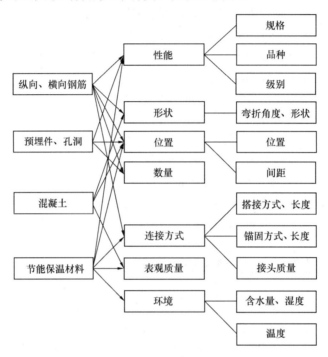

图1 施工质量监制目标分类

3.2 隐蔽工程施工质量监控项目

为了达到以上施工质量监控目标，应对施工过程中的施工方案选定、施工工艺操作、施工材料进场和施工环境等进行控制。混凝土建筑结构施工过程中主要是混凝土浇筑分项工程与钢筋分项工程。

依据《混凝土结构工程施工规范》[24]，混凝土浇筑分项工程应对如下施工过程进行质量控制：(1)表面干燥的地基、垫层、模板上应洒水湿润；现场环境温度高于35℃时宜对金属模板进行洒水降温，但洒水后不得留有积水；(2)混凝土浇筑应保证混凝土的均匀性和密实性。混凝土宜一次连续浇筑；当不能一次连续浇筑时，可留设施工缝或后浇带分块浇筑；(3)混凝土浇筑过程应分层进行，上层混凝土应在下层混凝土初凝之前浇筑完毕。

混凝土振捣是混凝土浇筑过程中的核心质量保障环节，其施工质量需要满足的要求为：(1)混凝土振捣应能使模板内各个部位混凝土密实、均匀，不应漏振、欠振、过振。(2)混凝土振捣应采用插入式振动棒、平板振动器或附着振动器，必要时可采用人工辅助振捣。(3)当使用插入式振捣棒时，应按分层浇筑厚度分别进行振捣，振动棒的前端应插入前一层混凝土中，插入深度不应小于50mm；振动棒应垂直于混凝土表面并快插慢拔均匀振捣；当混凝土表面无明显塌陷、有水泥浆出现、不再冒气泡时，可结束该部位振捣；振动棒与模板的距离不应大于振动棒作用半径的0.5倍；振捣插点间距不应大于振动棒的作用半径的1.4倍。当使用表面振动器时，应覆盖振捣平面边角；移动间距应覆盖已振实部分混凝土边缘；倾斜表面振捣时，应由低处向高处进行振捣。当使用附着式振捣器时，应与模板紧密连接，设置间距应通过试验确定；根据混凝土浇筑高度和浇筑速度，依次从下往上振捣；模板上同时使用多台附着振动器时应使各振动器的频率一致，并应交错设置在相对面的模板上。(4)混凝土分层振捣的最大厚度应遵循，振捣棒不超过作用部分的1.25倍，表面振捣器不超过200mm，附着式振捣器通过实验确定。

在钢筋分项工程中，应检查包括钢筋屈服强度、抗拉强度、伸长率、单位长度重量偏差等，在预应力工程中，应检查预应力钢丝下料长度，钢丝墩头外观、尺寸，挤压锚具成型后锚具外钢绞线的外露长度，预应力筋的品种、级别、规格、数量和位置等，预留孔道的规格、数量、位置、形状及灌浆孔、排气兼泌水孔等，锚垫板和局部加强钢筋的品种、级别、数量和位置等。

3.3 隐蔽工程施工质量监控要素

为了实时对隐蔽工程施工质量监控项目进行过程控制，应根据目前技术水平确定合理的质量监控要素，包括定量参数和定性参数，其中定量参数又可以分成静态参数和动态参数。对于混凝土建筑结构工程，其可测参数如表1所列。

混凝土建筑结构工程施工过程可测参数　　　　　　　　　表1

施工过程可测参数	定量或定性	静态或动态
环境湿度	定量	动态
环境温度	定量	动态
混凝土振捣深度	定量	动态
混凝土振捣速度	定量	动态
混凝土振捣时表面现象	定性	动态
混凝土振捣位置	定量	动态
混凝土振捣姿态	定量	动态
混凝土振捣时间	定量	动态
混凝土浇筑高度	定量	动态
混凝土浇筑速度	定量	动态
钢筋下料长度	定量	动态
钢筋品种	定性	静态
钢筋级别	定性	静态
钢筋(预留孔道)间距	定量	动态
钢筋(预留孔道)数量	定量	动态
预留孔道形状	定性	静态

由于静态参数在隐蔽工程的施工过程中并不变化，则在施工过程中无需对其进行实时的监控与记录。依据目前在建筑业中应用的实时监控技术，目前少有低成本的监控解决方案，因此钢筋、混凝土浇筑等使用标签数量大且难以重用的参数控制与管理在本文中不作讨论。本文基于目前技术水平确定了核心混凝土浇筑分项工程和钢筋分项工程的监控要素见表 2。

隐蔽工程施工质量监控要素　　　　　　　　　　　表 2

施工过程监控要素	可用技术手段
环境湿度	湿度传感器
环境温度	温度传感器
混凝土振捣深度	惯性测量单元
混凝土振捣速度	惯性测量单元
混凝土振捣器位置	定位系统或惯性测量单元
混凝土振捣器姿态	惯性测量单元
混凝土振捣时间	计时器
施工人员操作动作	惯性测量单元

4. 基于 IMU 的施工过程质量实时监控系统

根据隐蔽工程施工过程质量实时监控的原理与方法，本文采用分层设计的思想和原则，终端的开发环境 IDE 为 PyCharm，语言为 Python，可视化工具为 Blender，蓝牙配对采用 Windows 配对环境，数据传输及解码采用 C♯上机位和 Pyserial 库调试。基于惯性导航的无线定位系统的具体架构见图 2。

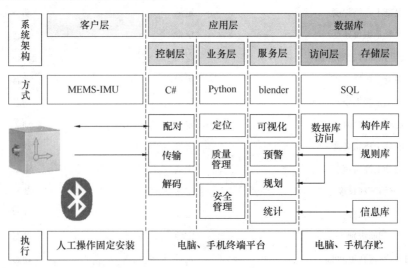

图 2　基于 IMU 的施工过程质量实时监控系统架构

4.1 惯性导航定位系统

惯性导航系统是一种不依赖任何外部信息,同时也无需向外发射信号的自主式导航系统,具有可靠、稳定、便携的特点,适合在复杂环境中对物体进行无线定位。基于该原理构建的建筑结构施工过程质量实时监控系统应能够对项目环境中的施工人员、建筑构件、机械设备或者施工工具进行高精度的追踪与定位。用于实现惯性技术的基本元器件称之为惯性测量单元(Inertial measurement unit,简称IMU),可用来测量物体的三轴姿态角、三轴角速率、三轴加速度等。在本研究中,惯性测量单元包括沿着正交直角坐标三边的三个单轴加速度计,三个单轴陀螺仪,三个地磁角测量仪、一个温度测量仪和一个湿度测量仪。其中,加速度计用于测量物体局部坐标系中独立的三个正交轴上的加速度;陀螺仪是用于测量在环境全局坐标系中物体的相对角速度信号;地磁角测量仪是用于测量惯性测量单元在地球磁场中的相对位置和重力方向;温度测量仪是用于测量观星测量单元周围环境的温度。惯性测量单元的工作原理是对比力(去除重力影响后的加速度)结合目标对象的姿态进行积分与二次积分,以求得目标对象的位置和姿态,其任务流程见图3,并在随后对每一个传感器进行说明。

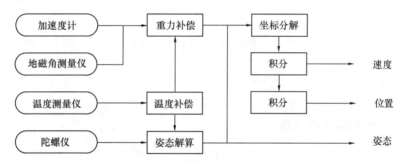

图3 惯性测量单元任务流程

4.1.1 MEMS加速度计

本文主要采用电容式MEMS加速度计。以双轴电容式MEMS加速度计为例,其结构如图4。当敏感质量块因惯性力产生位移时,四个弹性支撑对其进行水平与竖直方向的解耦,引起可变电极与固定电极之间的相对平移,通过测量可变电极与固定电极的电容变化推算出相应的加速度。

双轴电容式MEMS加速度计的力学原理如图5所示,设敏感质量块的质量为m,质量块与基片之间通过弹簧支撑连接的弹性系数为k,在体硅填充物中质量块运动的黏性阻力系数为b,则当载体在水平或者竖直方向上产生加速度时,质量块所受到的反向惯性力为F,同时产生相应的位移为x。

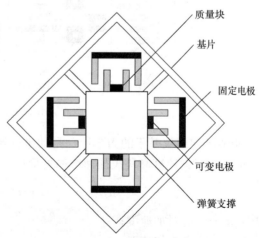

图4 双轴电容式MEMS加速度计结构示意

若弹簧支撑的弹力与质量块的位移呈线性变化关系，同时黏滞阻力与质量块的速度呈线性变化关系，根据质量块的力平衡条件，可以得到动力学方程：

$$m\frac{d^2x}{dt^2} + b\frac{dx}{dt} + kx = F = ma \tag{1}$$

式中：m——敏感质量块的质量；
$\quad\quad b$——阻尼系数；
$\quad\quad k$——弹性系数；
$\quad\quad x$——敏感质量块的位移；
$\quad\quad F$——外力；
$\quad\quad a$——外力加速度。

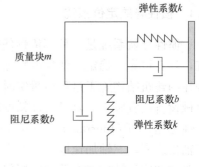

图 5 双轴电容式 MEMS 加速度计力学原理

令 $n = \dfrac{c}{2m}$ 为衰减系数，$\omega^2 = \dfrac{k}{m}$ 为无阻尼时的加速度计固有频率，进一步令 $\zeta = \dfrac{n}{\omega} = \dfrac{c}{2k}$ 为相对阻尼系数，则式(1)可以表示成：

$$\frac{d^2x}{dt^2} + 2\zeta\omega\frac{dx}{dt} + \omega^2 x = ma \tag{2}$$

4.1.2 MEMS 陀螺仪

本文主要采用 MEMS 陀螺仪对载体进行角运动的测量。以单轴 MEMS 陀螺仪为例，其结构如图 6 所示。静电驱动使质量块/音叉在一定频率下产生一个方向的简谐振动，当外部施加一个角速度时，由于科氏效应将产生一个垂直于质量块/音叉运动方向的力，科氏力将导致质量块/音叉产生位移，类似 MEMS 加速度计，通过测量可变电极与固定电极的电容变化推算出相应的角运动。

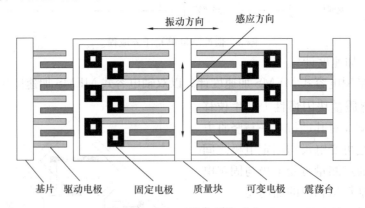

图 6 MEMS 陀螺仪结构示意

MEMS 陀螺仪基于的力学原理是动坐标系为定轴转动时，物体的绝对加速度时牵连加速度、相对加速度和科氏加速度的矢量和，即：

$$a_a = a_e + a_r + a_c = R\omega^2 + \frac{v^2}{R} + 2\omega v \tag{3}$$

式中：a_a——绝对加速度；
$\quad\quad a_e$——牵连加速度；
$\quad\quad a_r$——相对加速度；

a_c——科氏加速度；

R——物体在动坐标系内与转动定轴的距离；

ω——动态坐标系定轴转动的角速度；

v——物体的相对平移速度。

其中的科氏加速度为转动角速度与相对速度之积的两倍，方向垂直于相对运动的方向，与转动同方向。MEMS的电学原理图与MEMS加速度计近似，但是采用了一对相互反向振动的敏感质量块。可通过测量电容的变化量估计科里奥利力，推算载体的转动情况。

4.2 惯性导航的航位推算

惯性测量单元的定位算法是已知当前时刻下的位置条件，通过测量自身的位移和角度变化，来推算下一时刻的位置，属于航位推算法（dead reckoning）。航位推算起源于车辆、船舶航行过程中的航行定位，随着MEMS惯性测量单元的成本、重量和尺寸大大下降，航位推算法得以推广到日常的导航系统之中。以平面行位推算为例，如图7所示，当已知起始点、前进的方向和距离后，通过推算即可得知下一个位置。

一般来说，通过惯性测量单元，可以得到目标对象的三轴加速度以及其前进的方向，对加速度进行积分可以得到速度，对速度再次进行积分可以得到目标对象在单位时间内的位移变化，获得其移动的轨迹信息；对陀螺仪输出的角速度进行一次积分得到角度变化，与上一阶段的角度相结合，可以得到当前运动方向。但是在实际应用过程中，由于需要进行一次或二次积分，若测量的加速度存在误差，则使用积分获取的短时间内误差被放大，可能

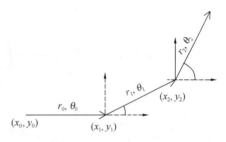

图7 航位推算原理示意

高达十几米。因此通过搜集目标对象的姿态等额外信息来调整并修正测量结果，是更为合理并常用的定位策略。当定位对象为人体时，根据人体行走的生理特征，可以直接利用加速度等运动参数的差异性来进行步态检测，通过统计行走的步数，利用步数乘以步长等于路程的原理可以推算相应时间内人体的步行距离，结合航位推算信息可得到相应的轨迹信息。

4.2.1 一般航位推算法

在t_0时刻，载体的初始位置为(x_0, y_0, z_0)，经过一段时间后到达t_n时刻，则载体的位置可以按照如下公式进行累计计算：

$$\begin{cases} x_n = x_0 + \sum_{i=0}^{n-1} d_i \cos \alpha_i \\ y_n = y_0 + \sum_{i=0}^{n-1} d_i \cos \beta_i \\ z_n = z_0 + \sum_{i=0}^{n-1} d_i \cos \gamma_i \end{cases} \tag{4}$$

式中：d_i——载体在t_{i-1}到t_i时刻之间移动的距离；

α_i、β_i、γ_i——载体在t_{i-1}到t_i时刻之间移动方向变化的角度（以欧拉角表示）。

当采样周期较短时，可以认为在较短的时间内，载体的运动速度、加速度保持恒定，此时的位移可以采用积分推算。由航位推算的定义可知，航位推算的本质是对路径的迭代求解，当某一计算步骤出现误差时，误差会持续累积并难以消除，因而从原理上来说，航位推算定位算法不应长时间使用。

4.2.2 人体航位推算法

人体航位推算法也被称为运动步长模型 PDR，基本原理与公式(4)相同，不同的是其过程中利用了人体的步态特征，其流程如图8所示。

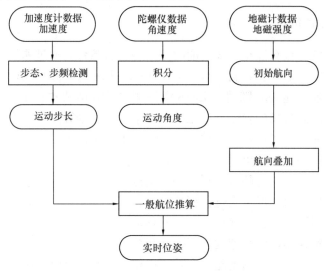

图 8 人体航位推算流程

4.3 坐标系转换

由于惯性导航系统采集到的数据是载体坐标系下的数据，而航位推算所需要的坐标系为全局坐标系，可视化系统使用的是相机坐标系，通过二维屏幕显示的是屏幕坐标系，因此需要建立多个坐标系之间的相互转换关系（图9）来实现实时的运动可视化监测。

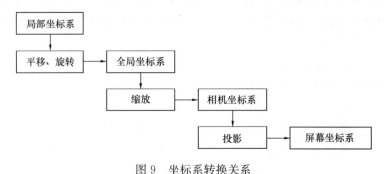

图 9 坐标系转换关系

4.3.1 相关坐标系
4.3.1.1 局部坐标系

局部坐标系也指载体坐标系，其原点位于载体的质心，其轴向分布可以通过地磁调整为全局坐标系。

4.3.1.2 全局坐标系

全局坐标系(世界坐标系)在本文中指东北天坐标系,其原点在载体初始位置的质心,x轴沿着纬度线指向正东方,y轴沿着经度线指向正北方,z轴沿着地球的切线指向空中,这三轴构成常用的右手坐标系,其中x轴和y轴构成的平面平行于水平地面(图10)。

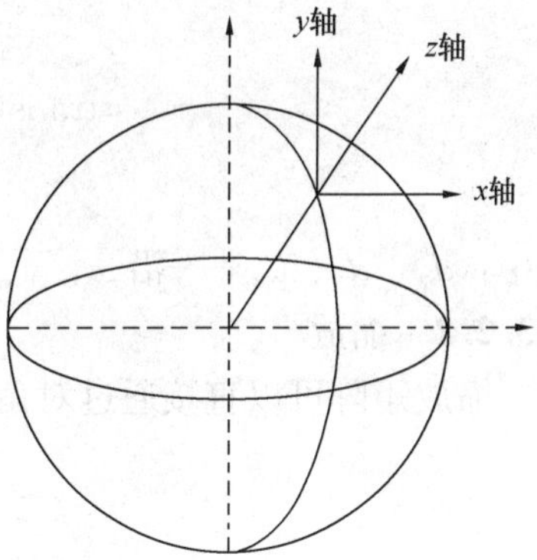

4.3.1.3 相机坐标系

相机坐标系也被称为观察坐标系,以相机的光心为坐标系原点,x轴和y轴与全局坐标系平行,z轴沿着相机光轴,见图11。

4.3.1.4 屏幕坐标系

屏幕坐标系是电脑屏幕显示的坐标系,其原点为屏幕左上角点,x轴沿着水平线指向右边,y轴沿着竖直线指向下方,如图12所示。

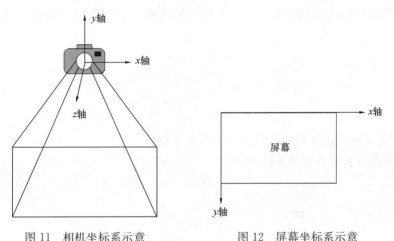

图10 全局坐标系示意

图11 相机坐标系示意　　图12 屏幕坐标系示意

4.3.2 局部坐标系、全局坐标系与相机坐标系的坐标转换

相关坐标系的坐标转换都可以通过一次平移和一到三次旋转实现。一次平移是将被转换坐标系的坐标原点移动到转换坐标系的原点,有限次的旋转是将被转换坐标系的坐标轴转动到与转换坐标系的坐标轴同向。

局部坐标系、全局坐标系与相机坐标系的转换体现在三维模型的运动过程中,本文中涉及的刚体运动可以分成三种:平移、缩放和旋转,其中缩放主要用于世界坐标系与相机坐标系之间的转换。考虑到计算机 GPU 能够进行较为强大的平行计算,当物体含有较多的三维坐标时,采用矩阵运算效率更高,因此本文中的所有坐标转换都采用矩阵形式。为了统一所有的刚体运动形式,在三维坐标的基础上添加一个辅助维度,其值为1。因此所有点的坐标存储格式变成$(x,y,z,1)$,当产生刚体运动转换时,可以直接使用矩阵点积的形式进行批量运算。下面针对各个转换矩阵进行说明:

4.3.2.1 平移

平移运算是最为简单的运动计算,其变化矩阵为:

$$\text{trans}(d_x, d_y, d_z) = \begin{bmatrix} 1 & 0 & 0 & 0 \\ 0 & 1 & 0 & 0 \\ 0 & 0 & 1 & 0 \\ d_x & d_y & d_z & 1 \end{bmatrix} \tag{5}$$

式中：d_x、d_y、d_z——沿 x、y、z 轴的位移。

4.3.2.2 缩放

缩放矩阵可以直接通过对角线对坐标进行缩放：

$$\text{scale}(s_x, s_y, s_z) = \begin{bmatrix} s_x & 0 & 0 & 0 \\ 0 & s_y & 0 & 0 \\ 0 & 0 & s_z & 0 \\ 0 & 0 & 0 & 1 \end{bmatrix} \tag{6}$$

式中：s_x、s_y、s_z——沿 x、y、z 轴的缩放系数。

4.3.2.3 旋转

旋转矩阵需要按照旋转顺序依次进行坐标的变换，其中按照 x 轴进行转动的变化矩阵为：

$$\text{rotate}_x(\alpha) = \begin{bmatrix} 1 & 0 & 0 & 0 \\ 0 & \cos\alpha & -\sin\alpha & 0 \\ 0 & \sin\alpha & \cos\alpha & 0 \\ 0 & 0 & 0 & 1 \end{bmatrix} \tag{7}$$

式中：α——沿 x 轴的旋转角度，按照右手螺旋法则。

按照 y 轴进行转动的变化矩阵为：

$$\text{rotate}_y(\beta) = \begin{bmatrix} \cos\beta & 0 & \sin\beta & 0 \\ 0 & 1 & 0 & 0 \\ -\sin\beta & 0 & \cos\alpha & 0 \\ 0 & 0 & 0 & 1 \end{bmatrix} \tag{8}$$

式中：β——沿 y 轴的旋转角度，按照右手螺旋法则。

按照 z 轴进行转动的变化矩阵为：

$$\text{rotate}_z(\gamma) = \begin{bmatrix} \cos\beta & -\sin\gamma & 0 & 0 \\ \sin\gamma & \cos\gamma & 0 & 0 \\ 0 & 0 & 1 & 0 \\ 0 & 0 & 0 & 1 \end{bmatrix} \tag{9}$$

式中：γ——沿 z 轴的旋转角度，按照右手螺旋法则。

当转动轴为局部坐标系时，变化矩阵左边连乘，当转动轴为全局坐标系时，变化矩阵右边连乘。本文的定位系统的旋转顺序为 xyz，相应的旋转矩阵为：

$$\text{rotate}(\alpha, \beta, \gamma) = \text{rotate}_x(\alpha)\text{rotate}_y(\beta)\text{rotate}_z(\gamma) \tag{10}$$

4.3.3 相机坐标系与屏幕坐标系的转换

相机坐标系与屏幕坐标系的转换就是将三维模型的坐标点投影到二维的平面上，平面投影一般可以分成两种，平行投影和透视投影。其中平行投影主要用于建筑与工程领域，投影的长度与真实长度一致，而透视投影，考虑了远处的物体较小近处的物体较大的物理现象，在本文中根据计算难度分成弱透视和全透视。

4.3.3.1 正交投影

当使用(x, y, z)表示三维模型的真实坐标而(x', y')表示其投影坐标，正交投影的关系表达式为：

$$\begin{bmatrix} x' \\ y' \end{bmatrix} = \begin{bmatrix} s_x & 0 & 0 \\ 0 & 0 & s_z \end{bmatrix} \begin{bmatrix} x \\ y \\ z \end{bmatrix} + \begin{bmatrix} c_x \\ c_z \end{bmatrix} \tag{11}$$

式中：s_x、s_z——任意缩放系数，在本文中取视域宽度与可视距离的比；

c_x、c_z——任意偏移系数，在本文中取三维物体初始轴的相应坐标。

通常正交投影用于工程表达，但是并不符合人类观察的现状，因此仅在系统调试时使用。

4.3.3.2 弱透视

弱透视假设三维物体的z轴坐标均等，在此基础上确定一个缩放因子后进行正交投影运算。首先计算目标对象某平面的平均z轴坐标，并使之取代相关平面内各个节点的深度，相比于全透视图，弱透视图计算简便，效率高，可以获得较为真实的近似。

4.3.3.3 全透视

全透视图考虑了人眼的真实情况，即相同尺寸的物体在远处看起来较小在近处看起来较大的现象，因此全透视图相比而言最接近人眼的情况。在全透视图中，视域宽度、相机的位置、相机的角度都会影响三维物体在屏幕之上的投影。因此，全透视图的计算过程首先是将物体在全局坐标系中的左边转化成为相机坐标系，后从相机坐标系转化成为屏幕坐标系。从全局坐标系到相机坐标的坐标运算形式如下：

$$\begin{bmatrix} x'' \\ y'' \\ z'' \end{bmatrix} = \text{view}_x \cdot \text{view}_y \cdot \text{view}_z \begin{bmatrix} x - c(x) \\ y - c(y) \\ z - c(z) \end{bmatrix}$$

$$\text{view}_x = \begin{bmatrix} 1 & 0 & 0 \\ 0 & \cos\alpha & \sin\alpha \\ 0 & -\sin\alpha & \cos\alpha \end{bmatrix}$$

$$\text{view}_y = \begin{bmatrix} \cos\beta & 0 & -\sin\beta \\ 0 & 1 & 0 \\ \sin\beta & 0 & \cos\beta \end{bmatrix}$$

$$\text{view}_z = \begin{bmatrix} \cos\gamma & \sin\gamma & 0 \\ -\sin\gamma & \cos\gamma & 0 \\ 0 & 0 & 1 \end{bmatrix} \tag{12}$$

式中：(x'', y'', z'')——物体三维坐标在相机坐标系中的坐标；

α、β、γ——相机视角，即相机轴与全局坐标系坐标轴的欧拉角；

$(c(x), c(y), c(z))$——相机在全局坐标系中的位置。

若采用简化c代表$\cos$，s代表$\sin$，则公式可以简化为：

$$x'' = c_y(s_z\Delta y + c_y\Delta x) - s_y\Delta z$$
$$y'' = s_x(c_y\Delta z + s_y(s_z\Delta y + c_z\Delta x)) + c_x(c_z\Delta y - s_z\Delta x)$$
$$z'' = c_x(c_y\Delta z + s_y(s_z\Delta y + c_z\Delta x)) - s_x(c_z\Delta y - s_z\Delta x) \tag{13}$$

式中：$(\Delta x, \Delta y, \Delta z)$——三维物体坐标到相机位置坐标的距离。

从相机坐标到屏幕坐标的转换如下：

$$x' = \frac{e_z}{z''}x'' - e_x$$

$$y' = \frac{e_z}{z''}y'' - e_y \tag{14}$$

式中：e_x、e_y、e_z——景深，相机到目标物体的视距；

(x', y')——三维物体在屏幕坐标系中的投影坐标。

5. 实时监控系统的精度提高与定位优化

完整的无线定位系统样机需要在平稳的环境中达到精准的定位精度，然而工程项目现场不仅需要覆盖的暴露面域较广，同时在周围动态环境的影响下，无线定位系统极易受到干扰导致难以提供可靠的定位数据。

5.1 误差分析

惯性测量单元的误差可以分成两类：系统性误差和随机性误差。系统性误差具有规律，可以通过实时补偿减小甚至消除其影响，而随机性误差一般难以使用合适的函数关系进行描述，因此而难以建立相应的数学模型并进行有针对性的补偿。

从误差来源角度 MEMS 惯性测量单元的误差可以分成随机误差、系统漂移、刻度因子误差和安装误差。随机误差包括量化噪声、零偏不稳定、角度随机游走、速率随机游走等；系统偏移包括运行偏差、环境敏感漂移等；刻度因子误差则来源于刻度因子的标定；安装误差包括轴失准误差、交叉耦合误差等。其中刻度因子误差和安装误差是确定性误差，是由于安装、制造 MEMS 惯性测量单元或者 MEMS 惯性测量单元敏感于环境变化而造成的，可以通过调整硬件安装或者硬件实验标定来消除或减小。随机误差和系统偏差是随机性误差，是指由无规律因素导致而引起的随机误差，很多还在探索之中，目前认为运行偏差在运行后较为稳定，可以通过启动后的短暂分析对随后的测量进行补偿，而随机误差则具有随机特征，只能通过建立准确的噪声模型通过预测噪声补偿来减小其影响。

为了对噪声成分进行识别，本文采用 Allan 方差（Allan variance，AVAR），通过相关时间及时间分布来区分不同噪声的特征从而识别出噪声来源，进而采用时间序列的噪声建模方法来消除或者减小噪声的影响。

Allan 方差是一种时域分析方法，能够对误差来源的统计特性进行细致的表征和识别，是仪器噪声分析的一种主流分析方法。Allan 方差基于一个基本假设：惯性测量单元噪声的方差可表示为相关时间的级数，即：

$$\sigma^2(\tau) = \sum_{n=-2}^{2} C_n \tau^n \tag{15}$$

式中：σ——惯性测量单元的噪声标准差；

τ——相邻时间差；

n——相邻时间级数。

在频域中，Allan 方差具有如下表达式，其值与噪声的总能量成正比，不同的随机过程造成可以通过调整 τ 来检验。

$$\sigma_y^2(\tau) = 4\int_0^{+\infty} S_\Delta(f)\frac{\sin^4(\pi f\tau)}{(\pi f\tau)^2}\mathrm{d}(f) \tag{16}$$

式中：$S_\Delta(f)$——随机过程误差的功率谱密度函数。

因此根据历史数据，建立噪声与相关时间的关系如图 13 所示。则根据相关时间与 Allan 方差之间的关系，例如斜率，就能够对噪声进行区分与辨识，并有针对性地削弱或去除相应噪声声源的影响。

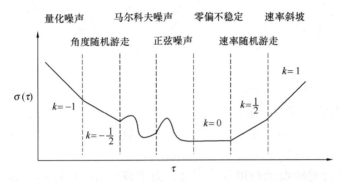

图 13　不同噪声声源与相关时间的关系示意

当惯性测量单元的噪声样本为 M 时，其 M-sample 方差定义为：

$$\sigma_y^2(M,T,\tau) = \frac{1}{M-1}\left\{\sum_{i=0}^{M-1}\overline{y}_i^2 - \frac{1}{M}\left[\sum_{i=0}^{M-1}\overline{y}_i\right]^2\right\}$$

$$\overline{y}_i = \frac{x(iT+\tau)-x(iT)}{\tau} \tag{17}$$

式中：$\sigma_y^2(M,T,\tau)$——M-sample 方差；

　　　$x(t)$——惯性测量单元在 t 时刻输出的加速度、角速度或角度；

　　　$\overline{y}(t,\tau)$——平均分组频率，定义为 $\overline{y}(t,\tau)=\frac{1}{\tau}\int_0^\tau x(t+t_v)\mathrm{d}t_v$，在离散的惯性测量单元的信号输出时，简化为 $\overline{y}(t,\tau)=\frac{x(t+\tau)-x(t)}{\tau}$；

　　　M——样本总数；

　　　T——采样时间，相邻采样之间的时间间隔；

　　　τ——分组时间长度，一般情况下 $\tau=m_0T$，当采样含有死区（dead-time）时，$\tau > m_0T$。

Allan 方差是 M-sample 方差的一种特例形式，定义为：

$$\sigma_y^2(\tau) = \langle\sigma_y^2(2,\tau,\tau)\rangle = \frac{1}{2}\langle(\overline{y}_{n+1}-\overline{y}_n)^2\rangle = \frac{1}{2}\langle(x_{n+2}-2x_{n+1}+x_n)^2\rangle \tag{18}$$

式中：$\sigma_y^2(\tau)$——Allan 方差；

　　　$\langle\cdots\rangle$——期望，总体平均；

　　　$\overline{y}_n$——以时间 τ 分组的第 n 个分组频率。

由于 Allan 方差的原始定义是基于无穷的噪声数据序列而得到的无偏期望，而在实际工程中难以获得并处理无穷多的噪声数据，因此本文采用以下形式来对 Allan 方差进行估计：

5.1.1 固定估计（fixed τ estimator）

固定估计是最为简单的估计形式，即 $\tau=T$，可采用以下表达式计算：

$$\sigma_y^2(\tau) = AVAR(\tau, N) = \frac{1}{2\tau^2(N-2)} \sum_{i=0}^{N-3} (x_{i+2} - 2x_{i+1} + x_i)^2 \tag{19}$$

5.1.2 无重叠估计（non-overlapped variable estimator）

将固定估计的间隔时间扩展，可以得到不重叠估计，具有如下的形式：

$$\sigma_y^2(\tau) = \frac{1}{2\tau^2\left(\frac{N-1}{n}-1\right)} \sum_{i=0}^{\frac{N-1}{n}-2} (x_{ni+2n} - 2x_{ni+n} + x_{ni})^2 \tag{20}$$

$$n \leqslant \frac{N-1}{2}$$

式中：τ——分组时间长度，$\tau=nT$；

n——每一个分组中所含的连续噪声数据的个数。

5.1.3 重叠估计（overlapped variable τ estimator）

由于无重叠估计对数据的利用效率不高，为了减少所需的噪声数据，将数据重叠使用，得到重叠估计的 Allan 方差形式：

$$\sigma_y^2(\tau) = \frac{1}{2\tau^2(N-2n)} \sum_{i=0}^{N-2n-1} (x_{i+2n} - 2x_{i+n} + x_i)^2 \tag{21}$$

$$n \leqslant \frac{N-1}{2}$$

重叠估计能够使用较少的噪声数据对 Allan 方差进行较为准确的估计，因此 IEEE 等国际标准也采用的这种噪声估计方法对 Allan 方差进行估计。

5.2 误差辨识

应用 Allan 方差测量并辨识 MEMS 惯性测量单元在工作状态时的噪声成分，本文采用直流稳压电源为 MEMS 惯性测量单元供电，实时采集并在计算机中记录传出的数据信号，在测试过程中，温度和湿度恒定。当采样频率为 100Hz 时，MEMS 惯性测量单元静止采集 3600s 的零漂数据。由于 MEMS 惯性测量设备在 x 和 y 轴上静止且无外力，因此其标准数值应为零，而在 z 轴上存在重力，其标准数据应为 1.0 倍的标准重力加速度。由结果可知，在 x 轴上，总的零漂误差随着时间不断累积增加，在 y 轴上保持稳定，但是在 z 轴上反而出现了下降，说明其噪声的成分和来源不用，具有不同的噪声特征。本文所使用 MEMS 惯性测量单元在角速度测量方面十分准确，在静止状态下，角速度标准数值应为零，而测量曲线完全贴合水平轴，基本不存在零漂误差。MEMS 惯性测量单元的输出角度是角速度的一次积分，在静止状态下的零漂数据，其标准数值应为零，但是在 x、y 和 z 轴上均有较为明显随时间变化的误差，其中 x 和 y 轴上的误差随时间不断下降，而 z 轴上的误差保持了较长时间的缓慢增长。MEMS 惯性测量单元对于地磁强度的测量存在较为明显的偏差，在 x、y 和 z 轴上其波动的幅值较其他测量变量更大，其不稳定的原因可能来源于周围的磁场干扰。

首先应用基本的 Allan 方差固定估计对零漂数据进行分析,其结果见表 3。可见 MEMS 惯性测量单元对于角速度的测量十分准确,对于加速度的测量较为准确,对于角度的测量存在误差,而对于地磁强度的测量存在较大的误差。此外,MEMS 惯性测量单元对于加速度、角速度、地磁强度的测量在不同轴向上表现相近,而在角度测量方面,水平面内的 x 和 y 轴的误差较小,而竖直方向的 z 轴误差较大。

MEMS 惯性单元 Allan 方差的固定估计　　　　　　　　　表 3

	x 轴	y 轴	z 轴
加速度	0.0224	0.0252	0.0282
角速度	0.0000	0.0000	0.0000
角度	0.1407	0.1304	0.8015
地磁强度	63.6115	63.5647	64.3674

由于角速度的测量实时准确,因此随后的分析缺省该项测量指标,仅针对加速度、角度以及地磁进行 Allan 方差的进一步估计。同时,由于无重叠估计需要大量的噪声数据,难以在实践中稳定获取,在本研究中应用更加准确,效率更高的重叠估计来对绘制 Allan 方差与相关时间之间的双对数关系图,如图 14 所示。设 MEMS 惯性测量单元的噪声声源相互独立,Allan 方差是各噪声导致误差的平方和。既有的实验数据表明,由于不同误差分项出现在不同的相关时间区域内,具有统计意义上的独立性,因此,通过对比图 13,可以将 Allan 方差中的各个误差项进行分离。

从图 14(a)中可以观察到,大部分 $\sigma\text{-}\tau$ 曲线上点的斜率 $k=-1$,因此 MEMS 惯性测量单元对加速度的测量误差主要有量化噪声和角度随机游走,在 x 轴方向有局部的斜率 $k=0$,说明误差中也存在少量的零偏不稳定。相比加速度测量误差,MEMS 惯性测量单元对于角度的测量存在更复杂的噪声来源。如图 14(b)所示,在 x 轴和 z 轴方向,角度的测量误差主要体现为量化噪声和角度随机游走,而在 y 轴方向上,出现了长期的下降段、短暂的水平段与局部的上升段,说明在该方向上的噪声以量化噪声和角度随机游走为主,同时混有零偏不稳定、速率随机游走等。而在图 14(c)中,地磁强度的测量噪声仅包含量化噪声和角度随机游走。

MEMS 惯性测量单元的噪声成分主要应包括量化噪声、角度随机游走、马尔科夫噪声、正弦噪声、零偏不稳定、速率随机游走和速率斜坡。由于各噪声成分相互独立,则 Allan 方差可以表示为:

$$\sigma_y^2(\tau) = \sigma_{QN}^2 + \sigma_{ARW}^2 + \sigma_{MN}^2 + \sigma_{SN}^2 + \sigma_{BI}^2 + \sigma_{RRW}^2 + \sigma_{RR}^2 \tag{22}$$

式中:σ_{QN}^2——量化噪声;

σ_{ARW}^2——角度随机游走;

σ_{MN}^2——马尔科夫噪声;

σ_{SN}^2——正弦噪声;

σ_{BI}^2——零偏不稳定;

σ_{RRW}^2——速率随机游走;

σ_{RR}^2——速率斜坡。

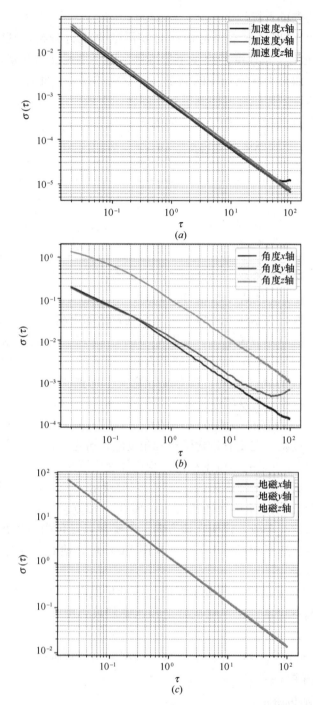

图 14　MEMS 惯性测量单元 Allan 方差与相关时间(σ-τ)双对数图
(a)加速度测量 Allan 方差与相关时间对数图；(b)角度测量 Allan 方差与相关时间对数图；
(c)地磁测量 Allan 方差与相关时间对数图

每一种误差所对应的功率谱密度及其 Allan 方差如表 4 所示。由于正弦噪声在环境中呈现出周期性的特性，可以使用其他噪声替代，同时马尔科夫噪声在不同相关时间内可以

分别使用角度随机游走和速率随机游走代替,因此 Allan 方程可以表达为:

$$\sigma_y^2(\tau) = \frac{3C_{QN}^2}{\tau^2} + \frac{C_{ARW}^2}{\tau} + \frac{2\ln 2 C_{BI}^2}{\pi} + \frac{C_{RRW}^2 \tau}{3} + \frac{C_{RR}^2 \tau^2}{2} = \sum_{n=-2}^{2} C_n \tau^n \quad (23)$$

式中:C_{QN}——量化噪声系数

C_{ARW}——角度随机游走系数

C_{BI}——零偏不稳定系数

C_{RRW}——速率随机游走系数

C_{RR}——速率斜坡系数

随机过程误差的功率谱密度与 Allan 方差系数　　　　　　表 4

随机过程误差	功率谱密度$S_\Delta(f)$	Allan 方差σ^2	$\sigma\tau$ 斜率
量化噪声	$\frac{4C_{QN}^2}{\pi}\sin^2(\pi f \tau_0), f \leq \frac{1}{2\tau_0}$	$\frac{3C_{QN}^2}{\tau^2}$	$k_{QN}=-1$
角度随机游走	C_{ARW}^2	$\frac{C_{ARW}^2}{\tau}$	$k_{ARW}=-\frac{1}{2}$
马尔科夫噪声	$\frac{q^2\tau_c^2}{1+4\pi^2 f^2 \tau_c^2}$	$\begin{cases}\frac{C_{MN}^2}{\tau}, \tau \gg \tau_c \\ \frac{C_{MN}^2}{3}, \tau \ll \tau_c\end{cases}$	$\begin{cases}-\frac{1}{2}, \tau \gg \tau_c \\ \frac{1}{2}, \tau \ll \tau_c\end{cases}$
零偏不稳定	$\begin{cases}\frac{C_{BI}^2}{2\pi f}, f \leq f_0 \\ 0, f > f_0\end{cases}$	$\frac{2\ln 2 C_{BI}^2}{\pi}$	$k_{BI}=0$
速率随机游走	$\frac{C_{RRW}^2}{2\pi f^2}$	$\frac{C_{RRW}^2 \tau}{3}$	$k_{RRW}=\frac{1}{2}$
速率斜坡	$\frac{C_{RR}^2}{8\pi^3 f^3}$	$\frac{C_{RR}^2 \tau^2}{2}$	$k_{RR}=1$

采用最小二乘拟合,对 MEMS 惯性测量的 Allan 方差进行拟合,可以得到相应的噪声系数如表 5。

随机过程误差的 Allan 方差系数　　　　　　表 5

测量变量	C_{QN}	C_{ARW}	C_{BI}	C_{RRW}	C_{RR}
	(g)	(g/h$^{0.5}$)	(g/h)	(g/h$^{1.5}$)	(g/h^2)
加速度 x 轴	−0.0003	0.0000	0.0000	0.0000	1.8582
加速度 y 轴	0.0004	0.0008	0.0000	0.0000	0.1237
加速度 z 轴	−0.0004	0.0004	0.0000	0.0457	0.2663
	(deg)	(deg/h$^{0.5}$)	(deg/h)	(deg/h$^{1.5}$)	(deg/h^2)
角度 x 轴	0.0050	0.0457	−0.0002	−0.0017	−7.5361
角度 y 轴	0.0069	−0.1147	−0.0001	0.0076	−103.2606
角度 z 轴	0.0578	0.0524	0.0000	0.0000	0.0002
	(mG)	(mG/h$^{0.5}$)	(mG/h)	(mG/h$^{1.5}$)	(mG/h^2)
地磁 x 轴	0.7940	0.7101	0.0016	−0.0005	0.0004
地磁 y 轴	0.7767	0.6118	0.0028	0.0612	−0.0005
地磁 z 轴	0.7975	1.7058	−0.0031	−0.0003	0.0144

5.3 卡尔曼滤波去噪

卡尔曼滤波的思想是使用前一时刻的位置为基准，根据移动速度、移动加速度、转动角速度、转动角加速度等测量值，通过计算或者给定可靠系数，将测量值与估计值进行融合，从而得到对下一时刻载体目标位置的最优预测，进而修正和平滑移动数据。

由于定位算法是典型的离散时间系统且不具备控制系统，因此采用无控制离散型的卡尔曼滤波，包括模型估算过程和测量过程。设目标物体的真实位置坐标为一个含有多个变量的状态向量，则模型的估算过程为：

$$x_t = \Phi x_{t-1} + \Gamma w_{t-1} \tag{24}$$

式中：x_t——t 时刻的系统状态；
$\quad\quad\Phi$——状态转移矩阵；
$\quad\quad w_t$——t 时刻的过程噪声；
$\quad\quad\Gamma$——噪声驱动矩阵。

同时其测量过程为：

$$z_t = H x_t + v_t \tag{25}$$

式中：z_t——t 时刻通过惯性测量单元直接测量的物理量；
$\quad\quad H$——测量矩阵；
$\quad\quad v_t$——t 时刻的测量噪声。

在实际项目工程中，一般测量噪声与过程噪声相互独立，满足以下条件：

$$\begin{aligned} E[w_t] &= 0 \\ E[v_t] &= 0 \\ Cov[w_i, w_j] &= P_i \delta_{i,j} \\ Cov[v_i, v_j] &= Q_i \delta_{i,j} \end{aligned} \tag{26}$$

式中：P_i——过程噪声的协方差，为非负定阵；
$\quad\quad Q_i$——测量噪声的协方差，为正定阵。

其中过程噪声的协方差是主要的系统误差来源，当其扩大时，系统趋向于不稳定，而测量噪声的协方差决定了测量数据的被信任程度，当其值较大时，滤波趋向于忽略测量值。卡尔曼滤波的应用包含以下五个步骤：

(1) 状态预测

根据上一步位置状态 x_{t-1} 预测下一步位置状态 $\hat{x}_t$，其表达式为：

$$\hat{x}_t = \Phi x_{t-1} \tag{27}$$

(2) 误差预测

根据上一步均方误差 R_{t-1} 预测下一步均方误差 $\hat{R}_t$，其表达式为：

$$\hat{R}_t = \Phi R_{t-1} \Phi^T + \Gamma P_{t-1} \Gamma^T \tag{28}$$

(3) 滤波增益

滤波增益过程，计算测量过程与估计过程的权重 Ψ_t，其表达式为：

$$\Psi_t = \hat{R}_t H^T (H \hat{R}_t H^T + Q_t)^{-1} \tag{29}$$

(4) 滤波估计

滤波估计方程，计算最优位置状态 x_t：

$$x_t = \hat{x}_t + \Psi_t(z_t - H\hat{x}_t) \tag{30}$$

(5) 均方误差更新

更新均方误差，计算最优均方误差 R_t：

$$R_t = (I - \Psi_t H)\hat{R}_t(I - \Psi_t H)^T + \Psi_t \hat{R}_t \Psi_t^T \tag{31}$$

当为了提高计算速率，适当容许累积误差时，也可以使用其简化形式：

$$R_t = (I - \Psi_t H)\hat{R}_t \tag{32}$$

这五个步骤构成了卡尔曼滤波的核心流程，通过滤波可以以最优的位置状态更新测量结果，并同时存储最优的误差估计，其结构如图 15 所示。其中 $x_{k|k-1}$ 表示在 $k-1$ 时刻对 k 时刻的状态预测值，R 表示相应的估计误差。

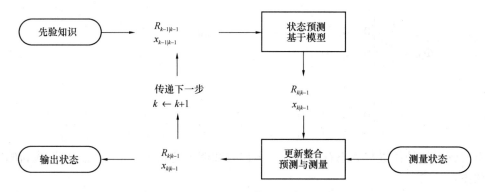

图 15　卡尔曼滤波的基本结构

在惯性测量单元的定位过程中，关注的主要变量有位置和姿态，其中位置状态包括位移坐标、运动速度和运动加速度，即：

$$x_t = (x_t, y_t, z_t, \dot{x}_t, \dot{y}_t, \dot{z}_t, \ddot{x}_t, \ddot{y}_t, \ddot{z}_t) \tag{33}$$

式中：(x_t, y_t, z_t)——t 时刻的目标位置坐标；

$(\dot{x}_t, \dot{y}_t, \dot{z}_t)$——$t$ 时刻的目标运动速度；

$(\ddot{x}_t, \ddot{y}_t, \ddot{z}_t)$——$t$ 时刻的目标运动加速度。

相对应的模型状态估算过程中的状态转移矩阵根据牛顿定律为：

$$\Phi_{\text{location}} = \begin{bmatrix} 1 & 0 & 0 & \Delta t & 0 & 0 & 0.5\Delta t^2 & 0 & 0 \\ 0 & 1 & 0 & 0 & \Delta t & 0 & 0 & 0.5\Delta t^2 & 0 \\ 0 & 0 & 1 & 0 & 0 & \Delta t & 0 & 0 & 0.5\Delta t^2 \\ 0 & 0 & 0 & 1 & 0 & 0 & \Delta t & 0 & 0 \\ 0 & 0 & 0 & 0 & 1 & 0 & 0 & \Delta t & 0 \\ 0 & 0 & 0 & 0 & 0 & 1 & 0 & 0 & \Delta t \\ 0 & 0 & 0 & 0 & 0 & 0 & 1 & 0 & 0 \\ 0 & 0 & 0 & 0 & 0 & 0 & 0 & 1 & 0 \\ 0 & 0 & 0 & 0 & 0 & 0 & 0 & 0 & 1 \end{bmatrix} \tag{34}$$

式中：Δt——滤波中每一步的时间步长，若为实时滤波，则等于采样周期。

而目标姿态的状态包括姿态角度、运动角速度和运动角加速度,即:
$$x_t = (\alpha_t, \beta_t, \gamma_t, \dot{\alpha}_t, \dot{\beta}_t, \dot{\gamma}_t, \ddot{\alpha}_t, \ddot{\beta}_t, \ddot{\gamma}_t) \tag{35}$$

式中:$(\alpha_t, \beta_t, \gamma_t)$——$t$ 时刻的目标姿态角度;

$(\dot{\alpha}_t, \dot{\beta}_t, \dot{\gamma}_t)$——$t$ 时刻的目标运动角速度;

$(\ddot{\alpha}_t, \ddot{\beta}_t, \ddot{\gamma}_t)$——$t$ 时刻的目标运动角加速度。

在本文中姿态的角度使用欧拉角表示,包括章动角(Pitch)、旋进角(Yaw)和自转角(Roll),也被称为俯仰角、偏航角和滚转角,分别对应于绕 y 轴、z 轴和 x 轴的旋转。其相对应的模型估算过程中的状态转移矩阵为:

$$\Phi_{\text{posture}} = \begin{bmatrix} 1 & 0 & 0 & \Delta t & 0 & 0 & 0.5\Delta t^2 & 0 & 0 \\ 0 & 1 & 0 & 0 & \Delta t & 0 & 0 & 0.5\Delta t^2 & 0 \\ 0 & 0 & 1 & 0 & 0 & \Delta t & 0 & 0 & 0.5\Delta t^2 \\ 0 & 0 & 0 & 1 & 0 & 0 & \Delta t & 0 & 0 \\ 0 & 0 & 0 & 0 & 1 & 0 & 0 & \Delta t & 0 \\ 0 & 0 & 0 & 0 & 0 & 1 & 0 & 0 & \Delta t \\ 0 & 0 & 0 & 0 & 0 & 0 & 1 & 0 & 0 \\ 0 & 0 & 0 & 0 & 0 & 0 & 0 & 1 & 0 \\ 0 & 0 & 0 & 0 & 0 & 0 & 0 & 0 & 1 \end{bmatrix} \tag{36}$$

考虑到目标的位置与姿态存在耦合,将位置和姿态的状态变量相互连接形成新的完整位姿状态:

$$x_t = (x_t, y_t, z_t, \dot{x}_t, \dot{y}_t, \dot{z}_t, \ddot{x}_t, \ddot{y}_t, \ddot{z}_t, \alpha_t, \beta_t, \gamma_t, \dot{\alpha}_t, \dot{\beta}_t, \dot{\gamma}_t, \ddot{\alpha}_t, \ddot{\beta}_t, \ddot{\gamma}_t) \tag{37}$$

则卡尔曼滤波的模型估算过程具有如下形式:

$$x_t = \begin{bmatrix} \Phi_{\text{location}} & 0 \\ 0 & \Phi_{\text{posture}} \end{bmatrix} x_{t-1} + \Gamma w_{t-1} \tag{38}$$

式中:Φ_{location}——位置状态转换矩阵;

Φ_{posture}——姿态状态转换矩阵;

Γ——噪声驱动矩阵;

w_{t-1}——过程噪声。

测量过程具有如下形式:

$$z_t = \begin{bmatrix} 0 & 0 & I & 0 & 0 & 0 \\ 0 & 0 & 0 & I & 0 & 0 \end{bmatrix} x_t + v_t \tag{39}$$

式中:I——单位阵;

v_t——测量噪声。

5.4 扩展卡尔曼滤波去噪

在工程项目现场,由于存在大量的遮挡、反射以及干扰信号,会导致基于线性的卡尔曼滤波的最优状态预测不精确,主要体现在模型估计以及观测过程中的噪声信号往往并非标准高斯噪声,在通过长期积累以后,对误差的估计将出现较大的偏差,从而导致对位置坐标的估计精度逐步下降。此外由于多种测量定位系统的测量周期并不相同且目标对象的移动是连续过程，为了适应非线性连续的环境，离散线性的卡尔曼被扩展到连续非线性区域，其状态转移矩阵和测量矩阵就等效成为状态转移函数和测量函数，其模型估算过程为：

$$x_{t+\Delta t} = \Phi(x_t,t) + \Gamma(x_t,t)w_t \tag{40}$$

式中：x_t——t 时刻的系统状态；

　　　Φ——状态转移函数；

　　　w_t——t 时刻的过程噪声；

　　　Γ——噪声驱动函数。

测量过程为：

$$z_t = H(x_t,t) + v_t \tag{41}$$

式中：z_t——t 时刻通过惯性测量单元直接测量的物理量；

　　　H——测量函数；

　　　v_t——t 时刻的测量噪声。

若在 t 时刻系统状态的估计为 $\hat{x}_t$，则将状态转移函数与测量函数在估计值处根据泰勒级数进行展开，根据精度需求去除高阶项，实现在估计值附近处的线性化。以忽略二阶以上的高阶项为例，扩展卡尔曼滤波的模型估计过程为：

$$x_{t+\Delta t} = \Phi(\hat{x}_t,t) + \frac{\partial \Phi(\hat{x}_t,t)}{\partial \hat{x}_t}(x_t - \hat{x}_t) + \Gamma(\hat{x}_t,t)w_t \tag{42}$$

式中：x_t——t 时刻的系统状态；

　　　Φ——状态转移函数；

　　　w_t——t 时刻的过程噪声；

　　　Γ——噪声驱动函数。

测量过程为：

$$z_t = H(\hat{x}_t,t) + \frac{\partial H(\hat{x}_t,t)}{\partial \hat{x}_t}(x_t - \hat{x}_t) + v_t \tag{43}$$

式中：z_t——t 时刻通过惯性测量单元直接测量的物理量；

　　　H——测量函数；

　　　v_t——t 时刻的测量噪声。

通过整理并简化，对于非线性系统线性化后的模型估计过程为：

$$x_{t+\Delta t} = F_t x_t + U_t + L_t w_t$$

$$F_t = \frac{\partial \Phi(\hat{x}_t,t)}{\partial \hat{x}_t}$$

$$U_t = \Phi(\hat{x}_t,t) - \frac{\partial \Phi(\hat{x}_t,t)}{\partial \hat{x}_t}\hat{x}_t$$

$$L_t = \Gamma(\hat{x}_t,t) \tag{44}$$

测量过程为：

$$z_t = G_t x_t + V_t + v_t$$

$$G_t = \frac{\partial H(\hat{x}_t,t)}{\partial \hat{x}_t} \tag{45}$$

$$V_t = H(\hat{x}_t, t) - \frac{\partial H(\hat{x}_t, t)}{\partial \hat{x}_t} \hat{x}_t$$

由于使用惯性测量单元进行定位的位置状态向量长度较大，在使用扩展卡尔曼滤波时，对过程函数、观测函数难以求导，因此并不适合在实时定位系统的每一次测量周期中应用，因此本文仅使用扩展卡尔曼滤波进行周期性校验。

5.5 应用案例——混凝土工程振捣过程质量实时监控

在浇筑混凝土建筑构件时，使用混凝土振捣可以使混凝土密实结合，排除其中混杂的气泡，消除混凝土构件表面蜂窝麻面等现象，保证混凝土强度和构件的质量。其原理是在混凝土振动的过程中，混凝土内部颗粒物之间的内摩擦力和黏着力减小，呈现出液体状态，骨料之间相互滑动并重新排列，使骨料缝隙之间的气体被挤出，从而达到密实的效果。

然而由于混凝土工程的隐蔽性，在实际工程中混凝土振捣的过程往往难以直接观测，导致混凝土的建造过程难以进行有效的质量过程控制。可能引起的质量安全隐患包括混凝土的漏振、少振和过振，降低了建筑混凝土构件的承载能力和整体性，同时削弱了混凝土构件正常工作的功能需求，造成建筑使用缺陷。

因此将基于IMU的施工过程质量实时监控传感器安置在混凝土振捣棒尾部即可实时提供可视化的振捣棒方位，据此实现对于隐蔽工程的施工过程质量控制与管理。

6. 结论与展望

本文基于惯性导航系统建立针对混凝土结构隐蔽工程施工质量的过程监控原理与方法和实时监控系统，类似黑匣子，实时反馈并记录了隐蔽工程在建造过程中的相关动态参数，主要包括混凝土振捣棒的位置、深度、施工环境的温度、湿度等。一方面，提供了实时的混凝土建筑结构隐蔽工程的可视化施工过程，施工人员和管理人员可以根据常规工程来对隐蔽工程进行项目质量过程控制，另一方面，记录了隐蔽工程的完整施工过程，一旦隐蔽工程出现了质量安全问题，可以通过追查相应的隐蔽工程黑匣子来追溯相关的施工过程，确定质量安全问题的原因和形成过程。

参考文献

[1] 戴波．建筑工程质量管理新技术推广策略研究[D]．City：重庆大学，2008．
[2] 韩国波．基于全寿命周期的建筑工程质量监管模式及方法研究[D]．City：中国矿业大学（北京），2013．
[3] 周福新，黄莹，李清立．基于流程化视角的建筑工程项目质量管理研究[J]．工程管理学报，2016，(01)：98-102．
[4] 蒋勤俭，黄清杰，常双九，et al. 装配式混凝土结构工程质量管理与验收[J]．工程质量，2016，(04)：5-13．
[5] Bell L. C., McCullouch B. G. Bar code applications in construction[J]. Journal of construction engineering and management, 1988, 114(2): 263-278.
[6] Rasdorf W. J., Herbert M. J. Bar coding in construction engineering[J]. Journal of construction engi-

neering and management, 1990, 116(2): 261-280.

[7] Cheng M.-Y., Chen J.-C. Integrating barcode and GIS for monitoring construction progress[J]. Automation in Construction, 2002, 11(1): 23-33.

[8] Jaselskis E. J., Anderson M. R., Jahren C. T., et al. Radio-frequency identification applications in construction industry[J]. Journal of construction engineering and management, 1995, 121(2): 189-196.

[9] Song J., Haas C. T., Caldas C. H. Tracking the location of materials on construction job sites[J]. Journal of construction engineering and management, 2006, 132(9): 911-918.

[10] Goodrum P. M., McLaren M. A., Durfee A. The application of active radio frequency identification technology for tool tracking on construction job sites[J]. Automation in Construction, 2006, 15(3): 292-302.

[11] Teizer J., Venugopal M., Walia A. Ultrawideband for automated real-time three-dimensional location sensing for workforce, equipment, and material positioning and tracking[J]. Transportation Research Record: Journal of the Transportation Research Board, 2008, (2081): 56-64.

[12] Cheng T., Venugopal M., Teizer J., et al. Performance evaluation of ultra wideband technology for construction resource location tracking in harsh environments[J]. Automation in Construction, 2011, 20(8): 1173-1184.

[13] Li H., Yang X., Wang F., et al. Stochastic state sequence model to predict construction site safety states through Real-Time Location Systems[J]. Safety science, 2016, 84: 78-87.

[14] 刘文平. 基于BIM与定位技术的施工事故预警机制研究[D]. City: 清华大学, 2015.

[15] 孙凌云. 施工自动定位跟踪技术选择的决策支持研究[D]. City: 大连理工大学, 2011.

[16] 杨晓波. 公路高瓦斯隧道施工人员定位及管理系统研究[D]. City: 重庆交通大学, 2014.

[17] 金雷鸣. 港口工程施工定位技术应用研究[D]. City: 大连海事大学, 2012.

[18] Bohn J. S., Teizer J. Benefits and barriers of construction project monitoring using high-resolution automated cameras[J]. Journal of construction engineering and management, 2009, 136(6): 632-640.

[19] Teizer J., Vela P. A. Personnel tracking on construction sites using video cameras[J]. Advanced Engineering Informatics, 2009, 23(4): 452-462.

[20] Yang J., Vela P., Teizer J., et al. Vision-based tower crane tracking for understanding construction activity[J]. Journal of Computing in Civil Engineering, 2012, 28(1): 103-112.

[21] Brilakis I., Park M.-W., Jog G. Automated vision tracking of project related entities[J]. Advanced Engineering Informatics, 2011, 25(4): 713-724.

[22] GB 50300—2013. 建筑工程施工质量验收统一标准[S]. 2013.

[23] GB 50204—2015. 混凝土结构工程施工质量验收规范[S]. 2015.

[24] GB 506666—2011. 混凝土结构工程施工规范[S]. 2011.

8 钢管再生混凝土构件力学性能研究（摘要）

韩林海，吕晚晴

（清华大学土木工程系）

随着我国城镇化建设的快速推进，新老建筑物频繁更替扩大了对建筑原材料的需求，同时又会产生大量的建筑废弃物，以天然骨料为主的建筑原材料短缺问题和以废弃混凝土为主的建筑垃圾剧增问题日益突出。将废弃混凝土通过破碎、加工等处理工序形成再生混凝土并用于工程结构是建筑业可持续发展方向之一。而以往的研究成果表明，再生混凝土的徐变较大，且破碎过程中可能产生微裂缝。因此与普通混凝土相比，再生混凝土徐变大、强度低、耐久性差、并带有一定的性能离散性。目前再生混凝土结构工程应用尚局限于非承重结构或者荷载相对较小的多层建筑等。

钢管再生混凝土是指在钢管中填充再生混凝土而形成且钢管及其核心再生混凝土能共同承受外荷载作用的结构构件。其中，钢管可采用普通强度（钢材屈曲强度＜460N/mm^2）和高强（钢材屈曲强度≥460N/mm^2）碳素钢管或不锈钢管，再生混凝土强度等级不超过 60 N/mm^2，再生粗骨料替代率为 0~100%。

已开展的普通强度钢管再生混凝土柱的研究结果初步表明，外包钢管的约束作用可有效地改善再生混凝土的脆性，提高其塑性和韧性性能[1]。钢管再生混凝土结构在受力过程中，管内再生混凝土的存在则可有效地延缓或避免钢管的局部屈曲，从而使结构具有承载力高、抗震和抗火性能好等特点。因此，再生混凝土和钢管有效组合形成钢管再生混凝土，能够有效促进再生骨料的高效高层次利用，开辟再生骨料在结构工程中应用新前景，符合可持续土木工程结构发展的趋势，可以预计，钢管再生混凝土是钢-混凝土组合结构中的一种性能优的新形式，将是垃圾资源化的有效途径之一。

以往，研究者们对钢管再生混凝土结构的力学性能已开展了一些研究，包括对钢管-再生混凝土界面力学性能的研究以及对钢管再生混凝土构件力学性能的研究等，如文献[2]开展了普通钢管再生混凝土构件的推出滑移试验，分析了截面形式、再生粗骨料取代率、混凝土强度等级等参数对其粘结性能的影响，文献[3-5]开展了圆形和方形钢管再生混凝土轴压构件、纯弯构件和压弯构件的试验研究与有限元模拟分析，文献[6-8]研究了地震作用下不同再生粗骨料取代率对钢管再生混凝土构件破坏形式、滞回特性和承载性能的影响规律，文献[9]研究了不同再生粗骨料取代率的钢管再生混凝土构件在撞击作用下的力学性能，文献[10-12]研究了高温后钢管再生混凝土构件的承载能力和高温作用的影响等。然而，目前在钢管再生混凝土结构领域仍缺乏一套成熟的理论体系及设计方法。

为此，本文第一作者领导的课题组对钢管再生混凝土结构进行了一系列的试验研究和理论分析，研究了钢管-再生混凝土界面的工作机理、钢管再生混凝土构件的静力性能、抗震性能与抗火性能。具体研究工作包括：

（1）针对钢管-再生混凝土结构界面的工作机理，进行了钢管再生混凝土构件的推出滑移试验，分析了钢管-再生混凝土构件的粘结机理，研究了截面形式、截面尺寸、再生粗骨料取代率、再生混凝土强度等参数对粘结强度的影响，给出了适用于计算钢管再生混凝土结构设计粘结强度的实用计算方法。

（2）对钢管再生混凝土构件的在一次加载下的静力性能，进行了钢管再生混凝土轴压短柱、纯弯构件和压弯构件的试验研究[13,14]，研究结果表明，在所研究参数范围内，钢管再生混凝土轴压、纯弯和偏压构件的承载力和刚度均低于相应钢管普通混凝土试件，但差异并不十分显著；为了研究再生混凝土收缩导致的脱空对其轴压及偏压性能的影响，进行了带脱空缺陷的圆形和方形钢管再生混凝土轴压构件和带脱空缺陷的圆形和方形钢管再生混凝土偏压构件的试验，研究了钢管再生混凝土构件轴压及偏压承载力的计算方法；进行了钢管再生混凝土构件的抗剪试验研究，研究了不同剪跨比、轴压比、混凝土强度和含钢率对钢管再生混凝土柱抗剪性能及抗剪承载力的影响，试验过程中可观察到构件出现了剪切型破坏与弯曲型破坏两种破坏形态，随再生粗骨料替代率的提高，试件破坏后内部混凝土损伤逐渐加重；对于复合受力情况，进行了方形和圆形钢管再生混凝土构件在压扭和弯扭作用下的试验研究，分析了截面形式、轴压比与再生粗骨料替代率等参数的影响，提出了方形和圆形钢管再生混凝土柱压扭及弯扭承载力的计算公式。

（3）针对钢管再生混凝土构件在长期荷载下的力学性能，对钢管再生混凝土构件进行了长期变形（收缩变形和徐变变形）的测试[15]，同时对徐变变形测试结束后试件的轴压力学性能进行了测试。测试结果显示，与相应钢管普通混凝土试件相比，钢管再生混凝土试件核心再生混凝土的收缩和徐变变形高约6%～23%。考虑长细比、约束效应系数、荷载偏心率和长期荷载比的影响，给出了长期荷载作用下钢管再生混凝土构件承载力影响系数k_{cr}的计算公式。

（4）针对钢管再生混凝土柱的抗震性能，进行了恒定轴压力和低周往复荷载作用下钢管再生混凝土柱的力学性能试验研究[16]，分析比较了钢管再生混凝土柱的滞回特性，并在系统参数分析结果的基础上，提出了钢管再生混凝土压弯构件的M-Φ和P-Δ滞回模型。

（5）针对钢管再生混凝土柱的抗火性能，进行了钢管再生混凝土构件的抗火性能试验，研究了再生骨料取代率、混凝土材料强度、偏心距和荷载比等参数对构件抗火性能的影响。研究结果表明，在试验参数范围内，矩形钢管再生混凝土柱的耐火极限随偏心率和荷载比的增大而减小，对于C40混凝土，再生粗骨料取代率为50%时相应构件的耐火极限最高，钢管再生混凝土柱耐火极限普遍高于钢管普通混凝土柱；对于C60混凝土，柱的耐火极限随着取代率的增加而降低，但总体上，再生粗骨料取代率对偏压柱的耐火极限的影响较小。

上述研究工作为进一步深入研究钢管再生混凝土构件的力学性能和设计方法创造了条件。

课题组的研究工作得到苏州-清华创新引领行动专项项目的支持（项目号：No.2016SZ02122）。

关键词：钢管再生混凝土；界面工作机理；静力性能；抗震性能；抗火性能

参考文献

[1] 韩林海,李威,陶忠,王文达. 现代组合结构和混合结构-试验、理论和方法(第二版). 北京:科学出版社,2017.

[2] 陈宗平,徐金俊,薛建阳,苏益声. 钢管再生混凝土粘结滑移推出试验及黏结强度计算. 土木工程学报,2013,46(3):49-58.

[3] 肖建庄,杨洁,黄一杰,王正平. 钢管约束再生混凝土轴压试验研究. 建筑结构学报,2011,32(6):92-98.

[4] 陈宗平,李启良,张向冈,薛建阳,陈宝春. 钢管再生混凝土偏压柱受力性能及承载力计算. 土木工程学报,2012,45(10):72-80.

[5] 杨有福. 钢管再生混凝土构件受力机理研究. 工业建筑,2007(12):7-12.

[6] Ou YC, Chen HH. Cyclic behavior of reinforced concrete beams with corroded transverse steel reinforcement. Journal of Structural Engineering, 2014, 140(9):4014050.

[7] 黄一杰,肖建庄. 钢管再生混凝土柱抗震性能与损伤评价. 同济大学学报(自然科学版),2013,41(3):330-335.

[8] He A, Cai J, Chen QJ, Liu X, Huang P, Tang XL. Seismic behaviour of steel-jacket retrofitted reinforced concrete columns with recycled aggregate concrete. Construction and Building Materials, 2018, 158:624-639.

[9] Li W, Luo Z, Wu C, Tam VWY, Duan WH, Shah SP. Experimental and numerical studies on impact behaviors of recycled aggregate concrete-filled steel tube after exposure to elevated temperature. Materials & Design, 2017, 136:103-118.

[10] Yang YF, Hou R. Experimental behaviour of RACFST stub columns after exposed to high temperatures. Thin-Walled Structures, 2012, 59:1-10.

[11] 王兵,刘晓,赵磊,高璐. 高温后方钢管再生混凝土短柱轴压受力性能分析. 工程力学,2015,S1:153-158.

[12] 陈宗平,经承贵,薛建阳,张超荣. 高温后圆钢管再生混凝土偏压柱力学性能研究. 工业建筑,2014,44(11):19-24,181.

[13] Yang You-Fu and Han Lin-Hai. 2006. Experimental behaviour of recycled aggregate concrete filled steel tubular columns. Journal of Constructional Steel Research, 62(12):1310-1324.

[14] Yang You-Fu and Han Lin-Hai. 2006. Compressive 2 and flexural behaviour of recycled aggregate concrete filled steel tubes (RACFST) under short-term loadings. Steel and Composite Structures, 6(3):257-284.

[15] Yang You-Fu, Han Lin-Hai and Wu Xin. 2008. Concrete shrinkage and creep in recycled aggregate concrete-filled steel tubes. Advances in Structural Engineering, 11(4):383-396.

[16] Yang You-Fu, Han Lin-Hai and Zhu Lin-Tao. 2009. Experimental performance of recycled aggregate concrete-filled circular steel tubular columns subjected to cyclic flexural loadings. Advances in Structural Engineering, 12(2):183-194.

9 预应力正交胶合木剪力墙抗侧性能分析

何敏娟，孙晓峰，李 征

（同济大学土木工程学院）

摘 要：正交胶合木（Cross-laminated timber，简称CLT）是一种新型的工程木产品，由于具有强度高、承压刚度大、整体性优、面外抗弯和抗剪性能好等优点，越来越多地被用于建造多高层木结构建筑。论文首先基于该种新型工程木产品，试验研究了其面外抗剪、抗弯及面内承压力学性能；然后开发出耗能件及用于普通CLT剪力墙的金属连接件，并通过节点试验获取耗能件及金属连接件的力学性能。在获取相关材性参数及节点性能参数后，设计了规格基本相同的无耗能件预应力CLT剪力墙、含耗能件预应力CLT剪力墙及普通CLT剪力墙试件，基于试验对比了这三类CLT剪力墙的抗侧性能。试验结果表明：与普通CLT剪力墙相比，预应力CLT剪力墙的初试抗侧刚度及极限承载力显著提高，并且通过施加预应力，改善了侧向荷载下普通CLT墙体节点区域较早被撕裂的脆性破坏模式；此外，可通过添加耗能件来提高预应力CLT剪力墙的耗能能力。

关键词：正交胶合木；多高层木结构；材性试验；耗能件；预应力CLT剪力墙

The Lateral Performance of Post-Tensioned Cross-Laminated Timber (CLT) Shear Walls

Minjuan He, Xiaofeng Sun, Zheng Li

(School of Civil Engineering, Tongji University)

Abstract: Cross-laminated timber is an innovative engineering timber product. It is used in mid-rise and high-rise timber building more and more widely due to its high strength, large compressive stiffness, good integrity, and excellent out-of-plane shear and bending properties. In this study, the out-of-plane properties in bending and shear, and the in-plane compressive properties of CLT panels were tested. Then the energy-dissipating connections and the metal connections (i.e. hold down and angle bracket) used in conventional CLT walls were developed. The mechanical properties of these connections were tested. Based on the experimental properties of the CLT panels and the developed connections, the specimens of post-tensioned single-panel CLT wall, post-tensioned double-panel CLT wall with energy-dissipating connections and conventional single-panel CLT wall were designed and tested. The lateral performance of the three types of CLT walls were studied and compared. It was found that the lateral stiffness and ultimate load resisting capacity of post-tensioned CLT walls were larger that of conventional CLT walls significantly. The failure mode of the conventional CLT walls, namely, the brittle failure in the area of the connections, was avoided by employing the post-tensioned steel strands. In addition, the energy-dissipating capacity of post-tensioned single-panel CLT walls could be increased significantly using the developed energy-dissipating connection.

Keywords: Cross-laminated timber, multi-storey timber structures, material tests, energy-dissipating connection, post-tensioned CLT shear wall

1. 引言

20世纪末德国和澳洲的一些住宅建筑最先使用正交胶合木材料。正交胶合木（Cross-laminated timber，简称CLT）产品是一种至少由三层实木锯材组坯，采用结构胶黏剂压制而成的工厂预制工程木产品，其具有承压强度高、正交垂直方向材性较接近、工厂预制化程度高等优点，越来越多地被运用于现代木结构尤其是多高层木结构建筑的建造中[1]。当前北美及欧洲的CLT产品主要以云杉-松-冷杉（Spruce-pine-fir，简称SPF）及落叶松树种为主，并且规范PRG 320[2]对该类CLT的强度分级标准做了相关规定。

当前国外针对CLT力学理论的研究已涵盖了各个方面，在节点方面，Gavric等[2,3]详细研究了用于CLT结构的不同类型节点的滞回特性，计算了诸如强度、刚度、延性比、耗能能力等力学指标参数。Schneider等[4-6]试验研究了两种角钢类型和五种钉类型组合成的六种角钢节点连接形式的力学性能。在普通CLT剪力墙方面，Dujic[7]等试验研究了CLT墙面开有门窗洞口对其抗侧性能的影响，研究表明普通CLT剪力墙的破坏主要集中在墙肢底部容易发生脆性破坏金属连接区域。Ceccotti[8,9]等试验研究了竖向荷载、墙面开洞、连接件中的紧固件数目、连接件数量和布置位置等因素对CLT墙体抗侧性能影响。Popovski等[10]等通过试验研究了32片CLT墙体的抗侧性能，分析了竖向荷载、加载制度、墙板开洞、墙体平面比等对墙体抗侧性能的影响。在结构体系方面，Ceccotti等[11]针对七层CLT坡屋顶结构开展了足尺振动台试验，试验结果表明CLT剪力墙结构具有良好的抗震性能，并建议了CLT剪力墙结构的水平地震力折减系数（Reduction factor）。此外，最近，Akbas等[12]和Ganey等[13]通过试验研究了竖向连续的预应力CLT剪力墙的抗侧力性能，并建立了理论分析模型。

普通CLT剪力墙在侧向荷载下容易较早产生节点附近墙板撕裂、金属连接件变形过大、节点区域紧固件被拔出等"弱节点、强构件"的破坏模式，墙板自身强度得不到充分发挥。为了解决上述问题，本文将预应力筋引入普通CLT剪力墙结构体系，设计了具有自复位功能的预应力CLT剪力墙结构，并通过安装耗能件来提高该种墙体的耗能件。试验总共分为三个阶段：（1）通过试验研究所用CLT板的面外抗弯、抗剪及面内受压力学性能；（2）开发预应力CLT剪力墙所用耗能件及普通CLT剪力墙所用金属连接节点，通过试验获取其力学性能参数；（3）设计预应力CLT剪力墙、普通CLT剪力墙及带有耗能件的预应力CLT剪力墙，通过试验对比其抗侧力性能。下面将分别介绍各阶段试验研究及成果。

2. CLT材料性能试验

2.1 CLT板面外抗弯试验

2.1.1 试验装置

试验共测试了20组CLT板弯曲试件（图1），单层层板厚度35mm，总厚度均为

175mm，CLT 所用铁杉锯材（截面尺寸：140mm×35mm）仅上下面涂胶。在这 20 组试件中，10 组 CLT 试件用于测试 CLT 板强轴方向的弯曲性能，其 1、3、5 层层板为顺纹层，2、4 层为横纹层；另外 10 组 CLT 试件用于测试 CLT 板弱轴方向弯曲性能，其 1、3、5 层层板为横纹层，2、4 层为顺纹层。根据 PRG 320[14]中抗弯试件的长厚比约 30 的规定，所有 20 组抗弯 CLT 板的长度取为 5000mm（试件长厚比 28.6），试件宽度根据规范取 310mm。根据 GB/T 50329[15]，采用对称三分点加载

图 1 CLT 板抗弯试验

装置（图 2），两个加载点之间距离为梁截面高度 6 倍，梁的纯弯段竖向挠度通过 U 形挠度测量装置获得，U 形装置的两端（水平间距为梁截面高度 5 倍）钉在梁的中性轴上。试验加载速度 0.64 cm/min。

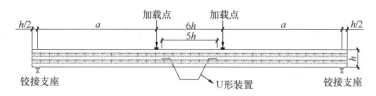

图 2 CLT 板抗弯试验加载装置

2.1.2 面外抗弯力学特性

强轴方向 10 组受弯 CLT 板的荷载-跨中挠度曲线如图 3 所示，位移为固定于前后两个 U 形装置跨中的位移计读数平均值，10 组受弯 CLT 板在加载阶段几乎均处于弹性阶段，在达到极限承载力后发生比较突然的脆性破坏，并且承载力降至极限承载力的 20%～30%。试件 9 和试件 10 在加载过程中发生脱胶破后承载能力下降，该两组试件极限承载力较低。除去脱胶的试件不计，其余 8 组试件的极限承载力范围为 42.07～54.35kN，初始弹性刚度范围为 518.9～657.9kN/m。图 4 为弱轴方向 10 组受弯 CLT 板的荷载-跨中挠度曲线。与强轴方向的受弯试件相比，弱轴方向受弯试件的塑性要稍好，在接近于极限承载力时，试件的抗弯刚度略有下降。弱轴受弯试件在极限承载力处发生脆性破坏后，其承载力降至接近于零，试验中也发现弱轴方向受弯试件的脆性破坏要更突然、更剧烈。试件 10 最终发生脱胶破坏，但其承载力并未因此显著降低，这是由于试件 10 的锯材抗拉强度要高于其他试件，当其跨中挠度较大时，由变形造成的内力通过层板间的胶合面脱落来释放。除去脱胶的试件不计，其余 9 组试件的极限承载力范围为 20.66～26.01kN，初始弹性刚度范围为 140.4～174.6kN/m。此时，承载力和弹性刚度分别降至强轴方向的 50% 和 30% 左右，这是由于弱轴方向受弯试件的底部第 5 层层板（横纹层）对截面抗弯刚度几乎没有贡献，导致截面有效高度折减所致。

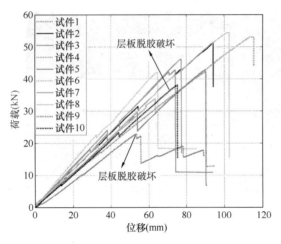
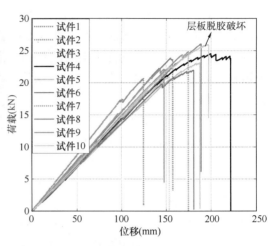

图 3 受弯试件荷载-挠度曲线（强轴方向）　　　图 4 受弯试件荷载-挠度曲线（强轴方向）

2.1.3 破坏模式

强轴方向 CLT 板受弯试件的破坏模式（图 5）包括：顺纹层受拉破坏、滚剪破坏、横纹层剪切破坏。一般情况下滚剪破坏模式伴随着其他破坏模式共同发生。弱轴方向 CLT 板受弯试件的破坏模式（图 6）包括：顺纹层受拉破坏、顺纹层受拉及撕裂破坏弱轴方向受弯 CLT 板主要以第四层层板（顺纹层）受拉破坏为主，若干试件除第四层发生受拉破坏外，第二层层板（顺纹层）发生撕裂破坏。弱轴方向 CLT 板受弯试件破坏地更彻底、剧烈。

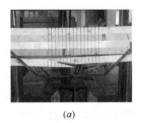
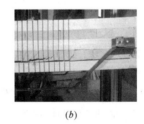

(a)　　　　　　　　　(b)　　　　　　　　　(c)

图 5 受弯试件破坏模式（强轴方向）

（a）顺纹层受拉破坏；（b）滚剪破坏；（c）横纹层剪切破坏

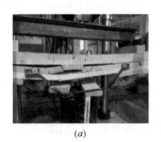
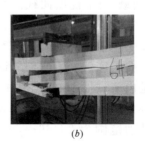

(a)　　　　　　　　　(b)

图 6 受弯试件破坏模式（弱轴方向）

（a）受拉破坏；（b）受拉及撕裂破坏

2.2 CLT板面外抗剪试验

2.2.1 试验装置

共测试了20组CLT板剪切试件（图7），试件所用锯材规格与抗弯试件的相同。类似的，10组用于强轴方向抗剪性能测试的试件采用1、3、5层顺纹组胚，另10组用于弱轴方向抗剪性能测试的试件采用2、4层顺纹组胚。根据PRG 320[14]中抗剪试件的长厚比5~6的规定，所有20组抗剪CLT板的长度取为1000 mm（试件长厚比5.7），试件宽度310 mm。采用跨中单点加载（图8），记录跨中竖向挠度，试验试验加载速度1.27 mm/min。

 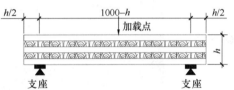

图7 CLT板抗剪试验　　　　图8 CLT板抗剪试验加载装置

2.2.2 面外抗剪力学特性

强轴方向10组受剪CLT板的荷载-跨中挠度曲线如图9所示，受剪试件在荷载接近于极限承载力时，曲线斜率有所降低，CLT板表现出一定的塑性变形。10组强轴抗剪试件中有两组最终由于发生脱胶破坏而承载力降低，除去该两组不计，强轴方向受剪试件的极限承载力范围为71.25~86.34 kN，初始弹性刚度范围为12031.8~15048.1 kN。弱轴方向10组受剪CLT板的荷载-跨中挠度曲线如图10所示，几乎所有曲线的竖向荷载都存

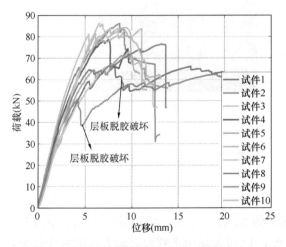 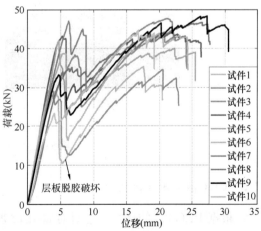

图9 受剪试件荷载-挠度曲线（强轴方向）　　图10 受剪试件荷载-挠度曲线（弱轴方向）

在两个极值点，试件在加载至第一极值点处发生破坏，承载力下降，继续加载后，试件的承载力先降后升至第二极值点，最终于第二极值点处试件发生整体脱胶破坏而彻底丧失承载能力。弱轴方向抗剪试件的极限承载力范围为 31.52~46.23 kN，初始弹性刚度范围为 7698.8~10125.1 kN/m。此外，弱轴方向 CLT 板的极限承载力和初试刚度约为强轴方向的 50% 和 65%，这是由于弱轴方向受剪试件的底部第 5 层层板（横纹层）对截面抗剪刚度几乎没有贡献所致。

2.2.3 破坏模式

强轴方向受剪试件的破坏模式（图 11）包括：横纹层滚剪破坏、横纹层剪切破坏。弱轴方向受剪试件的破坏模式（图 12）包括：顺纹层受拉（撕裂）破坏、横纹层受拉破坏。

 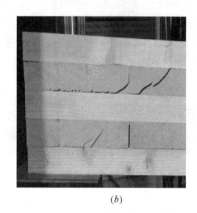

(a) (b)

图 11 受剪试件破坏模式（强轴）
(a) 横纹层滚剪破坏；(b) 横纹层剪切破坏

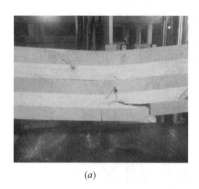 

(a) (b)

图 12 受剪试件破坏模式（弱轴）
(a) 顺纹层受拉破坏 (b) 横纹层受拉破坏

2.3 CLT 板内承压试验

根据 PRG320 的规定，通过对如图 13 所示的 20 组 80 mm×80 mm×175 mm 的 CLT 木块（10 组木块用于获取强轴方向的抗压性能，10 组木块用于获取弱轴方向的抗压性能）进行承压试验，来获取 CLT 板面内强轴方向和弱轴方向的抗压性能。面内受压试件的应力及应变关系曲线见图 14、图 15。所用 CLT 板面内强轴方向抗压强度及刚度为 18.3MPa 及 6851.4MPa，弱轴方向的抗压强度及刚度为 14.4MPa 及 4893.9MPa。

图 13 面内受压试件

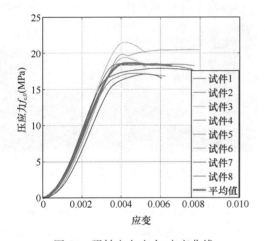

图 14 强轴方向应力-应变曲线

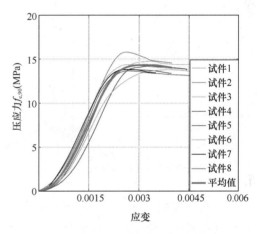

图 15 弱轴方向应力-应变曲线

3. 节点性能试验

3.1 耗能件力学性能试验

3.1.1 试件设计及加载制度

本文基于 UFP（U-shaped Flexural Plate，U 型弯板）开发出新型软钢阻尼器 OFP（O-shaped Flexural Plate，O 型弯板），试件成品和安装过程见图 16。OFP 由 Q235 钢制成，上下各有一段弯曲段，抗震时墙体间错动而产生位移差，OFP 剪切变形，弯曲段受弯屈服，实现能量耗散。OFP 的尺寸如图 17 所示，弯曲段直径为 30mm，厚度为 5mm，宽度为 60mm，钢插板长度为 150mm，高度为 100mm，厚度为 5mm，与 OFP 所连接的 CLT 木墙板端面间距为 35mm，螺栓孔规格均为 M16，螺栓规格为 8.8 级。OFP 剪切试验加载装置如图 18 所示。

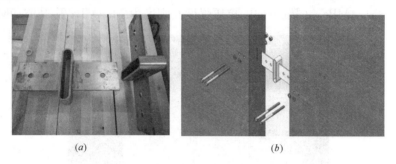

图 16　OFP 构造和安装过程
(*a*) OFP 构造；(*b*) OFP 安装过程

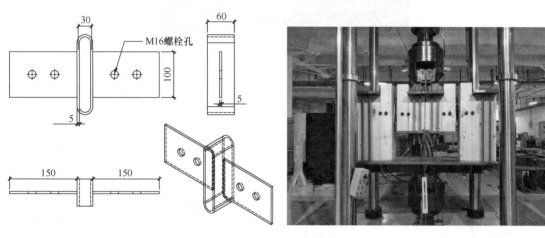

图 17　OFP 的尺寸　　　　　　　图 18　OFP 剪切试验加载装置

参考欧洲规范 Eurocode BS-EN-12512-2001[16]，往复加载采用短程三循环（或短循环）和全程循环（全循环）两种制度。短循环按单调加载得到的破坏位移为目标值，循环往复加载三周，如图 19 所示。全循环按单调加载得到的屈服位移为目标值，先以位移目标值的 25%、50%、75%、100% 往复循环各加载 1 次，然后按位移目标值的 2 倍、4 倍、6 倍等各循环 3 次，直至位移达到 30mm，如图 20 所示。

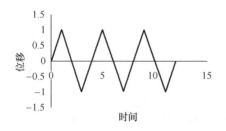 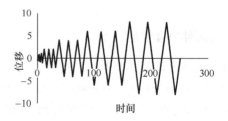

图 19　短程三循环（或短循环）加载制度　　图 20　全程循环（或全循环）加载制度

3.1.2　OFP 耗能件力学特性

对 OFP 节点分别进行了 1 组单调加载、3 组全循环加载和 1 组短循环加载，得到了相应的力-位移曲线和各项性能指标。OFP平接节点单调加载试验结果显示，OFP延性较好，其力学性能如表1所示。OFP平接节点往复加载试验结果显示，OFP延性较好；持力性能较好，承载力逐圈变化小；耗能性能稳定，等效粘滞阻尼比逐圈下降幅度小，其力学性能如表2所示。OFP耗能节点滞回曲线如图21所示，其首圈形状较饱满，后两圈不如首圈。

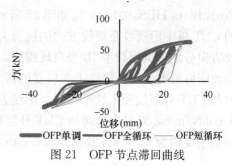

图 21 OFP 节点滞回曲线

单个 OFP 节点单调加载力学性能   表 1

项目	物理量	值
破坏位移	V_u (mm)	30.00
承载力峰值	F_{max} (kN)	31.77
屈服位移	V_y (mm)	5.74
屈服力	F_y (kN)	20.48
延性系数	D	5.22

单个 OFP 节点往复加载力学性能   表 2

项目	物理量	值	
		正向	反向
破坏位移	V_u (mm)	27.91	
第1圈承载力峰值	F_{max1} (kN)	31.09	
弹性刚度	k_{el} (kN·mm^{-1})	2.60	
塑性刚度	k_{pl} (kN·mm^{-1})	0.43	
屈服力	F_y (kN)	19.39	
屈服位移	V_y (mm)	9.01	
延性系数	D	3.10	
第2圈承载力峰值	F_{max2} (kN)	29.60	−22.19
1-2承载力降低比	ΔF_{1-2}	4.79%	−0.68%
第3圈承载力峰值	F_{max3} (kN)	28.52	−22.01
1-3承载力降低比	ΔF_{1-3}	8.27%	0.11%
第1圈滞回环面积	E_{d1} (kN·mm)	455.27	241.76
第2圈滞回环面积	E_{d2} (kN·mm)	285.24	167.19
第3圈滞回环面积	E_{d3} (kN·mm)	179.25	142.93
第1圈阻尼比	v_{eq1}	16.71%	11.78%
第2圈阻尼比	v_{eq2}	11.16%	8.12%
第3圈阻尼比	v_{eq3}	7.29%	7.04%

3.2 金属连接件性能试验（用于普通 CLT 剪力墙）

3.2.1 自攻螺钉试验

所有金属连接件均采用规格为 $\phi5\times80$mm 的意大利产 CLT 专用自攻螺钉（型号：

RothoBlass HBS580），单钉如图 22 所示。通过试验研究了单根自攻螺钉在厚度为 175mm 的 CLT 板中沿顺纹及横纹方向的抗剪力学性能，试验装置如图 23 所示。沿木纹横纹及顺纹方向分别单调加载了 10 个自攻螺钉，将试验所得的剪力-剪切位移曲线取平均值，单根自攻螺钉沿着顺纹及横纹方向所受的剪力及剪切变形关系如图 24、图 25 所示，当剪力方向与 CLT 外层层板木纹平行时（顺纹方向），单钉极限承载力及剪切刚度约为 5.53kN 及 0.60kN/mm，当剪力方向与 CLT 外层层板木纹垂直时（横纹方向），单钉极限承载力及剪切刚度约为 5.89 kN 及 0.59kN/mm。

图 22 意大利 RothoBlass 产自攻螺钉 HBS580

图 23 单根自攻螺钉抗剪试验

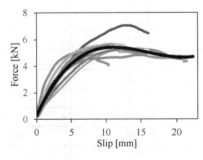

图 24 单根自攻钉顺纹方向抗剪试验曲线

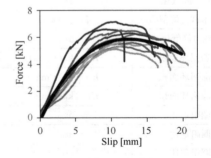

图 25 单根自攻钉横纹方向抗剪试验曲线

3.2.2 角钢连接件及抗拔件力学性能试验

（1）试件设计

共开发了两种规格的抗拔件及一种规格的角钢连接件，如图 26 所示，均采用厚度 6mm 的钢板焊接而成。抗拔件 1 竖向肢尺寸为 110mm×369mm，水平肢尺寸为 110mm×80mm，抗拔件 2 竖向肢高度 319mm，宽度 150mm，水平肢尺寸为 150mm×80mm。角钢连接件竖向肢尺寸为 134mm×140mm，水平肢尺寸为 140mm×80mm。分别通用 9 根 $\phi 5 \times 80$ mm 的自攻螺钉将抗拔件 1 及抗拔件 2 的竖向肢锚固于墙板上，通用 7 根 $\phi 5 \times 80$mm 的自攻螺钉将角钢连接件的竖向肢锚固于 CLT 墙板上，抗拔件 1、抗拔件 2 及角钢连接件的水平肢均通过两根 $\phi 16$ 的 8.8 级螺栓与钢板锚固。每一种类型的节点在受拉及受剪方向均进行单调及往复加载试验，单调加载速度取 0.1mm/s，参考欧洲规范 Eurocode BS-EN-12512-2001[16]，往复加载制度如图 27 所示，加载速度取 0.5mm/s，加载试件限值在 30min 以内。

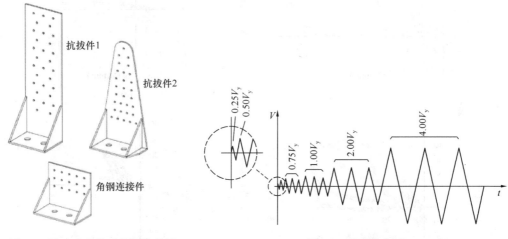

图 26 用于试验的两种抗拔件及一种角钢连接件

图 27 节点往复试验加载制度

图 28 所示为抗拔件 1 抗拉试验装置，图 29 所示为角钢连接件抗剪试验装置，每种金属连接件（抗拔件 1、抗拔件 2 和角钢连接件）在抗拉及抗剪方向上均进行一组单调试验及三种往复试验。

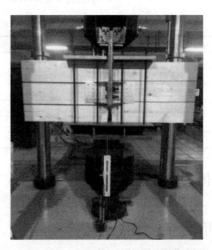

图 28 抗拉试验装置（抗拔件 1）　　图 29 抗剪试验装置（角钢连接件）

（2）节点力学性能

部分节点在往复荷载下的滞回曲线如图 30 所示。通过试验发现所有金属连接件均具有较好的耗能能力，抗拔件 1 和 2 及角钢连接件的力学性能见表 3 所示，除了在极限抗拉承载力方面，抗拔件 1（51.6kN）要略高于抗拔件 2（52.59kN）以外，其他力学性能参数方面，抗拔件 1 均要优于抗拔件 2，最终选取抗拔件 1 及角钢连接件用于普通 CLT 剪力墙试验中连接墙板与基础及楼板的金属连接件。

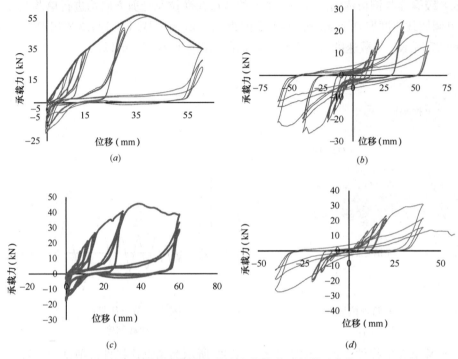

图 30 部分金属连接件在往复荷载下的滞回曲线
(a) 抗拔件 1 抗拉滞回曲线；(b) 抗拔件 1 抗剪滞回曲线
(c) 抗拔件 2 抗拉滞回曲线；(d) 角钢连接件抗剪滞回曲线

金属连接件在抗拉及抗剪方向上的力学性能　　表 3

力学性能参数	抗拔件 1	抗拔件 2	角钢连接件
等效屈服剪力（kN）	24.57	22.71	20.97
等效屈服拉力（kN）	45.16	43.96	38.00
极限抗剪承载力（kN）	26.70	24.34	33.37
极限抗拉承载力（kN）	51.60	52.59	40.91
剪切初始刚度（kN/mm）	1.30	1.24	1.34
抗拉初始刚度（kN/mm）	2.71	2.47	1.62

4. 墙体力学性能试验

4.1 试件设计

4.1.1 普通 CLT 剪力墙

普通 CLT 剪力墙试件及所采用的抗拔件及角钢连接件如图 31 所示，该两层墙体试件的缩尺比为0.5，宽度1.2m，总高达到了2.2m，一层墙板与钢基础通过抗拔件与角钢连接件连接，楼板与一层墙板及二层墙板均通过抗拔件及角钢连接件连接，墙板顶部与水平加载的300kN液压千斤顶铰接连接，二层墙板顶部及CLT楼板位置处均受到平面外约束，该平面外约束由两对安装在墙体面外约束框架上的滑轮提供。

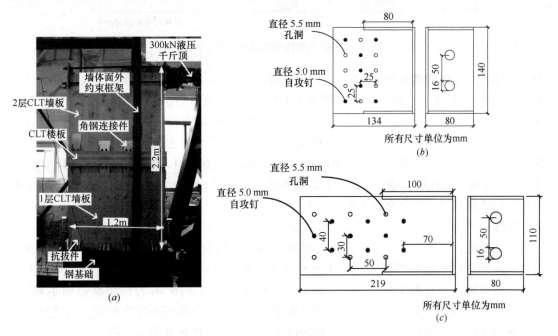

图31 普通CLT剪力墙构造示意
(a) 墙体；(b) 角钢连接件；(c) 抗拔件

4.1.2 预应力CLT剪力墙

预应力CLT剪力墙构造如图32所示，该两层预应力CLT剪力墙体试件的缩尺比为0.5，每面墙体中设置了两根直径15.2mm的预应力钢绞线，为了防止墙体水平滑移，在一层CLT墙板两端各设置一个与基础通过螺栓锚固的抗剪键，CLT墙板的截面构造如图33所示，在CLT板交错组胚的过程中，在该CLT板的第三层层板中沿纵向各抽取两列截面为35mm×120mm的规格材，从而形成贯穿墙板横截面的纵向孔洞，孔洞位置沿CLT板横截面对称布置，孔洞边缘至墙板端部140mm，竖向贯通的孔洞用于墙体内置预应力筋，每根预应力筋的截面形心至CLT墙板端部距离为200mm。

在竖向预应力作用下，CLT楼板受到横纹承压作用，在横纹压力下楼板产生的较大变形将引起预应力钢绞线中张拉力的损失。为了减小横纹压力作用下楼板的大变形，对CLT楼板进行特殊

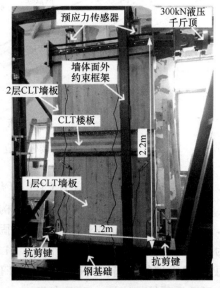

图32 预应力CLT剪力墙构造

143

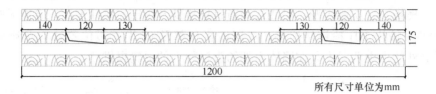

图 33 预应力墙体所用墙板的截面构造

构造设计,其具体构造如图34所示。CLT楼板板面开有孔洞,竖向荷载传递钢支撑(图35)的钢管穿过CLT楼板,钢管长度等于CLT楼板厚度(175mm),竖向荷载传递钢支撑的翼缘上开有直径5.5mm的孔洞,两个竖向荷载传递钢支撑通过三块钢盖板拼接成一个C形槽口,通过规格 $\phi5\times80$mm 的自攻钉与2层CLT墙板连接。在竖向荷载传递钢支撑(图35)中,竖向钢管的下端焊接有直径16mm抗剪销钉,抗剪销钉穿过C形槽钢(图34)腹板上的孔洞,插入下层CLT墙板的上端部截面中,抗剪销钉主要用于防止楼板与下层墙板接触面发生水平滑移(当接触面的摩擦力不足以抵抗水平剪力时),C形槽钢的翼缘上开有直径5.5mm的孔,并且通过规格 $\phi5\times80$mm 的自攻钉与下部墙板连接。为了保证预应力钢绞线能够自上至下贯穿所有墙板及楼板,在相邻的CLT楼板间设有宽度35mm的间隙,在C形槽钢腹板上开有两个直径30mm的孔洞,并且在竖向荷载传递钢支撑的腹板上开有两个尺寸为17.5mm×35mm的缺口。图36为楼板横截面示意图,图37为拼装好的CLT楼板与竖向荷载传递钢支撑。试验中发现预应力张拉力通过竖向荷载传递钢支撑中的焊接钢管由二层墙板传递至一层墙板,楼板几乎不受横纹承压作用。

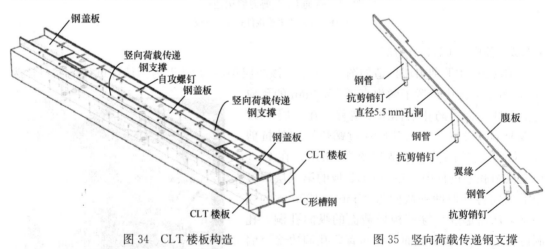

图 34 CLT楼板构造　　图 35 竖向荷载传递钢支撑

4.1.3 带有耗能件的预应力CLT剪力墙

带有预应力的CLT剪力墙试件构造如图38所示,采用缩尺比0.5,单层墙板宽度0.6m,总高2.2m,通过与钢基础锚固的抗剪键来约束墙体可能发生的水平滑移,同样,为了避免楼板收到横纹承压作用,CLT楼板的构造与4.1.2节中所述的楼板构造相同,每层CLT墙板与墙板间通过3个OFP耗能件拼接。墙体顶端与施加水平荷载的300kN液压千斤顶铰接连接。两根直径15.2mm的预应力钢绞线自上而下贯穿每面带有耗能件的预应力 CLT 剪力墙试件。4.1.2 节及 4.1.3 节中所述墙体所用预应力钢绞线均为直径 15.2mm，横截面积 140 mm² 的高强冷拉钢绞线，其生产工艺符合规范《预应力用混凝土钢绞线》GB 5224—2015[17]，屈服强度 f_{py} 达到 1070MPa。

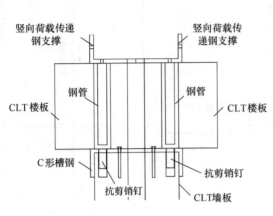

图 36 楼板横截面示意

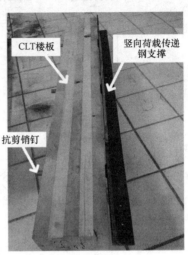

图 37 拼装好的 CLT 楼板与竖向荷载传递钢支撑

图 38 带有耗能件的预应力 CLT 剪力墙

4.1.4 剪力墙传感器布置

预应力 CLT 剪力墙的传感器布置如图 39 所示，其中位移计 1 号～4 号用于采集侧向荷载下墙板的面内剪切机弯曲变形，位移计 5 号～6 号用于采集侧向荷载下墙角竖向提升位移，位移计 7 号～8 号用于采集侧向荷载下一层墙板与二层墙板之间张口边缘的竖向距离，位移计 9 号～11 号用于采集侧向荷载下墙体试件不同高度处的侧向位移，力传感器 1

号~2号用于采集预应力钢绞线的内力变化。带有耗能件的预应力 CLT 剪力墙的传感器布置图如图 40 所示，其中位移计 10 号用于采集耗能件处墙板与墙板间的竖向错动位移。

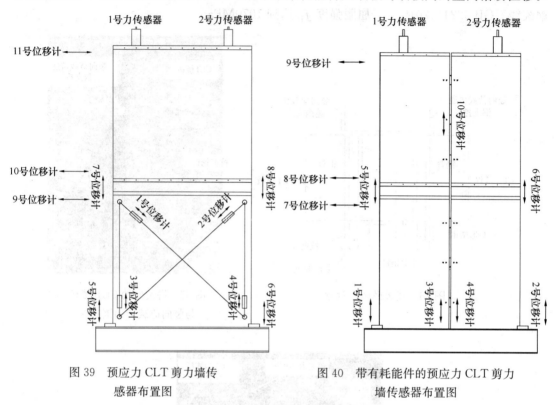

图 39　预应力 CLT 剪力墙传感器布置图　　　图 40　带有耗能件的预应力 CLT 剪力墙传感器布置图

4.2　试验分组及加载制度

墙体试验分组信息见表 4，总共测试了 7 面 CLT 剪力墙，包括 3 面无耗能件的预应力 CLT 剪力墙，2 面普通 CLT 剪力墙，2 面带有耗能件的预应力 CLT 剪力墙。对于 3 面无耗能件的预应力 CLT 剪力墙，单根预应力筋的初试张拉力分别取为 23kN，46kN 和 70kN，对于 2 面带有耗能件的预应力 CLT 剪力墙，单根预应力筋的初试张拉力分别取为 23kN 和 70kN。2 面普通 CLT 剪力墙的构造弯曲相同，1 面用于往复加载，1 面用于单调加载。

墙体试验分组　　表 4

墙体分类	试件编号	初试张拉力	初试张拉应力比	是否有耗能件	加载制度
预应力 CLT 剪力墙	PT23	23kN	15.4%	无	往复
	PT46	46kN	30.7%	无	往复
	PT70	70 kN	46.1%	无	往复
普通 CLT 剪力墙	T1	—		无	单调
	T2	—		无	往复
带有耗能件的预应力 CLT 剪力墙	PTO23	23kN		有	往复
	PTO70	70 kN		有	往复

所有往复加载试验采用的加载制度取自美国规范 ACI Innovative Task Group 5 (2008)[18] 中的 ACI ITG-5.1-07 加载制度,如图 41 所示,记载速度 40mm/min,对于预应力 CLT 剪力墙,往复加载的目标位移角取 2.5%。对于普通 CLT 剪力墙往复加载试验,其目标位移角由普通 CLT 剪力墙单调加载试验决定,将承载力下降至极限承载力 80% 时所对应的侧向位移角定为普通 CLT 剪力墙往复加载的目标位移角。

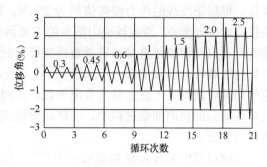

图 41 往复加载制度

4.3 试验结果

4.3.1 墙体力学性能

(1) 无耗能件的预应力 CLT 剪力墙

无耗能件的预应力 CLT 剪力墙的滞回曲线对比如图 42（a）所示,随着初试张拉力的提高,墙体的极限承载力及初试抗侧刚度均有所提高,预应力筋的内力随着墙体侧移的变化曲线如图 42（b）~（d）所示,虽然不同剪力墙钢绞线初始张拉力不同,但加载结

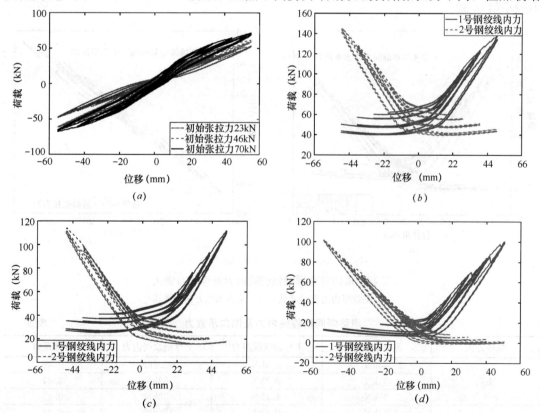

图 42 无耗能件的预应力 CLT 剪力墙

（a）滞回曲线；（b）PT70 钢绞线内力；（c）PT46 钢绞线内力；（d）PT23 钢绞线内力

束后，每根钢绞线的内力损失值约为 20kN，其中墙体 PT23 加载结束后，钢绞线中的残余张拉力接近于 0，造成预应力损失的主要因素包括：木材的局部受压变形、钢绞线锚固端变形，以及钢绞线自身的轻微塑性变形。此外，正向加载时，试件的极限承载力要略大于负向加载时的极限承载力，如图 40（a）所示，这是由于，在加载制度中，每一等级幅值的加载阶段均先正向加载再负向加载，正向加载的过程中造成钢绞线一定程度的松弛，因此在随后的负向加载过程中，墙体在侧推至与正向加载时相同幅度的位移角时，其极限承载力要较小。

试验过程中发现，在加载墙体 PT23（初试张拉力 23kN）的过程中，墙体与基础之间发生了轻微的滑移。在发生滑移的瞬间，钢绞线内力及墙体侧向承载力均有较为突然的下降，图 43 中的红点表示墙体与基础滑移瞬间钢绞线内力及承载力的突然降低，在整个加载过程中，在正向加载及负向加载过程中各发生 3 次滑移。六次滑移瞬间钢绞线内力及侧向承载力见表 5 所示，墙体自重约为 4.4kN，正向加载方向平均摩擦系数 0.42，负向加载方向平均摩擦系数 0.32。试验发现，正向加载时的平均摩擦系数要大于负向加载时的平均摩擦系数，这是因为依据加载制度，墙板先正向加载随后负向加载，在正向加载至一定位移角时墙体发生水平滑动，已经发生过滑移的 CLT 墙板与钢基础之间的接触面要相对更光滑，因此在随后的负向加载过程中，墙板与基础之间相对更容易滑移，负向加载时的平均摩擦系数要更小。出于保守因素考虑，建议 CLT 墙板及钢基础试件的摩擦系数取较小值 0.32。

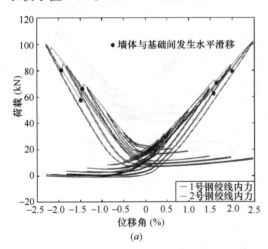
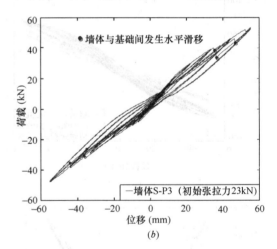

(a)   (b)

图 43 墙体滑移瞬间钢绞线内力及承载力变化
(a) 钢绞线内力突然降低；(b) 墙体承载力突然降低

滑移瞬间钢绞线内力及侧向承载力　　　　　　　　　　　表 5

加载方向	侧向承载力	1号钢绞线内力	2号钢绞线内力	摩擦系数
正向	40kN	5.3kN	82.2kN	0.42
	35kN	8.1kN	69kN	0.42
	30kN	6.1kN	58kN	0.43
负向	35kN	83.0kN	15.8kN	0.34
	30kN	71.0kN	17.2kN	0.32
	26kN	63.7kN	18.0kN	0.30

（2）普通 CLT 剪力墙

普通 CLT 剪力墙加载曲线如图 44 所示，其中单调加载至承载力下降至普通 CLT 剪力墙 T1 极限承载力 80％时停止，其对应位移角约为 8.1％，以此位移角为目标位移角对普通 CLT 剪力墙 T2 进行往复加载，但是当往复加载至位移角 5.0％时，由于固定于约束框架上的用于墙体面外约束的滑轮及约束框架本身变形过大，已经对后续往复加载产生了影响，因此往复加载被终止。通过对比往复加载试验 T2 及单调记载试验 T1 的荷载位移曲线，发现单调加载时的荷载位移曲线基本上与往复加载时所得滞回曲线的外包络线吻合。

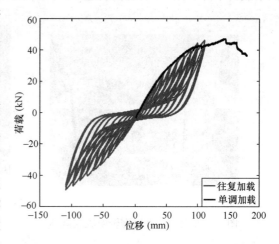

图 44　普通 CLT 剪力墙单调及往复加载曲线

（3）力学性能及破坏模式对比

无耗能件预应力 CLT 剪力墙与普通 CLT 剪力墙在侧向荷载下的力学性能见表 6 所示，其中虽然预应力 CLT 剪力墙 PT23 与普通 CLT 剪力墙 T1 的极限承载力几乎差不多，但是普通 CLT 剪力墙达到极限承载力时对应的位移角高达 6.2％，当侧推至 2.5％位移角时，普通 CLT 剪力墙的侧向承载力为 33.4kN，仅仅为墙体 PT23 极限承载力约一半左右。普通 CLT 剪力墙的初试抗侧刚度仅仅约为预应力墙体 PT23 的 40％。此外，随着初始张拉力每提高 23kN，预应力 CLT 剪力墙的极限承载力及初试抗侧刚度均显著提高 10kN 及 600kN/mm 左右。

侧向荷载下的力学性能对比　　　表 6

墙体分类	墙体编号	初试张拉力（kN）	极限承载力（kN）	初试刚度（kN/mm）	加载制度
预应力 CLT 剪力墙	PT23	23	52.5（2.5％位移角）	2835.6	往复
	PT46	46	60.8（2.5％位移角）	2218.8	往复
	PT70	70	71.7（2.5％位移角）	1654.0	往复
普通 CLT 剪力墙	T1	—	51.2（6.2％位移角）	702.0	单调
	T2		33.4（2.5％位移角）	700.6	往复

此外，在破坏模式方面，加载结束后预应力 CLT 剪力墙基本上没有发生破坏，只有墙体 PT70 加载结束后墙角 CLT 板发生轻微脱胶破坏（图 45），但这并未对预应力 CLT 剪力墙的抗侧性能产生太大影响。而在普通 CLT 剪力墙加载过程中，角钢、抗拔件自身钢板弯曲变形，节点区所用自攻螺钉被从 CLT 墙板中拔出，并且对于普通 CLT 剪力墙而言，在加载过程中，节点区域的破坏将导致其侧向承载力上升速率显著降低甚至导致其侧

向承载力下降。图 46 所示为普通 CLT 剪力墙加载过程中发生的节点区域破坏。

图 45 墙体 PT70 墙脚轻微脱胶破坏

图 46 普通 CLT 剪力墙节点区域破坏

(4) 变形模式及耗能能力对比

预应力 CLT 剪力墙的主要以转动变现为主，在侧向荷载下，其一层墙板的墙脚最大竖向抬升距离为 14.2 mm，并且其上下层墙板之间几乎没有张开（上下墙板间张口边缘最大竖向距离 0.45 mm），侧向荷载下，上下墙板之间作为一个整体共同转动，图 47 所示为预应力剪力墙 PT70 中侧向荷载与底层墙脚竖向抬升距离之间的关系，对于预应力墙体，由于设置了抗剪键，其水平滑移几乎为零。图 48~图 50 为普通 CLT 剪力墙侧向荷载下水平滑移及墙角竖向抬升距离关系图，对于普通 CLT 剪力墙，在加载到 2.5% 位移角时，其一层墙板的墙脚最大竖向抬升距离为 15.1 mm，此外在侧推普通 CLT 剪力墙时，其二层墙板底与一层墙板顶张开，在加载到 2.5% 位移角时，该张口边缘最大竖向距离为 7.3 mm，由于没有设置抗剪键，墙板水平滑移距离约为 2.7 mm。

普通 CLT 剪力墙 T2 与预应力 CLT 剪力墙 PT23 的耗能曲线对比如图 51 所示，普通 CLT 剪力墙的耗能能力要稍优于预应力 CLT 剪力墙，这是由于侧向荷载下普通 CLT 剪力墙的耗能主要来源于节点区域角钢连接件自身钢板的变形，及性塑性铰的自攻钉的塑性变形。而预应力 CLT 剪力墙在加载至 2.5% 位移角时，几乎没有太多的塑性变形，所以其耗能能力一般。

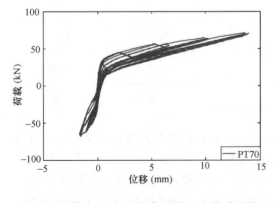 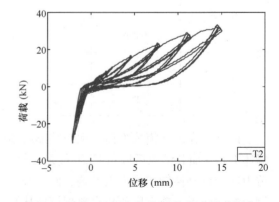
图 47 墙体 PT70 侧向荷载-墙脚竖向抬升距离　　图 48 墙体 T2 侧向荷载-墙脚竖向抬升距离

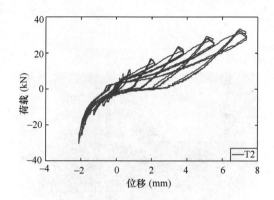

图49　墙体T2侧向荷载-上下墙板间张口边缘竖向距离

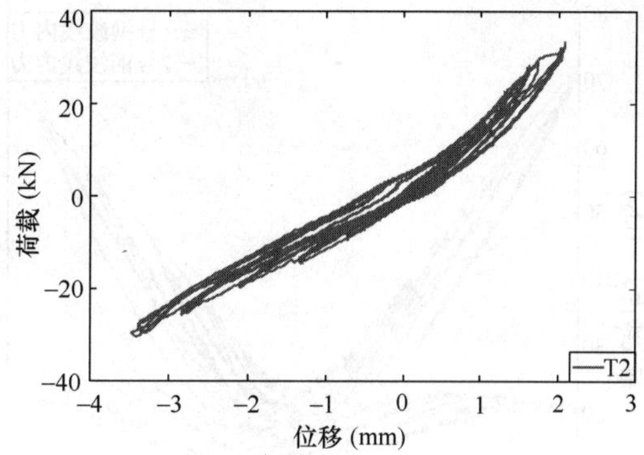

图50　墙体T2侧向荷载-墙体水平滑移

4.3.2　带有耗能件的预应力CLT剪力墙

为了提高预应力CLT剪力墙的耗能能力，在其中加入了OFP耗能节点，并对代有耗能件的预应力CLT剪力墙PTO23和PTO70做往复记载试验，其滞回曲线及预应力钢绞线内力变化曲线如图52~图54所示。图55为加载结束后产生剪切变现的耗能件OFP。从图53、图54可以看出，对于带有耗能件的预应力CLT剪力墙，加载结束后，其每根钢绞线预应力损失值约为10kN，在加载至最大位移角2.5%时，墙体PTO70的钢绞线最大张拉力达到了将近100kN。此外，通过试验结束后从墙体上拆

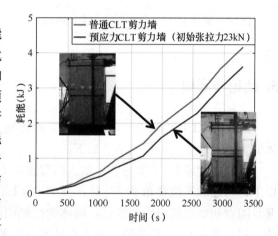

图51　耗能能力对比

下的OFP耗能件来看，安装于墙体二层的耗能件承受的剪切变形相对较大，安装于二层墙板最顶部的产生剪切变形的OFP耗能件如图55所示。

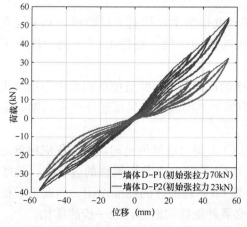

图52　带有耗能件的预应力墙体滞回曲线

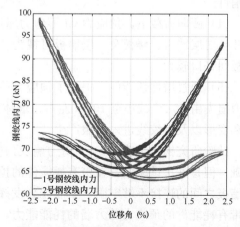

图53　墙体PTO70钢绞线内力变化曲线

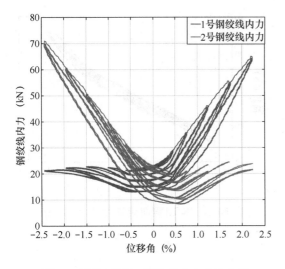
图 54 墙体 PTO23 钢绞线内力变化曲线

图 55 剪切变形的耗能件 OFP

5. 结论与展望

通过对 CLT 材料特性、用于 CLT 墙体的金属连接节点、预应力 CLT 和普通 CLT 墙体的大量试验研究，得到结论如下：

（1）在普通 CLT 剪力墙中加入内置预应力钢绞线，墙体的初始抗侧刚度提高至 2 倍以上；当加载至位移角 2.5% 时，预应力 CLT 剪力墙的承载力是普通墙体的 2 倍以上，并且随着初始张拉应力的提高，墙体的抗侧刚度及极限承载力也显著提高。

（2）普通 CLT 墙体较早在节点区域发生破坏，而预应力 CLT 剪力墙直至加载结束也几乎没有任何破坏。

（3）预应力 CLT 墙板与基础之间的摩擦系数建议取 0.32，墙体初试张拉应力比建议取 30%～50%，此时墙板与基础之间的摩擦力足以抵抗水平剪力，可以不设抗剪键及抗剪销钉。

（4）侧向荷载下，预应力 CLT 剪力墙以转动变形为主，其上下层墙板间几乎没有脱开并且二层墙板、楼板及一层墙板可视为整体发生侧向转动；而普通 CLT 剪力墙上下层墙板间张口较大。

（5）通过增加耗能件，提高了预应力剪力墙的耗能能力，并且使单根预应力钢绞线的内力损失由 20kN 减小至 10kN 左右；墙体内力损失主要由木板局部受压变形、锚固端松弛和变形以及钢绞线进入屈服产生。

试验研究揭示了普通 CLT 剪力墙及预应力 CLT 剪力墙抗侧机理，为工程应用奠定了基础。但是本次试验中，在带有耗能件的预应力 CLT 剪力墙发生侧移的过程中，耗能件与墙板之间的连接不够牢固，不能充分发挥 O 形 OFP 耗能件的耗能作用，因此为了提高该带有耗能件的预应力剪力墙的耗能能力，有必要耗能件构造进行进一步的优化。

参考文献

[1] 何敏娟,孙晓峰,李征. 正交胶合木结构在地震作用下的层间位移角研究[J]. 特种结构,2017,34(1):1-7.

[2] Gavric, I., Fragiacomo, M., and Ceccotti, A. Cyclic behaviour of typical metal connectors for cross-laminated (CLT) structures [J]. *Materials and Structures*, 2015, 48(6), 1841-1857.

[3] Gavric, I., Fragiacomo, M., and Ceccotti, A. Cyclic behavior of typical screwed connections for cross-laminated (CLT) structures [J]. *European Journal of Wood and Wood Products*, 2015, 73(2), 179-191.

[4] Schneider, J., Stiemer, S. F., Tesfamariam, S., Karacabeyli, E., and Popovski, M. (2012, July). Damage assessment of cross laminated timber connections subjected to simulated earthquake loads. In *Proceedings of the 12th World Conference on Timber Engineering*, Auckland, New Zealand.

[5] Schneider, J., Karacabeyli, E., Popovski, M., Stiemer, S. F., and Tesfamariam, S. Damage assessment of connections used in cross-laminated timber subject to cyclic loads [J]. *Journal of Performance of Constructed Facilities*, 2013, 28(6), A4014008.

[6] Schneider, J., Shen, Y., Stiemer, S. F., and Tesfamariam, S. Assessment and comparison of experimental and numerical model studies of cross-laminated timber mechanical connections under cyclic loading [J]. *Construction and Building Materials*, 2015, 77, 197-212.

[7] Dujic, B., Klobcar, S., and Zarnic, R. (2008). Shear capacity of cross-laminated wooden walls. In *Proceedings of the 10th World Conference on Timber Engineering*, Miyazaki, Japan.

[8] Ceccotti, A., Follesa, M., Lauriola, M. P., and Sandhaas, C. (2006, September). SOFIE project – test results on the lateral resistance of cross-laminated wooden panels. In *Proceedings of the First European Conference on Earthquake Engineering and Seismicity* (Vol. 3).

[9] Ceccotti, A., Follesa, M., Lauriola, M., & Sandhaas, C. (2006). SOFIE Project – Test Results on the Lateral Resistance of Cross-Laminated Wooden Panels. 1st ECEES 2006. *Geneva, Switzerland*.

[10] Popovski, M., Schneider, J., and Schweinsteiger, M. (2010). Lateral load resistance of cross-laminated wood panels. In World Conference on Timber Engineering (pp. 20-24).

[11] Ceccotti, A., Sandhaas, C., Okabe, M., Yasumura, M., Minowa, C., and Kawai, N. SOFIE project – 3D shaking table test on a seven-storey full-scale cross-laminated timber building [J]. *Earthquake Engineering & Structural Dynamics*, 2013, 42(13), 2003-2021.

[12] Akbas T, Sause R, Ricles J M, et al. Analytical and experimental lateral-load response of self-centering post-tensioned CLT walls [J]. Journal of Structural Engineering, 2017, 143(6): 04017019.

[13] Ganey R, Berman J, Akbas T, et al. Experimental investigation of self-centering Cross-laminated timber walls [J]. Journal of Structural Engineering, 2017, 143(10): 04017135.

[14] ANSI/APA PRG 320, Standard for Performance-rated Cross-laminated Timber, American National Standards Institute/APA-The Engineering Wood Association, APA, Tacoma, WA, 2012.

[15] GB/T 50329-2012, 木结构试验方法标准, 中国建筑工业出版社, 北京, 2012.

[16] BS EN 12512: 2001, Timber structures-Test methods-Cyclic testing of joints made with mechanical fasteners [S].

[17] GB/T 5224-2015, 预应力混凝土用钢绞线, 国家质量监督检验检疫总局, 北京, 2015.

[18] ACI Innovation Task Group 5. (2008). "Acceptance criteria for special un-bonded post-tensioned precast structural walls based on validation testing and commentary: An ACI standard." American Concrete Institute, Farmington Hills, MI.

10 BIM-Based Bridge Performance Assessment

Weixiang Shi, Jian Chai, Peng Wang, Xiangyu Wang*
Australasian Joint Research Centre for Building Information Modelling (BIM),
Curtin University, Perth, Australia.
* Corresponding author (Xiangyu. wang@curtin. edu. au)

Abstract: With increasing collapses of bridge structures and popularized utilization of long-span bridges, worldwide bridge performance assessment is of importance in emerging and future bridge management industry. A reliable and visualised bridge performance assessment method can promote serviceability and safety of bridges because bridge management supervisors usually take it as guideline to make decisions. Although new technologies, building information modelling (BIM), laser scanning by point clouds and structural health monitoring, have gained rapid development in recent years, present bridge performance assessment are still mainly based on practical experience and on-site inspection. On the other hand, the collected data obtained from SHM system needs a smart data interpretation approach due to the usage of data is still limited in most of the current bridge performance assessment methods. In this regard, this paper presents a framework of BIM-based bridge performance assessment method. Firstly, the existing BIM implementation in Australia is introduced in Section. 2. Secondly, automated as-built bridge modelling by laser scanning technology is introduced in Section. 3. Finally, a framework of BIM-based bridge damage detection and localization method is introduced in Section. 4. We demonstrate the advantages and potential future of BIM-based bridge performance assessment on bridge structures.
keywords: Building Information Modelling; As-built Modelling; Laser Scanning; Damage Detection and Localization

1. Introduction

Optimal utilisation, maintenance and management of the existing bridge infrastructure have been identified as a critical factor in ensuring and raising the productive capacity of our economy. In Australia, over 60% of bridges for local roads are over 50 years old and approximately 55% of all highway bridges are over 20 years old [1] . There are around 22,500 bridges with a replacement value of about AUD $ 3 billion, and an annual maintenance expenditure of about AUD $ 300 million [2] . However, the current engineering practices of infrastructure performance assessment are based on visual inspections therefore require a tremendous labour cost and time. Those practices are also biased and subjected to the inspectors' experiences and judgements. Previous research works on non-model based and model based approaches [3] have been conducted on vibration-based system identification and condition assessment. Non-model based methods require measuring and analysing the vibration data from civil infrastructure to indicate the possible structural condition change. Model-based methods require an accurate initial finite element (FE) model to represent the true structural behaviour. Both approaches suffer from a high level of systematic modelling errors and uncertainties for the structural identification and performance assessment [4].

Building information modelling (BIM) has recently gained widespread developments in the architecture, engineering, and construction industry. With BIM technology, a virtual model of the structure can be digitally simulated. Therefore, BIM applications for planning, design, construction, and operation of the facility are emerging in recent years. Laser scanning is also a prominent technology can produce rich information regarding bridge infrastructures, and there has been lots of studies on extracting various road and bridge elements from point clouds. The automated 3D as-built model from laser scanning technology can be used as a baseline model to BIM platform which can integrate and visualize other information related to bridge performance assessment.

As a result, BIM-based bridge performance assessment method is proposed with the objective improving accuracy of as-built model for reliable bridge assessment and knowledge of present bridge performance assessment with effective maintenance planning and decision making through visualising

2. BIM Implementation in Australia

Building information modelling (BIM) is one of the most promising recent developments in the architecture, engineering, and construction industry. With BIM technology, an accurate virtual model of a building is digitally constructed. BIM can be used for planning, design, construction, and operation of the facility. It helps architects, engineers, and constructors visualize what is to be built in a simulated environment to identify any potential design, construction, or operational issues [5].

2.1 BIM Policies in Australia

Australian Government set a date of 1 July 2016 from which procurement for all its buildings will require full collaborative BIM based on open standards for information exchange. The further recommends that the Australian Government encourage the State and Territory Governments, through the Council of Australian Governments, to commit to a similar timeframe for full collaborative BIM based on open standards to be required for procurement of their buildings.

The National BIM initiative Implementation Plan requires execution of the following project work programs [6]:

1) Procurement: Manage risk, intellectual property, insurance and warranty require-

ments for clients, consultants and constructors through new forms of procurement contracts that support collaborative, model-based procurement processes.

2) BIM Guidelines: Provide industry and government clients, consultants and constructors with a set of Australian BIM Guidelines based on collaborative working, open standards and alignment with global best practice.

3) Education: Deliver a broad industry awareness and re-training program through a national BIM education taskforce based on core multi-disciplinary BIM curriculum, vocational training and professional development.

4) Product Data and BIM Libraries: Enable easy access to building product manufacturers' certified information for use in all types of model-based applications through an Australian on-line BIM Products Library.

5) Process and Data Exchange: Establish open standard data exchange protocols that will support collaboration and facilitate integration of the briefing, design, construction, manufacturing and maintenance supply chain throughout the entire life of a built facility.

6) Regulatory Framework: Establish a mechanism for planners, local government and government regulatory bodies with integrated data of building and service system elements, land, geospatial and definition of human and related activities to measure and analyse performance of built form.

Pilot Projects: Encourage pilot projects to be undertaken to demonstrate and verify the readiness of the outputs from the above six work programs to be deployed on an economy-wide basis.

2.2 BIM policies in each state

Building SMART Australasia, the Department of Industry, Innovation, Science, Research and Tertiary Education (DIISRTE) and others co funded a report Productivity in the Buildings Network: Assessing the Impacts of Building Information Models which found that accelerating the adoption of building information modelling (BIM) in the Australian built environment sector could improve productivity by between six to nine percent. It found that concerted government support for the use of BIM by the architects, engineers, builders, contractors, owners and facility managers involved in a building's lifecycle would increase BIM adoption in 2025 by six to sixteen percent and produce an economic benefit equivalent to $5 billion added to Australia's Gross Domestic Product [7].

The NSW Government actively encourage innovation that provides the best value-for-money, including in procurement models and methods. All Australian governments, including the NSW government, are faced with challenging fiscal environments. BIM is a money-saving concept that will improve the budget, drive productivity and make Australian businesses more efficient and competitive. Requiring BIM on government projects would deliver substantial productivity dividends for the budget of participating jurisdictions [8].

The adoption of BIM by industry permits clients and contractors to rehearse the construction of buildings on a computer before it is built. The benefits that this can lead to are far reaching, and extend not only to each and every player involved in the design, construction and maintenance of a building, but also to those who commission the work and use the building. As a significant player in the construction of new facilities for public use, the benefits to Government from using BIM in buildings that it commissions will be dramatic. By mandating BIM, the construction industry will be able to offer the Commonwealth certainty of project spend and drive significant savings for asset acquisition [8].

In a report released in 2015, the Queensland Department of Transport and Main Roads outlined how they expected professional engineers to evaluate risks and benefits of innovation that lead to 'value for money' solutions. building SMART submits that the NSW Government should adopt the same principle, including seeing innovation as an opportunity to improve Australia's future [8].

2.3 BIM Standards and Guidelines

2.3.1 NATSPEC National BIM Guide [7]

The National BIM Guide is a reference document to be read in conjunction with the Project BIM Brief which outlines the particular requirements for each project.

It is expected that the Project BIM Brief, whether developed using the NATSPEC Project BIM Brief or other means, is formulated by the client in consultation with the project team. The guide can also be used as a planning tool by consultants to clarify the services they propose to provide when preparing bids for projects.

The National BIM Guide is to assist clients, consultants and stakeholders to clarify their BIM requirements in a nationally consistent manner. This will reduce confusion and duplication of effort.

2.3.2 National Guidelines and case studies for Digital Modelling [9]

These National Guidelines and Case Studies for Digital Modelling are the outcomes from one of a number of Building Information Modelling (BIM) -related projects undertaken by the CRC for Construction Innovation.

The guidelines is to assist in and promote the adoption of BIM technologies in the Australian building and construction industry, and try to avoid the uncertainty and disparate approaches that created inefficiencies with the implementation of 2D CAD over the past three decades. The guidelines are also part of a larger CRC for Construction Innovation program that seeks to encourage increasing digital modelling practice in the whole building and construction industry.

The guidelines are supported by six case studies including a summary of lessons learnt about implementing BIM in Australian building projects. A key aspect of these publications is the identification of a number of important industry actions: the need for BIM-compatible product information and a national context for classifying product data; the need for an industry agreement and setting process-for-process definition; and finally, the need to ensure a national standard for sharing data between all of the participants in the facility-development process.

2.3.3 BIM-MEPAUS Practices [10]

These guidelines set out the BIM _ MEPAUS framework for the development and management of Industry Foundation Models and the BIM _ MEPAUS Master Shared Parameter Schedule.

Industry Foundation Models (IFMs) provide Autodesk®Revit® MEP plant, equipment and fitting families for designers and manufacturers. The models are sourced from the Autodesk Australian Content Pack and incorporate BIM _ MEPAUS Shared Parameters to provide families that can be used for design modelling as well as the basis for development of Manufacturer's Certified Models (MCMs).

The BIM _ MEPAUS Master Shared Parameter Schedule is a BIM _ MEPAUS controlled document that lists and classifies all shared parameters for use in BIM _ MEPAUS model content and workflows related to design, construction and supply chain integration. The shared parameters also include for facility and asset management requirements. Use of the shared parameters can extend to off model data applications including a range of BIM-to-Field construction and commissioning workflows and e-procurement.

3. Automated as-built bridge modelling

A properly functioning and reliable transport network is essential to Australia's economic activity. Bridges are a vital part of Australian transportation infrastructure. By 2013, there were approximately 2815 road bridges in Western Australia [11], and in 2014-2015, the expenditure of local governments on bridge maintenance was over 17 million dollars [12]. Effective asset management needs to be integrated into the asset lifecycle such that all practices and management strategies are aimed at achieving lowest long-term costs.

One of the fundamental success factors in bridge asset management is timely collection of qualified data on bridge performance. Inaccurate or incorrect data on bridge condition and performance would lead to flawed decisions on bridge management. For the past decades, laser scanning technique has been widely applied in measurement of transportation infrastructure.

As a remote sensing technique, laser scanners can be deployed in multiple platforms, including terrestrial laser scanning, mobile laser scanning, and UAV-borne laser scanning. Flexibility in platforms makes it applicable to bridges that are conventionally difficult to access. Figure 1 shows laser scanning data (e.g. point cloud) of a highway bridge, and there is a total of over 35 million points. As for particular bridge performance assessment method, as-built modelling by laser scanning technologies could be potentially used in improving accuracy of finite element model.

Figure 1 Point cloud of a bridge from laser scanning

3.1 Functionalities of laser scanning in bridge asset management

Laser scanning can facilitate asset management process in various tasks, including bridge infrastructure inventory [13], as-is bridge modelling [14], condition inspections, and information integration and virtualization [15].

Laser scanning can produce rich information regarding bridge infrastructures, and there has been studies on extracting various road and bridge elements from point clouds. Those exploit elements include road surface, edges, markings, traffic signs and roadside objects as tress and utility poles. For instance, Zheng et al. (2017) developed automatic lighting poles extraction using Gaussian-mixture-model, and achieved 90% true positive rate [16]. Figure 2 shows point cloud of roads and extracted traffic elements, where different elements are colour-coded. In general, a relative high accuracy rate can be achieved on those feature extraction algorithms, making it applicable for efficient bridge inventory. Those applications demonstrates high flexibility and efficiency of laser scanning as a results of its high collection speed, high geometric accuracy, no interference to traffic and high accessibility with various platforms.

Figure 2 laser scanning data of roads (Coloured with object classes)

Another major area of applications on bridge laser scanning is on bridge inspection. Laser scanning can be applied in different levels of bridge inspection tasks, including routine inspection, condition rating, and structural and loading assessment of a bridge [17]. Liu et al. (2010) proposed and validated methods of using laser scanning to detect mass loss volumes of concrete bridge girder deflection [18]. Park et al. (2007) applied laser scanning to measure displacement of a bridge and achieved an accuracy of sub millimetre [15]. Lubowiecka et al. (2009) used laser scanning to geometric models of historic bridges and then combined finite element analysis to test structures. Their experiments showed that laser scanning can generate geometric models accurate enough for structure analysis, and enable finite element analysis of complex structures without existing geometric models.

As for as-built modelling, research and practises in other areas as buildings show benefits of laser scanning. As-built models reflect actual and current conditions of a building compared to as-designed or previous models. Laser scanning can rapidly produce dense measurements of object surface, from which an as-built model can be manually created. In addition to manual modelling, there has also been research on developing automated modelling algorithms. The advancement in data processing can improve modelling efficiency and widen application areas. After the above steps, the point cloud is already segmented into components, each of which is recognised by certain types of primitives. We then will investigate the use of a graph matching method based on the connection relationship and structural rules to organise geometric primitives into civil infrastructure components, and add sematic information from prior knowledge, such as BIM at the planning stage or existing design documents. The graph will be derived from a topography graph that represents connectivity among detected primitives and be matched between BIM components and that of the point cloud. The produced 3D models can be used as a platform to integrate and visualize other information related to bridge assessment, where n-dimensional information system can be built. In the following, we present a case study using laser scanning to create as-built models of bridges.

3.2 Case Study: fully automated as-built modelling of highway bridges

Laser scanning is capable of rapidly capturing a large volume of as-built data. If not automated, manual data processing and interpretation is time-consuming and labour intensive. Those drawbacks limit efficiency and applications of laser scanning. Automated as-built modelling methods directly produce models from raw inputs and requires no user interaction. Automated as-is bridge modelling is significant since it can largely reduce time of data processing and potentially enable bridge modelling in nearly real time.

Even though there has been many research focusing on automated modelling of buildings and industrial plants, there is currently no available methods for bridges. Due to structure differences between buildings and bridges, existing as-built modelling methods could not be applied to bridges. To address the challenge of automated as-built modelling, a fully automated approach is developed. The developed approach is for concrete highway bridges and based on point density of bridge components. The developed approach can achieve component-level point cloud classification, and enable production of parametric models with semantic information. The overall process of the developed method is as follows:

1) Pre-processing to remove outliers and align point cloud with a predefined coordinate system;

2) Clustering points into different groups according to point density histogram in the vertical direction;

3) Refining point clusters into component level with point density and depth maps;

4) Fitting geometric primitives e.g. planes and cylinders to each clustered components;

5) Refining geometric models based on contextual constraints in a bridge.

Figure 3 shows an example of as-built modelling, where the left image is raw laser scanning data as input, the middle one represents classified points and the right one represents parametric models as final output. In the parametric model, every components are coded with different colours.

Figure 3 Process of automated as-built modelling

The developed approach is validated with laser scanning data of two highway concrete bridges. Figure 4 shows the output of the developed approach, where different components of the bridge are coded with colours. Experiments show that for a bridge with over 1.6 million points, the algorithm takes about 90 seconds to produce parametric models.

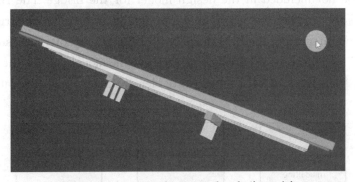

Figure 4 Automated generated as-built models
(different components are coded with colours)

4. BIM-based Bridge Damage Detection and Localization

In structural health monitoring (SHM) systems, a large number of sensors are usually deployed to collect vibration measurement data for damage detection and localization. These sensors, such as accelerometers which can measure vibration more accurately compared to displacement sensors, are used to collect raw data for extracting modal parameters of the structure. From the structural engineering domain knowledge, sensed data from poor locations of structure is not sufficient for accurately describing modal parameters of the structure. However, the modal parameter are most common parameters related to structural health state. As a result, sensor locations primarily determine ability of SHM systems for damage detection and localization. Finite-element model (FEM) is a mathematical model for dividing a complex structure into finite elements which contain enough structural information of the entire structure. In practical engineering, researchers mainly use elements and nodes obtained from FEM to decide locations available for sensor nodes. Therefore, for damage detection and localization, accuracy of predication will be improved from the accurate FEM. The BIM-based FEM not only can be seen in 3D with all accurate as-built information, but also includes sensed data which be used by further modal parameters estimation. The reminder of this section discusses the BIM-based FEM and the usage of sensed data in modal parameters estimation for damage detection and localization.

4.1 BIM-based FEM

Traditionally, an initial FEM is created based on existing documentary of design information to define both geometric and structural parameters of the model. Then, the FEM will be manually adjusted to include specific elements which are not typically included in a design model and as-built information. These components may influence the dynamic response of the infrastructure. For example, the additional elements diaphragms, sidewalks and curbs may be not included in the design model for the bridge. The BIM-based FEM precisely reflects the true behaviour and all minor details of the physical structure. By tackling a key issue on integrating the BIM model with FE analysis software, such as ANSYS, a BIM-based FE model can be established as the baseline model and for the further system identification. A framework of BIM-based finite element analysis is illustrated in Figure5.

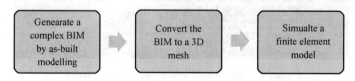

Figure 5　A framework of BIM-based finite element analysis

4.2 Data Interpretation

After sensor system deployed on structures, advance and robust data interpretation will be fed into the SHM system to improve damage diagnosis and infrastructure maintenance management process. Typically, sensed raw data such as acceleration needs to be processed to extract modal parameters and then diagnose the structure's health state with damage detection algorithms. A flowchart of data interpretation is illustrated in Figure6.

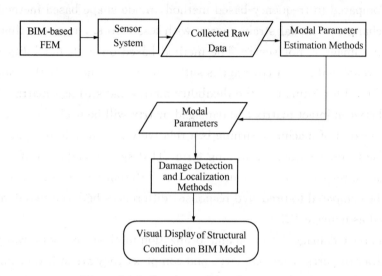

Figure 6 A flowchart of data interpretation

Extensive time-domain and frequency-domain modal parameters estimation methods have been proposed in last decades [19]. However, most of methods are based on traditional centralised algorithms which is very limited in promising wireless-based SHM system because wireless sensor network (WSN) prefer to transmit processed data rather than raw data to data processing and control system, also could be called as central control station. Distributed Eigen-system Realization Algorithm (DERA) has been proposed to achieve same accuracy of traditional modal parameter estimation methods while meets with WSN constraints, such as transmission bandwidth and energy consumption [20]. In DERA, the singular value decomposition (SVD) of Hankel data matrix is not calculated with data from all sensors in the data processing centre (centralised method) but in an incremental manner where the SVD is updated along specific paths.

Once modal parameters (natural frequency or mode shape) have been estimated for the structure from real data sets, some damage detection methods based on natural frequency and mode shape comparison potentially become feasible to diagnose damage in the structure. If the difference between identified frequency or mode shape are beyond a pre-defined threshold, then some parts of the structure are considered to damaged. Furthermore, it is still necessary to recognise exact location of damage for maintenance work. Various damage localization methods based on natural frequency, mode shape, flexibility matrix and stiffness matrix could be found in previous work [21, 22]. Detecting damage location using identified natural frequency and mode shape is simple but not reliable in some cases. Commonly the proposed frequency method uses frequency-change ratio in each element of the structure to determine damage location. However, in practical engineering, even visualised damage may not influence the natural frequency. Moreover, identified natural frequency could be seriously influenced by temperature and so in most practical cases it's not reliable. Compared to frequency-based method, mode shape based method is to locate damage from changes in modal strain energy with less errors. However, this method will be unreliable if damage is not severe. The method based on both flexibility and stiffness matrix is more reliable if sensor coverage is sufficient. In this method, the estimated modal parameters are used for estimating the flexibility matrix and stiffness matrix. Then identifying changes between intact matrix and damaged matrix will be useful to find damage locations. The application of machine learning is a relatively new way in damage detection and localization. Using collected data from long-term SHM system and a set of neural network to predict dynamic response of the structure under different scenarios. Then the measured response will be compared to predicted response, differences between two dynamic responses are detected as damage [23].

Although above damage detection and localization methods are not always unequivocal in practical infrastructures which are large and complex, they are still very important indicators to assess the condition of infrastructure (usually supplemented by visual inspections). Machine learning have been considered to be the promising method for damage detection and localization, particularly when lots of data are recorded in long-term SHM system.

Finally, different elements of the structure are indicated in 3D BIM model and near-real time structural condition of each element are visualised in the BIM model. The condition of structure visualised in the 3D BIM model have a small delay due to it takes time to process raw data and analysis by detection methods. This BIM based user interface also enables researchers to retrieve raw data in local storage system for further research.

5. Conclusion

The integration of bridge performance assessment with BIM provides a potential way to improve serviceability and safety of bridges. Proposed framework is not limited to complex bridges but also in other complex infrastructures. Visualizing the near real-time condition of structures in BIM based software help users to find possible locations corresponding to damage and raw sensor data stored in on-board data storage can also be retrieved for a limited duration by users. The integrated system is believed to give a more efficient and practical approach to improve present bridge assessment performance.

References

[1] Austroads, Investigating the Development of a Bridge Assessment Tool for Determining Access for High Productivity Freight Vehicles, Research Report, 2012.

[2] Austroads, Guidelines for Ensuring Specified Quality Performance in Bridge Construction, 2003.

[3] B. F. Spencer, H. K. Jo, K. A. Mechitov, J. Li, S. H. Sim, R. E. Kim, S. Cho, L. E. Linderman, P. Moinzadeh, R. K. Giles, G. Agha, Recent advances in wireless smart sensors for multi-scale monitroing and control of civil infrastructure. J Civil Struct Health Monit (2015).

[4] M. Z. A. Bhuiyan, G. Wang, J. Cao, J. Wu, Deploying wireless sensor networks with fault-tolerance for structural health monitoring. IEEE Transactions on Computers 64 (2015) 382-395.

[5] S. Azhar, Building information modeling (BIM): Trends, benefits, risks, and challenges for the AEC industry. Leadership and management in engineering 11 (2011) 241-252.

[6] B. S. Australia, National Building Information Modelling Initiative, 2012.

[7] NATSPEC National BIM Guide, 2011.

[8] J. Mitchell, Procurement of Government Infrastructure Projects" Building SMART Australasia, 2016.

[9] C. C. Innovation, National Guidelines for Digital Modelling, 2009.

[10] BIM-MEP, BIM-MEPAUS Practices Industry Foundation Models and Shared Parameter Guideli-nes 2014.

[11] Structures Inspection & Information Management Policy, 2013.

[12] Report on Local Government Road Assets & Expenditure 2014/15, 2015.

[13] Y. He, Z. Song, Z. Liu, Updating highway asset inventory using airborne LiDAR. Measurement 104 (2017) 132-141.

[14] J. Li, Y. Wang, X. Wang, H. Luo, S.-C. Kang, J. Wang, J. Guo, Y. Jiao, Benefits of building information modelling in the project lifecycle: construction projects in Asia. International Journal of Advanced Robotic Systems 11 (2014) 124.

[15] H. Park, H. Lee, H. Adeli, I. Lee, A new approach for health monitoring of structures: terrestrial laser scanning. Computer - Aided Civil and Infrastructure Engineering 22 (2007) 19-30.

[16] H. Zheng, R. Wang, S. Xu, Recognizing Street Lighting Poles From Mobile LiDAR Data. IEEE Transactions on Geoscience and Remote Sensing 55 (2017) 407-420.

[17] S.-E. Chen, Laser scanning technology for bridge monitoring, Laser Scanner Technology, InTech, 2012.

[18] W. Liu, S. Chen, E. Hauser, LiDAR - BASED BRIDGE STRUCTURE DEFECT DETECTION. Experimental Techniques 35 (2011) 27-34.

[19] B. Peeters, C. Ventura, Comparative study of modal analysis techniques for bridge dynamic characteristics. Mechanical Systems and Signal Processing 17 (2003) 965-988.

[20] X. F. Liu, J. N. Cao, W. Z. Song, P. Guo, Z. J. He, Distributed sensing for high-quality structural health monitoring using WSNs. IEEE Transactions on parallel and distributed systems 26 (2015).

[21] J.-T. Kim, Y.-S. Ryu, H.-M. Cho, N. Stubbs, Damage identification in beam-type structures: frequency-based method vs mode-shape-based method. Engineering structures 25 (2003) 57-67.

[22] J. Li, H. Hao, Z. Chen, Damage identification and optimal sensor placement for structures under unknown traffic-induced vibrations. Journal of Aerospace Engineering 30 (2015) B4015001.

[23] E. K. Chalouhi, I. Gonzalez, C. Gentile, R. Karoumi, Damage detection in railway bridges using Machine Learning: application to a historic structure. Procedia Engineering 199 (2017) 1931-1936.

11 土木工程结构生命周期可持续量化评价

王元丰[1]，章玉容[2]，王京京[1]，石程程[1]，刘胤杉[1]，梅生启[1]，周硕文[1]
(1. 北京交通大学土木建筑工程学院；2. 浙江工业大学建筑工程学院)

摘　要：土木工程相关的建筑和基础设施极大地改善了人们的生活质量，促进了我国社会和经济的发展。但是，它们的建设和运营消耗了大量的能源和资源，并产生了很大的环境污染。促进土木工程的可持续发展，已经成为相关行业的重要任务。可持续性评估是实现可持续管理的前提，非常有必要开展土木工程可持续性的量化研究。本文立足于可持续的环境、经济和社会三个维度，同时针对土木工程结构特有的工程特性，从生命周期的各个阶段，对工程结构的可持续性开展了定量化的评价研究。首先，采用生命周期评价方法（LCA），考虑终点和中点评价模型的影响，分别对土木工程结构设计、施工和维修加固阶段的环境影响进行分析；其次，考虑到工程结构在服役期内的性能退化，把可靠度理论作为桥梁加固方案优化的依据，并基于生命周期评价方法，优选环境影响最小的桥梁加固方案；再次，结合生命周期评价、生命周期成本（LCC）和结构时变可靠度，采用统计学和数值模拟的方法，得到桥梁运营阶段采用不同维护策略对桥梁全生命周期内环境和成本的影响，运用帕累托最优的方法进行结构时变可靠度、生命周期环境影响和成本的桥梁维护多目标优化；最后，将技术指标（可靠度）综合考虑到可持续性评价体系中，即通过衡量单位可靠度的环境影响、经济性能和社会影响来评价工程结构的可持续性。本文从工程结构生命周期各个阶段以及可持续性的不同维度，定量化地构建了土木工程结构可持续评价的方法，并通过不同的案例对相关问题进行说明，以期为在世界和中国重要性越来越凸显的土木工程可持续性评估，提供方法支持和实践示范。

关键词：土木工程结构；可持续性；生命周期环境影响；生命周期成本；社会影响；可靠度；定量化评价

引言

改革开放以来经济的高速发展，使中国已经成为世界上最大的能源、资源消耗和温室气体排放国。与土木工程相关的建筑和基础设施极大地改善了人们的生活质量，促进了我国社会和经济的发展。但是，它们的建设和运营消耗了大量的能源和资源，并产生了很大的环境污染。据统计，中国城市既有建筑面积达 430 亿 m^2，每年新建房屋面积高达 16~20 亿 m^2，约占世界新建建筑总量的 40%[1]。仅 2014 年，中国施工企业因钢材使用所产生的二氧化碳排放量高达 21 亿吨，废弃物高达 4.5 亿吨[2]。建筑在使用过程中产生的碳排放也是惊人的，约占中国碳排放总量的 25%~27%[3]。

随着城市化进程的推进和人民生活水平的提高，未来中国建筑能耗还将会有较大比例增长。清华大学课题组考虑建筑总面积、生活方式变化和技术水平提高的影响，对各种情境下 2030 年中国建筑的耗能情况进行了预测，结果显示如果节能措施不力，2030 年中国建筑耗能可达 15.1 亿吨标煤，排放 39.6 亿吨 CO_2，为 2006 年的 2.7 倍，占 IPCC《第四次评估报告》预测2030年世界经济发展较快情景建筑156亿吨CO_2的26%[4]。因此，中国建筑和基础设施等土木工程相关行业节能减排、保护环境任重而道远。土木工程可持续性已成为未来结构设计的重要目标。要提升土木工程可持续性，需要大力研发绿色节能的技术，同时也要有促进可持续工程的激励与约束机制。可持续性评估是实现可持续管理的前提，非常有必要开展土木工程可持续性的量化研究。这方面无论是国际还是中国都还没有充分开展，尚处于初期探索阶段。

可持续性包括经济、社会和环境影响这三个相互依存且相辅相成的部分。然而，要实现土木工程可持续发展的目标，不仅要对其经济、环境和社会性能进行评价，还应保证其技术与功能满足要求。本文将从全生命周期的角度，对开展土木工程环境影响、经济成本和社会影响的量化评价方法予以阐述，对北京交通大学土木工程创新课题组（ICCE）考虑工程技术参数（可靠度）对环境、经济和社会影响，建立的土木工程综合可持续评价和优化方法，予以介绍，并给出相应一些具体工程项目生命周期环境影响评价和综合可持续评价的结果。

1. 土木工程结构生命周期环境影响评

20世纪60年代，全生命周期环境评价（Life Cycle Assessment，LCA）的思想开始萌芽，并发展成迄今为止唯一被标准化的产品环境影响评价和提供环境决策支持的工具。1969年，美国可口可乐公司对生产不同类别饮料容器过程中消耗的资源和产生的环境排放进行评价研究，标志着LCA研究的开始[5]。20世纪80年代末至今，LCA研究进入了快速增长期。1990年，首届关于LCA的国际研讨会由国际环境毒理学和化学协会（International Society for Environmental Toxicology and Chemistry，SETAC）主持召开，会上首次明确了"生命周期评价"的术语[6]。国际标准化组织（International Organization for Standardization，ISO）也对LCA的发展起了很大的推动作用：1993年，"环境管理标准技术委员会（TC207）"正式成立，1997年，有关LCA的一系列标准开始颁布实施[7-9]，自此以后，LCA被正式纳入到ISO环境管理体系中。20世纪90年代中后期，联合国环境规划署（United Nations Environment Program，UNEP）也开始参与到了LCA的研究当中，在1996年和1999年分别发表了题为《LCA：概念和方法》[10]和《面向全世界的LCA应用》的报告[11]。

LCA的实施框架最初由SETAC确定，该框架以三角形模型为基础，将LCA方法分为4个组成部分：目标与范围的确定，清单分析，影响评价和改善分析[12]。ISO[13]在SETAC技术框架的基础上，对生命周期评价技术框架做出了改进，去掉了改善分析阶段，而增加了解释阶段，并指出解释阶段是对前三个相互关联的阶段进行解释，且该解释是双向的，其相互关系如图1所示。

生命周期评价的四个阶段，其主要研究内容可以概括如表1所示。

全生命周期评价各阶段主要内容　　　　　　　　　　表1

阶段	内容	说明
目标和范围的确定	目标确定	确定研究原因及未来结果的应用
	功能边界	指定生命周期的每个阶段及每个阶段考虑影响因素的数量

续表

阶段	内容	说明
目标和范围的确定	地域边界	确定影响因素作用的地理范围
	时间边界	确定影响因素作用的时间范围
	定义功能单位	用来作为基准单位的量化的产品系统性能
	数据质量评估	调查数据来源、时间跨度、空间范围及获取方式并评估其可靠性
清单分析	建立流程图	确定功能单位在每个阶段的经济和实物的各种因素的输入和输出
	量化因素	量化功能单位的每个阶段的各种影响因素
	建立清单	经过上述步骤形成了产品环境因素清单
环境影响评价	分类和特征化	根据每种因素的影响方式对其进行分类
	标准化	对不同类型的环境影响进行单位统一
	权重	根据每种因素对环境影响的大小进行赋权
评价、解释	对结果进行评价	根据之前的分析,提出产品在环境影响方面的改进方法和改进方向

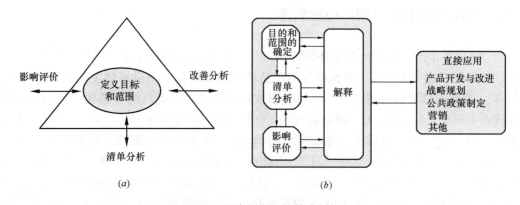

图 1 生命周期评价技术框架
(a) SETAC;(b) ISO

由于生命周期评价方法涉及的范围广泛,时间和空间的跨度较大,因此,不同的研究可能会选取不同的系统边界来对产品进行环境影响分析。由 PE-International 开发的生命周期评价专业软件——GaBi 软件说明书中[14]指出生命周期评价按系统边界的选取不同可分为以下四种类别:从摇篮到坟墓(Cradle to grave)、从摇篮到大门(Cradle to gate)、从大门到大门(Gate to gate)和从大门到坟墓(Gate to grave),如图 2 所示。

由于各国、各地区的生产及施工工艺选用及发展阶段的不同,以及自然条件的差异,导致各国生产相同产品所造成环境影响的差异。LCA 自 1990 年以来被用于建筑行业[15],许多发达国家的研究机构对建筑 LCA 便开始涉足并开展了不断深入的研究。而桥梁工程的 LCA 环境评价研究远远落后于民用建筑的研究,不管是国内还是国外更多的桥梁评估理论集中在桥梁工程本身的结构性能上,如安全性、耐久性及适用性等[16]。下面,本文以建筑和桥梁作为研究对象,对其全生命周期(材料生产、运输、设计、施工、运营和加固)不同阶段的环境影响评价做系统性的综合分析,结合具体工程案例给出量化分析结果。

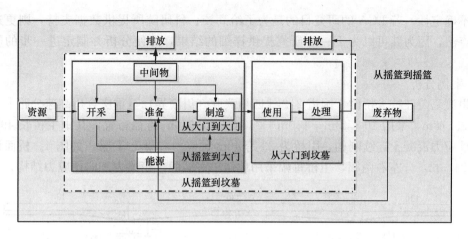

图 2 不同生命周期评价类别系统边界

1.1 设计阶段 LCA 分析

1.1.1 基本方法

生命周期环境影响评价是在完成目的和范围确定、清单分析后展开的数据计算和分析工作,需要根据清单分析所提供的资源、能源消耗数据以及各种排放数据对产品所造成的环境影响进行定性定量的评估,确定产品环境荷载。环境影响评价共包括三个步骤:分类及特征化、标准化、权重计算。

LCA 影响评价根据评价的深度,可以分为中点破坏环境影响评价和终点破坏环境影响评价。两种破坏评价模型如图 3 所示[17]。

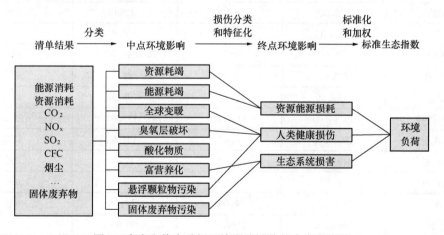

图 3 中点和终点破坏环境影响评价的内容和步骤

由图 3 可以看出,构建生命周期影响评价的模型方法是目前研究的主要方向,中点模型侧重于环境影响及其机理,对各种环境扰动因素采用当量因子转换来进行数据分析,终点模型则注重影响的后果[18]。Kägi 等[19]也指出,终点法与中点法相比,由于需要一个更完整的影响路径,因此会造成额外的不确定性。若 LCA 的研究目的是为了给政策制定提供支持,或便于企业的宏观了解,则选择终点法可以让其有更准确的了解,并降低其主观决策的不确定性。若 LCA 的研究目的是为了给高校、科研院所提供数据支持，则更适宜选择中点法，因为其可以为下一步研究提供详细的结果，方便分析并制定进一步的研究方案。

1.1.2 案例分析

白果渡嘉陵江特大桥位于国道 212 线四川武胜（川渝界）至重庆合川高速公路上，桥全长 1433.78m，采用 10×40m+130m+230m+13×40m 跨径布置，其中主桥长 490m，为三跨预应力混凝土连续刚构，引桥为 23 跨 40m 预应力混凝土 T 梁，如图 4。桥面设计宽度为 24.5m，分左右两幅，主桥每幅采用单箱单室截面，箱梁为三向预应力结构。

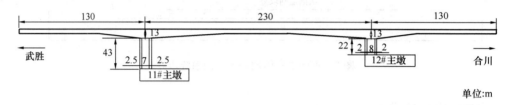

图 4 白果渡嘉陵江特大桥整体布置图

（1）桥梁环境影响中点模型分析

依据环境影响中点模型计算方法，可得出白果渡嘉陵江特大桥在物化阶段（主要包括材料生产阶段和施工阶段）内的环境影响潜值，见表 2。

白果渡嘉陵江特大桥在物化阶段的环境影响潜值　　　表 2

物质	材料生产阶段	施工阶段	总和	占比
GWP	1860.00	96.60	1956.6	4%
AP	202.00	867.00	1069	2%
EP	44.60	276.00	320.6	1%
POCP	151.00	88.70	239.7	0%
DUST	223.00	0.00	223	0%
固废	4790.00	6.98	4796.98	10%
化石燃料	10100.00	1130.00	11230	22%
石灰石	15600.00	0.00	15600	31%
铁矿石	4250.00	0.00	4250	8%
锰矿石	63.30	0.00	63.3	0%
水	161.00	0.06	161.06	0%
总和	47600.00	2460.00	50060	—
占比	95%	5%		

结果表明，在桥梁物化阶段内，材料生产阶段产生的环境影响占总环境影响的 95%，说明该阶段对环境的破坏最为严重。主要原因在于，材料生产阶段会消耗大量的矿石资源，如石灰石、铁矿石、锰矿石。不仅矿石需求量大，且由层次分析法得知，其重要性优先于其他所有环境影响。

在桥梁物化阶段内，石灰石影响潜值所占比重最大，为 31%，其次为化石燃料，占

22%。桥梁在材料生产阶段、施工阶段及运营维护阶段会消耗大量的资源、能源，因资源耗竭造成的环境影响远远大于由污染物排放对环境造成的破坏。

（2）桥梁环境影响终点模型分析

基于环境影响终点模型计算方法，可以得到白果渡嘉陵江特大桥的环境影响量化分析结果见表3。

白果渡嘉陵江特大桥在物化阶段的环境损伤权重值　　　　表3

阶段	人类健康 (capita·yr)	生态系统 (capita·yr)	资源能源 (capita·yr)	总和 (capita·yr)
材料生产阶段	4450.00	1200.00	21100.00	26800.00
施工阶段	3020.00	5530.00	714.00	9260.00
总和	7470.00	6730.00	21814.00	36060.00

终点破坏模型计算方法的最终目的是求得物化阶段内总环境荷载值，由表3可知，白果渡嘉陵江特大桥物化阶段的总环境影响为 3.61×10^4 capita·yr。综合以上数据可以得到桥梁物化阶段不同环境影响类型及不同阶段总环境荷载的对比，见图5。

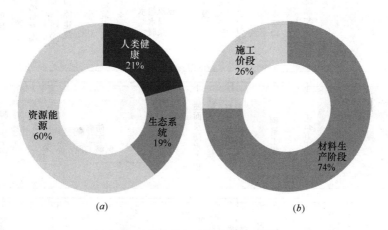

图5　环境损伤权重值的百分比组成
（a）不同类型；（b）不同阶段

图5（a）表示白果渡嘉陵江特大桥在物化阶段不同类型的环境损伤权重值。在三种环境影响类型中，以桥梁对资源能源的损伤最大，为60%，其次为人类健康受损和生态系统受损，分别占20%左右。图5（b）表示白果渡嘉陵江特大桥在不同阶段环境损伤权重值的百分比组成及贡献排序。其中材料生产阶段造成的环境影响最大，占物化阶段环境影响的74%，施工阶段占26%。

1.2 施工阶段LCA分析

在土木工程物化阶段（材料生产、运输、施工建造）考虑工程环境影响的计算过程中，由于以往数据来源往往是简略的分部分项的汇总，既不能给出详细的排放数据，又不能给设计人员提供针对墙、柱等结构构件的减排意见。因此，本节选择了建筑物化阶段的

CO_2 排放作为研究对象，采用生命周期评价方法对其进行计算和分析。利用工程概预算定额对建筑项目按照单位工程、分部工程、分项工程、单元工序的层级进行分解，建立了不同工序和不同施工阶段环境排放的计算模型，并通过实际案例进行了计算与分析环境影响。

1.2.1 基本方法

(1) 建筑施工过程分解与集成

建筑的项目分解与集成是根据建筑工程的施工单元过程，结合工程概预算定额来实现的，具体的分解和集成情况如图6所示[20]。

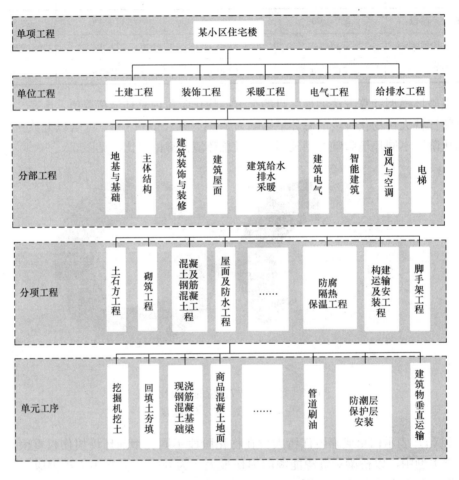

图6 建筑工程施工项目分级与集成

(2) 建筑物化阶段 CO_2 排放量计算

本文采用以工程量清单计算环境排放的模式，借鉴了工程概预算的方式，以建筑工程建设的单元工序为基本单位，将"环境排放（CO_2）"看作一种货币，构建各单元工序的环境排放清单和"综合环境排放系数"。利用单元工序清单组成分部分项工程清单，计算分部分项工程综合环境排放系数，再利用分部分项工程综合环境排放系数计算各单位工程物化阶段环境排放量，并最后得到整个单体建筑物化阶段的环境排放总量。

各单元工序的环境排放包括材料和机械两部分,本节考虑在建筑工程生产力平均水平的条件下,按照施工单位具有正常的施工条件、合理的劳动组织,使用恰当的材料和机械来完成单位合格建筑工程产品过程中对外界产生的环境排放量,作为单位工程的综合环境排放系数,即利用建筑工程概预算定额清单确定建筑工程环境排放定额清单。

各单元工序综合环境排放系数计算公式见式(1):

$$E_{m,n} = \Sigma p_i(W/w)EF_{mat,i,m} + \Sigma p_iL_iEF_{tra,m} + \Sigma(p_i\beta_iL_iEF_{tra,m} + p_i\beta_iEF_{dis,i,m} + q_iEF_{mach,j,m}) \quad (1)$$

各分项工程中包含不同的单元工序,完成单位工程量分项工程所造成的环境排放应包含完成相应工程量各单元工序过程中产生的全部环境排放量。因此,在单元工序定额清单的基础上,可建立分项工程定额清单,计算完成单位分项工程的综合环境排放系数。

各单位分项工程综合环境排放系数计算方法见公式(2):

$$E_{m,p} = \Sigma_n Q_n E_{m,n} \quad (2)$$

建筑工程可分解为诸多单位工程,每个单位工程中包含若干项分部工程,分部工程又是由各分项工程所构成的,由此按各层级分别求和,即可得到整个单体建筑工程物化阶段的环境排放。因此,建筑工程物化阶段总环境排放计算见式(3):

$$E_m = \Sigma_r \Sigma_q \Sigma_p E_{m,p} \quad (3)$$

除上述混凝土及钢筋混凝土工程等有明确工程资料和定额信息的施工过程之外,还有些工序虽然也耗能,但却无确切的定量信息。针对这部分施工过程,可根据单位面积的能耗统计值来估算工程的总耗能,并根据相应能耗计算环境排放量。这种方法的施工过程环境排放可按式(4)确定:

$$E_{cone,m} = \Sigma Q \times G \times EF_{ene,m} \quad (4)$$

式中各项符号含义见文献[20]。

1.2.2 案例分析

本节选用的项目位于河北省廊坊市大厂回族自治县,为某小区的一栋高层住宅。地上34层,地下1层,总建筑面积23886.87m^2,其中地上23212.79m^2,地下674.08m^2,建筑占地面积742.34m^2。整个小区的项目建设用地面积199331m^2。小区规划如图7所示。

根据工程量清单以及分部分项排放系数,按照公式(1)~公式(3),可计算出各分项工程CO_2排放量,进而求得分部工程CO_2排放量,最终得到单位工程CO_2排放量。对于每个单位工程,不同的施工阶段对其影响也有所不同,如图8所示。

根据上述对建筑物化阶段各施工阶段以及各单位工程CO_2排放量组成的分析可以看出:(1)若想减少一栋单体建筑物化阶段总体的CO_2排放量,可以主要考虑在

图7 案例项目规划

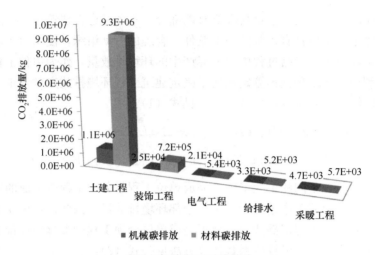

图8 各单位工程碳排放

材料生产阶段采取使用回收再利用原材料、增加生产流程中的环保措施、改变材料种类等减排措施,并辅以施工机械型号和所耗能源类别的改变、施工组织方案的改进;(2)若各单位工程想减少自身的 CO_2 排放,土建工程和装饰工程主要应考虑材料生产阶段进行减排,土建工程中也可以考虑施工机械方面的调整,装饰工程可以考虑调整运输方案、就近选择厂家、改变运输机械能耗类型等方面;电气工程需主要从机械使用方面考虑;给排水工程和采暖工程需要同时考虑材料生产和机械两个部分,才能达到良好的减排效果。

考虑施工组织进度的安排,施工建造阶段 CO_2 排放随时间变化的情况如图9所示。

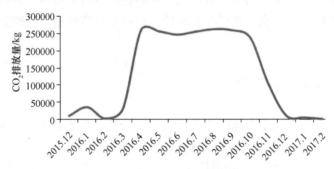

图9 CO_2 排放时间变化图

由图9可知,在土方开挖(2015.12~2016.1)、地下结构施工(2016.1~2016.3)和安装工程(2016.11~2017.2)阶段,CO_2 排放增速较慢,在主体结构施工(2016.4~2016.12)和初装修(2016.8~2017.2)阶段 CO_2 排放增速大。因为在主体结构施工和初装修阶段,会涉及起重机、施工电梯、起重机等大量大型设备的使用,而其余阶段所用机械单位台班的能耗较小。在土方开挖、地下结构施工和安装工程阶段,CO_2 排放总量较小,不足施工建造阶段 CO_2 排放总量的5%。主体结构施工(2016.4~2016.12)和初装修(2016.8~2017.2)阶段 CO_2 排放总量大,超过施工建造阶段 CO_2 排放总量的95%。因为主体结构施工和初装修阶段的持续时间长,所需的施工机械类型比其余阶段多。

1.3 维修加固阶段 LCA 分析

土木工程经常需要通过加固维修，才能达到其设计的寿命期。对于桥梁工程，随着我国交通事业的迅猛发展，运输量大幅提高，车辆载重日益增加，由于以前设计标准低，造成大量既有桥梁的通行能力不足；同时由于自然环境（大气侵蚀、温度湿度变化等）和使用环境（荷载作用频率增加，材料与结构的疲劳等）的长期作用，许多桥梁存在普遍老化和性能严重衰退的现象。目前绝大多数桥梁处于带病的工作状态，威胁着人民生命财产的安全，而拆除重建则会消耗大量资金与资源，所以对于桥梁发生的病害与缺陷要及时维修、加固补强，恢复和提高其承载能力。另外，由于桥梁具有较长的设计使用年限，人们发现桥梁的终身养护费用往往比最初的建造费用还要高。随着目前环境问题越来越严重，作为消耗加高的桥梁工程，加固方案的环境影响大小也应该成为与成本、性能同样重要的方案比选标准。

1.3.1 基本加固方法

目前旧桥的加固技术基本上有三种类型[21]：加强薄弱构件；增加或更换桥梁构件；改变结构受力体系。本文主要研究加强薄弱构件加固方法的环境影响。工程中最常用的加固方法有：增大截面加固法、粘贴钢板加固法、粘贴纤维增强复合材料加固法和体外预应力加固法等。

1.3.2 案例分析

某钢筋混凝土简支 T 梁桥，截面尺寸如图 10 所示。标准跨径 $L_b=20$m，计算跨径 $L=19.5$m，原设计荷载为公路-Ⅱ级。主梁采用 C30 混凝土，纵向主筋为 HRB335 级。跨中截面受拉钢筋为 $8\phi32+2\phi20$，钢筋横截面面积 $A_s=7062$mm²，保护层厚度 $c=35$mm。因交通量日益增加，需要提高该桥的承载力，加固设计按公路-Ⅰ级汽车荷载设计。结果重要性系数 $\gamma_0=1$。荷载等级提高后跨中截面弯矩组合设计值 $M_d=2519$kN·m。

对上述承载能力不足的 20m 简支梁桥的 T 形梁采用增大截面法、粘贴钢板法、粘贴碳纤维法和体外预应力四种常用加固方法进行加固计算，得到每种加固方法的加固方案。为了分析同一个构件采用不同加固方法其造成环境影响大小的不同，首先要根据每种加固方法的加固方案结合《公路桥梁维修与加固工程预算定额》确定每种加固方法的工程量清单，进而根据桥梁材料生产阶段和加固阶段各种材料、资源能源单位环境影响潜值计算四种加固方法的环境影响大小。本算例中，采取的具体的加固方案如下。

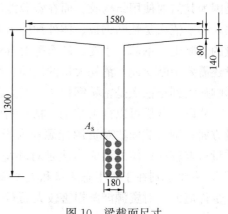

图 10　梁截面尺寸

（1）增大截面加固法（采用喷射混凝土）。
此加固方案所用混凝土体积与钢筋质量（$\phi25$ 钢筋单根理论质量 3.85kg/m[22]）：

混凝土体积：$V=0.25\times0.2\times20=1$m³

钢筋质量：$M=3\times3.85\times20=231$kg

（2）粘贴钢板加固法。此加固方案为宽度为 150mm，厚 5mm 的涂抹粘钢，粘钢面积

为 $S=0.15\times20=3\text{m}^2$。

(3) 粘贴碳纤维加固。此加固方案为粘贴一层1.2mm厚的Ⅰ级碳纤维板材，其单位面积质量 2kg/m^2，粘贴方案为梁底粘贴一层150mm宽的碳纤维板材，在梁肋两侧各粘贴一层50mm高的碳纤维板材。

(4) 体外预应力加固法。此加固方案为采用2束 $2\phi^s15.2$ 钢绞线，其理论质量为 1.101kg/m，所以钢绞线重量为 $2\times2\times1.101\times20=88\text{kg}$。

通过计算得到了桥梁加固最常用的四种加固方案：—增大截面加固法、粘贴钢板加固法、粘贴碳纤维加固法和体外预应力加固法，—在同一根梁的加固中的环境影响潜值与环境影响特征值，为了方便比较将四种加固方案的环境影响潜值和环境影响特征值分别列于表4。

不同加固方法的环境影响潜值　　　　　　　　　表4

影响类型	增大截面法	粘贴钢板法	粘贴CFRP法	体外预应力法
人类健康	0.095	0.127	0.048	0.006
生态质量	0.033	0.364	0.139	0.002
资源能源	0.292	0.011	0.004	0.008
总和	0.420	0.502	0.191	0.153×10^{-1}

对表4进行分析可知在四种常用的桥梁加固方案中对人类健康影响大小依次为：粘贴钢板法、增大截面法、粘贴CFRP法和体外预应力法。粘贴钢板加固法对人类健康和生态质量产生环境影响最大最主要的原因是粘贴钢板消耗大量的结构胶，在此两类型影响中生产结构胶产生的损伤占了90%以上，因为在结构胶的生产过程中会产生大量的氮氧化物会对人的呼吸系统和生态质量造成严重的损害。对生态质量影响的大小依次是：粘贴钢板法、粘贴CFRP法、增大截面法和体外预应力法。粘贴CFRP对生态质量影响较大也是因为其需要使用结构胶。而在资源能源方面，对其产生影响最大的加固方案是增大截面法，其余依次是粘贴钢板、体外预应力和粘贴CFRP。主要原因是增大截面法需要消耗大量的混凝土、钢筋、水泥、砂石等对自然资源需求量大的材料，而且其施工机械比较多同时还需要中断交通，消耗大量的汽油、柴油、电等能源。从环境影响总的潜值来分析，其影响从大到小依次是粘贴钢板法、增大截面法、粘贴CFRP法和体外预应力法。

从以上分析可以得出结论：粘贴钢板法对环境影响最大，尤其是在人类健康和生态质量方面；增大截面法对资源能源和人类健康影响较大，粘贴CFRP法对人类健康和生态质量影响较大；体外预应力法是对环境最友好的方法，其产生的环境影响仅为粘钢法的3%，主要差别在于对生态质量和人类健康方面。因为体外预应力法消耗的材料仅为钢材和少许能源，对资源能源损害较大而对于人类健康和生态质量则比其他三种方法小得多。

1.4　考虑不确定性的LCA分析

在土木工程全生命周期过程中存在大量不确定性，与确定性的LCA相比，采用考虑不确定性的LCA评价方法考虑各输入变量的变异性，以概率统计的方式来表示输入和输出变量，可使得LCA评价结果更加科学合理。因此，本节将分析清单数据不确定性产生的原因，以及误差在评价模型中的传递过程，并给出LCA不确定性的计算方法[23]。通过计算案例桥梁的环境影响值,分析了生命周期各个阶段、不同环境影响的特点以及形成的原因,通过统计方法计算出案例桥梁的LCA不确定性,得出了环境影响的概率分布。

1.4.1 基本方法

LCA是一种数据密集型的分析工具,数据的不确定性程度是影响其结果有效性的一个非常重要的因素。LCA的结果,只有在保证其可靠性的情况下,才能在实践中被接受和使用。而由于数据质量和模型算法的限制,LCA结果表示为确定的数值会带来较大的风险,因此有必要对LCA的结果进行不确定性分析。不确定性方法可以应用于清单分析(Life Cycle Inventories analysis,LCI)[24]和影响评价阶段(Life Cycle Impact Assessment,LCIA)[25],通过可信度的分析,便于更好的使用LCA结果。

LCA的分析基础来源于LCI,但在LCI阶段就存在着大量的不确定性,文献[26]中提到,导致全球变暖的清单中的不确定性就会占全部排放清单的百分之五到百分之二十,数据质量较好的情况下,不确定性仍然会达到百分之五左右。

对于LCI阶段,不确定性主要来源是由于使用统计方法或基于专家判断的方法来定量化清单数据,所带来的不确定性和可变性的累积效应[27]。对于影响评价阶段,由于使用了概率分布来描述清单数据的不确定性,可以用Monte-Carlo模拟用来评估统计的变异性或者不确定性传递的累积效应。

根据LCA不确定性的形成和来源,进行桥梁生命周期环境影响评价中不确定性分析计算时主要涉及两方面:①模型输入不确定性的量化分析;②不确定性在模型中的传播,可以用Monte-Carlo模拟的方法量化最终结果不确定性的大小。

1.4.2 案例分析

选取某大桥进行分析,该桥位于山西省岢岚至临县高速公路,桥梁全长568.00m,设计荷载为公路Ⅰ级;地震动峰值加速度:0.05g。梁体结构布置见图11。下部结构桥墩采用矩形墩、柱式墩,右幅桥台采用柱式台、肋板台,左幅桥台采用柱式台,基础采用灌注桩基础。

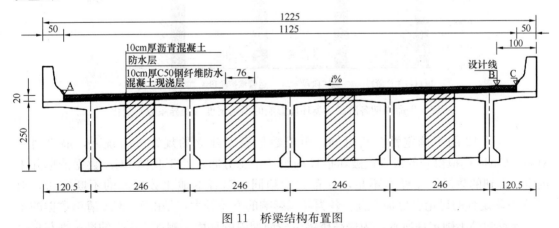

图11 桥梁结构布置图

根据国内外对环境影响潜值和清单数据的研究以及国内的调查和统计,可以对清单数据做出不同假设概率分布的χ^2拟合检验和K-S拟合检验。再结合1.1节中桥梁LCA的模型,使用Monte-Carlo模拟方法分析不确定性在模型中的传递,可以得到桥梁生命周期各个阶段的 3 类环境影响值的概率分布和统计特征,见表 5。图 12 为计算所得的生命周期各阶段 3 类环境影响的不确定度(变异系数)变化。

案例桥梁生命周期各阶段的环境影响值统计特征　　表 5

权重值 (capital·year)	分布类型	原材料生产阶段			施工阶段		
		均值	标准差	变异系数	均值	标准差	变异系数
人类健康	正态	379.2	151.1	39.84%	222.4	86.4	38.85%
生态环境	正态	79.4	26.2	33.04%	116.4	53.7	46.12%
资源能源	正态	1036.1	62.9	6.07%	460.2	28.8	6.26%
总和	正态	1494.7	165.7	11.09%	799	105.7	13.23%

权重值 (capital·year)	分布类型	运营维护阶段			拆除阶段		
		均值	标准差	变异系数	均值	标准差	变异系数
人类健康	正态	1084.6	434.1	40.02%	7.2	2.8	39.13%
生态环境	正态	1600.4	589.5	36.84%	9.8	4.2	43.26%
资源能源	正态	547.2	29.9	5.47%	40.7	2.4	5.99%
总和	正态	3232.2	732.7	22.67%	57.7	5.6	9.78%

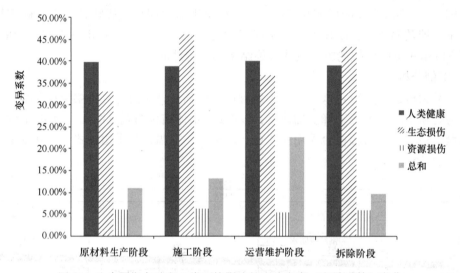

图 12　生命周期各阶段 3 类环境影响的不确定度(变异系数)变化

通过图 12 的不确定性分析可知,由于数据模型建立的规则较为统一,最终通过 Monte-Carlo 模拟得到的最终点破坏指标均服从正态分布,鉴于我国 LCA 数据清单并不健全,得到的数据往往样本不足,分布类型趋同,随着本地化数据库的不断完善,对 LCA 不确定性的讨论会更加充分。各类环境影响的方差较大,是由于 LCA 清单数据的不完善导致的高不确定性所致。为提高桥梁生命周期内的环境影响评价结果的准确性与可靠性,建立完备的中国基础建设工程主要建设材料、资源能源清单尤其重要,并且需要建立本地化的 LCA 数据库,给出中国的特征化、标准化因子,并对 LCA 结果的不确定性做出更为准确的判断。

2. 基于可靠度的土木工程 LCA 分析

土木工程技术参数的变化,会对工程的环境等方面的可持续性产生影响。如设计时结构的可靠性改变、工程寿命的长短,都直接导致工程的经济、环境和社会影响变化。因此,本文将分析该怎样综合工程结构技术参数变化,对工程的可持续进行评价。

本文将可靠度理论与 LCA 结合起来研究,把可靠度理论作为土木工程(桥梁)加固方案优化的依据,并基于 LCA 计算,寻找对环境的影响最小的方案。下面将以桥梁加固方案的综合评估为例,加以说明。

2.1 基本方法

(1) FRP 加固钢筋混凝土梁抗弯功能函数建立

在分析既有桥梁 FRP 加固后可靠度水平时,需要通过建立极限状态函数来对桥梁的不同状态进行划分。既有桥梁 FRP 加固后抗弯极限状态功能函数如式(5)所示:

$$Z = R - S = R - M_D - M_L = \begin{cases} > 0 & \text{安全状态} \\ = 0 & \text{极限状态} \\ < 0 & \text{失效状态} \end{cases} \quad (5)$$

在进行受弯可靠度分析时,需要根据不同的失效模式建立不同的极限状态函数。对于第 1 种失效模式,抗弯承载力表达式如式(6)所示:

$$R = \gamma_{\text{mfc}} Z1(1-\xi_e)\left[f_{\text{cd}}bx\left(h_0\sqrt{\xi_c} - \frac{1}{2}x\right) + f'_{\text{sd}}A'_s\xi_s(h_0\sqrt{\xi_c} - a'_s) + E_f\varepsilon_f A_f a_s\right] \quad (6)$$

对于第 2 种失效模式,抗弯承载力表达式如式(7)和(8)所示:

$$R = Z_1(1-\xi_e)\left[f_{\text{sd}}A_s\xi_s(h_0\sqrt{\xi_c} - 0.5\xi_{\text{fb}}h) + E_f[\varepsilon_f]A_f h(1-0.5\xi_{\text{fb}})\right] \quad (7)$$

$$R = Z_1(1-\xi_e)\left[f_{\text{sd}}A_s\xi_s(h_0\sqrt{\xi_c} - a'_s) + E_f[\varepsilon_f]A_f(h-a'_s)\right] \quad (8)$$

FRP 加固后抗弯极限状态功能函数中 M_D、M_L 为桥梁永久荷载及所受活荷载所产生的弯矩效应,如式(9)、式(10)所示,公式中参数见文献[28]。

$$M_D = \frac{l^2}{8}\left[(g_1 + g_2)\lambda_{\text{conc}} + g_3\lambda_{\text{asph}}\right] \quad (9)$$

$$M_L = \lambda_{\text{mtrk}}\left[I_{\text{beam}}\xi\left(m_{\text{cq}}q_k\frac{l^2}{8} + m_{\text{cq}}P_k\frac{l}{4}\right) + m_{\text{cr}}q_r\frac{l^2}{8}\right] \quad (10)$$

(2) 参数的随机分布及可靠度计算方法

在计算桥梁结构或者桥梁构件的可靠度指标时,将材料强度、构件几何尺寸作为随机变量来考虑的同时,也要考虑抗力模型以及荷载模型的不确定性。引入材料强度、结构尺寸、计算模型这三类随机变量,并且假定各个随机变量相互独立。可靠度指标计算过程中涉及到的随机变量统计特征见表 6。

随机参数统计特征[30]　　　　　　　表6

变量	平均值/名义值	变异系数	分布类型
抗弯模型不确定性因子	1.100	0.071	正态分布
钢筋屈服强度	1.160	0.069	正态分布
钢筋截面积不确定性因子	1.000	0.035	正态分布
C30混凝土28天抗压强度	1.250	0.146	正态分布
混凝土自重偏差系数	0.987	0.098	正态分布
沥青混凝土自重偏差系数	0.989	0.111	正态分布
汽车冲击系数	1.200	0.050	正态分布
活荷载产生的弯矩不确定性系数	1.350	0.162	极值Ⅰ型分布
梁板截面宽度	1.000	0.008	正态分布
梁截面有效高度	1.010	0.023	正态分布
汽车荷载不确定性系数	0.788	0.108	正态分布
人群荷载不确定性系数	0.579	0.391	极值Ⅰ型分布
CFRP厚度	0.825	0.040	正态分布
CFRP弹性模量	1.180	0.104	威布尔分布

采用Monte-Carlo抽样法与一次二阶矩法相结合的方法，能有效解决单独使用Monte-Carlo抽样法计算效率低和单独使用一次二阶矩法计算精度低的问题[29]。本文计算时，首先采用Monte-Carlo方法抽样来模拟功能函数Z的统计特征，然后采用二阶矩方法计算RC受弯T梁的抗弯可靠度指标。

2.2 案例分析

（1）工程概况

案例采用1.3.2节案例，桥梁信息见1.3.2节，截面尺寸见图10。

（2）可靠度计算

计算案例桥梁初始可靠度。在计算加固桥梁的可靠度时，首先计算桥梁的初始可靠度，建立桥梁初始受弯构件抗力模型如公式（11）：

$$R_M = \begin{cases} \gamma_{\mathrm{mfc}} f_c bx \left(h_0 - \dfrac{x}{2}\right) & \text{第一类T形梁} \\ \gamma_{\mathrm{mfc}} f_c \left[bx\left(h_0 - \dfrac{x}{2}\right) + (b'_f - b)h'_f\left(h_0 - \dfrac{h'_f}{2}\right)\right] & \text{第二类T形梁} \end{cases} \quad (11)$$

受压区高度值计算公式（12）：

$$x = \begin{cases} \dfrac{\lambda_{\mathrm{rebar}} f_s A_s}{f_c b} & \text{第一类T形梁} \\ \dfrac{\lambda_{\mathrm{rebar}} f_s A_s - f_c(b'_f - b)h'_f}{f_c b} & \text{第二类T形梁} \end{cases} \quad (12)$$

式中参数见文献［29］。

经计算得到梁的初始设计抗力值为$R_0 = 2266.549\mathrm{kN \cdot m}$，并根据规范，计算出荷载效应值$M = 2148.008\mathrm{kN \cdot m}$。将抗力表达式、截面受压区高度计算式以及荷载表达式中的参数都视作随机变量,采用 2.1 节所述的 Monte-Carlo 法与一次二阶矩法相结合的方法计算方法进行梁的初始可靠度指标计算,经计算得到梁初始可靠度指标 $\beta_0 = 5.44$。

下面,进一步计算案例桥退化后抗弯可靠度。建立既有桥梁加固前的功能函数,这里暂且认为荷载效应不随时间发生变化,只考虑桥梁自身性能的劣化。根据《公路桥梁承载能力检测评定规程》JTG/T H21—2011[31]引入 4 个桥梁计算系数,并建立既有受弯梁抗弯承载力计算模型如公式(13):

$$R_M = \begin{cases} Z_1(1-\xi_e)\gamma_{mfc}f_c bx\left(h_0\sqrt{\xi_c} - \dfrac{x}{2}\right) & \text{第一类 T 形梁} \\ Z_1(1-\xi_e)\gamma_{mfc}f_c\left[bx\left(h_0\sqrt{\xi_c} - \dfrac{x}{2}\right) + (b'_f - b)h'_f\left(h_0\sqrt{\xi_c} - \dfrac{h'_f}{2}\right)\right] & \text{第二类 T 形梁} \end{cases} \tag{13}$$

受压区高度值计算公式(14):

$$x = \begin{cases} \dfrac{\lambda_{rebar} f_s A_s \xi_s}{f_c b} & \text{第一类 T 形梁} \\ \dfrac{\lambda_{rebar} f_s A_s \xi_s - f_c(b'_f - b)h'_f}{f_c b} & \text{第二类 T 形梁} \end{cases} \tag{14}$$

式中参数及桥梁计算系数见文献[29]。

经计算得到不同环境条件下梁退化后的抗力值,荷载效应的计算结果不变,如表 7 所示。仍采用上述计算初始可靠度的方法,得到案例桥梁退化后的可靠度,如表 9 所示。

抗力及荷载效应计算结果 表 7

环境条件	Ⅰ类	Ⅱ类	Ⅲ类	Ⅳ类
退化后抗弯承载力(kN·m)	1932.809	1891.903	1832.589	1787.593
荷载效应(kN·m)	2148.008	2148.008	2148.008	2148.008

从表 7 可见,退化后的 RC 梁不论在何种环境条件下,它的承载能力已经不能满足荷载效应的要求。而且,随着条件恶化越来越严重,桥梁的承载能力退化的越快。那么就需要对 RC 梁进行抗弯加固来满足荷载效应的要求。另外,从表 9 中也可看出,各类环境条件状态下 RC 梁可靠度指标均有不同程度的退化,而且随着环境条件的恶化,可靠度指标减小越大。

由此,论证了该 RC 梁需要进行加固。依据《公路桥梁加固设计规范》JTG/T 122—2008[21],得出只考虑受外界环境影响自身性能退化情况下用 CFRP 布与 CFRP 板加固桥梁方案,如表 8 所示。

CFRP 布与 CFRP 板抗弯加固粘贴方案 表 8

材料类别 \ 环境条件	Ⅰ类	Ⅱ类	Ⅲ类	Ⅳ类
CFRP 布粘贴方案	梁底粘贴 3 层 160mm 宽 梁侧粘贴 3 层 160mm 宽	梁底粘贴 3 层 160mm 宽 梁侧粘贴 3 层 200mm 宽	梁底粘贴 3 层 160mm 宽 梁侧粘贴 3 层 250mm 宽	梁底粘贴 3 层 160mm 宽 梁侧粘贴 3 层 300mm 宽
CFRP 布截面积 A_f (mm²)	231	265	307	346

续表

材料类别＼环境条件	Ⅰ类	Ⅱ类	Ⅲ类	Ⅳ类
CFRP板粘贴方案	梁底粘贴1层160mm宽 梁侧粘贴1层80mm宽	梁底粘贴1层160mm宽 梁侧粘贴1层100mm宽	梁底粘贴1层160mm宽 梁侧粘贴1层120mm宽	梁底粘贴1层160mm宽 梁侧粘贴1层150mm宽
CFRP板截面积A_f（mm^2）	378	423	467	531

采用CFRP布与CFRP板进行梁抗弯加固后的抗弯可靠度指标计算以及RC梁初始可靠度、退化后抗弯可靠度对比结果列于表9中。

CFRP布与CFRP板加固后梁抗弯可靠度指标计算结果　　表9

环境条件	Ⅰ类	Ⅱ类	Ⅲ类	Ⅳ类
初始可靠度指标	5.44	5.44	5.44	5.44
退化后抗弯可靠度指标	4.51	4.39	4.18	4.02
CFRP布加固后抗弯可靠度指标	5.73	5.72	5.69	5.68
CFRP板加固后抗弯可靠度指标	5.92	5.91	5.89	5.87

从表9结果看，CFRP布和CFRP板加固对梁抗弯可靠度指标都有很大幅度的提升，采用CFRP板进行抗弯加固后的抗弯可靠度指标在5.9左右，采用CFRP布加固后的抗弯可靠度指标在5.7左右，采用CFRP板进行抗弯加固后的抗弯可靠度指标略高于采用CFRP布加固后的抗弯可靠度指标。

（3）LCA计算分析

由前计算结果，将不同环境条件下采用CFRP布及CFRP板进行RC梁的加固时材料使用量统计如表10所示。

CFRP布、CFRP板及相应结构胶使用量（kg）　　表10

项目＼环境条件	Ⅰ类	Ⅱ类	Ⅲ类	Ⅳ类
CFRP布	8.64	10.08	11.88	13.68
结构胶	33.12	38.64	45.54	52.44
CFRP板	13.44	15.12	16.8	19.32
结构胶	10.24	11.52	12.8	14.72

根据文献[32]，可知生产1kgCFRP布、1kgCFRP板以及生产1kg结构胶的环境影响潜值，由公式（15）及表10中所列材料用量便可算出采用CFRP布、CFRP板进行一次梁抗弯加固产生的环境影响潜值，计算结果如表11所示。

$$EI = EI_{CFRP} + EI_{结构胶} = Q_{CFRP} \times RM_{CFRP} + Q_{结构胶} \times RM_{结构胶} \quad (15)$$

式中，EI为环境影响潜值总和；EI_{CFRP}为CFRP材料的环境影响潜值；$EI_{结构胶}$为结构胶的环境影响潜值；Q_{CFRP}，$Q_{结构胶}$分别为CFRP材料和结构胶的使用量；RM_{CFRP}，$RM_{结构胶}$分别为CFRP材料与结构胶的环境影响系数。

CFRP 布和 CFRP 板单次加固环境影响潜值　　　　表 11

加固类型	影响类型	Ⅰ类环境条件	Ⅱ类环境条件	Ⅲ类环境条件	Ⅳ类环境条件
CFRP 布	人类健康	1.90E-01	2.22E-01	2.61E-01	3.01E-01
	生态质量	5.66E-01	6.60E-01	7.78E-01	8.96E-01
	资源能源	8.96E-03	1.04E-02	1.23E-02	1.42E-02
	总和	7.65E-01	8.92E-01	1.05E+00	1.21E+00
CFRP 板	人类健康	8.91E-02	1.00E-01	1.11E-01	1.28E-01
	生态质量	2.63E-01	2.96E-01	3.29E-01	3.79E-01
	资源能源	5.12E-03	5.76E-03	6.04E-03	7.36E-03
	总和	3.58E-01	4.02E-01	4.47E-01	5.14E-01

将不同环境条件下 CFRP 与结构胶分别产生的环境影响潜值用柱状图表示，对比结果如图 13 所示。

从图 13 对比结果可以看出，不论何种环境条件下，CFRP 布加固产生的环境影响要大于 CFRP 板加固产生的环境影响，而且 CFRP 布加固产生的环境影响大概是 CFRP 板加固产生环境影响的 2.3 倍左右。CFRP 布加固产生的环境影响潜值几乎都是来自结构胶，CFRP 布产生的环境影响只占很小的比例。CFRP 板加固的环境影响潜值中结构胶产生的环境影响潜值大约是 CFRP 板产生环境影响潜值的 2 倍。采用 CFRP 布加固需要粘贴三层，需要的结构胶多，CFRP 布加固所需的结构胶用量是 CFRP 板加固所需结构胶用量的 3.5 倍左右。因而，粘贴 CFRP 布加固产生的环境影响要比粘贴 CFRP 板加固产生的环境影响大。

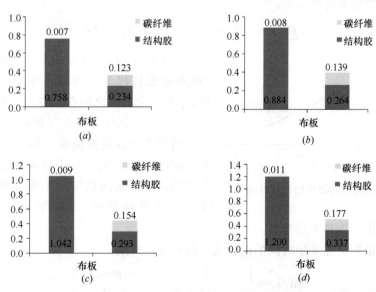

图 13　不同环境条件下 CFRP 布和 CFRP 板加固环境影响潜值
(a) Ⅰ类环境条件；(b) Ⅱ类环境条件；
(c) Ⅲ类环境条件；(d) Ⅳ类环境条件

3. 基于可靠度的土木工程结构经济—环境评价

上面综合考虑结构可靠度，进行了环境影响评价，本节将进一步综合考虑技术参数（可靠度）和经济参数（成本）的影响，对土木工程进行生命周期环境影响评价，并寻找综合的可持续最优解。

对桥梁维护策略与维护成本的研究由来已久，随着对维护后的桥梁可靠度进行随时间变化的考虑，基于生命周期成本的维护策略分析和优化的研究也日益增多[33]。但这些研究只是考虑了桥梁建设的成本，并没有考虑环境影响。本文结合 LCA、生命周期成本（Life Cycle Cost，LCC）和结构时变可靠度，采用统计学和数值模拟的方法得到桥梁运营阶段，采用不同维护策略对桥梁全生命周期内环境和成本的影响，运用帕累托最优化的方法进行结构时变可靠度、生命周期环境影响和成本的桥梁维护多目标优化。

3.1 基本方法

（1）桥梁结构时变可靠度模型

桥梁结构在建造施工及运营过程中存在着多种不确定性，这些不确定性需要在结构安全性分析中予以考虑，本章采用基于概率的时变可靠度方法，以失效概率作为桥梁结构安全性的衡量标准。首先建立结构抗力与荷载的功能函数 $Z(t)$：

$$Z(t) = R(t) - S(t) \tag{16}$$

式中，t 为时间（单位：年）；$R(t)$ 为结构抗力随时间演化的函数；$S(t)$ 为结构所承担的荷载效应随时间演化的函数。

则结构失效概率可表示为：

$$P_f(t) = P[Z(t)] = \int_{-\infty}^{0} f[Z(t)] \mathrm{d}Z \tag{17}$$

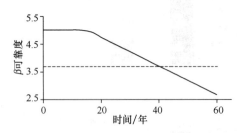

图 14　可靠度随时间退化的双线性模型

为了便于应用，Cornell[34] 采用可靠度指标 β 的概念来替代失效概率作为结构安全性的度量，β 与失效概率之间存在着一一对应的关系：

$$\beta(t) = -\Phi^{-1}[P_f(t)] \tag{18}$$

结构劣化效应被模拟为构件承载能力 $R(t)$ 随时间的不断折减，对于可靠度指标随时间演化规律，本文采用结构时变可靠度的双线性模型[35]，如图 14 所示，该模型基于大量的工程实例调研数据得出，更具实际意义。

模型假定结构可靠度随时间的退化为一条双线性折线：

$$\beta(t) = \begin{cases} \beta_0 & 0 \leqslant t \leqslant T_i \\ \beta_0 - \alpha(t - T_i) & t \geqslant T_i \end{cases} \tag{19}$$

式中，β_0 为桥梁结构的初始可靠度；α 为桥梁结构可靠度退化率，服从均匀分布（$\alpha \sim U(0.002, 0.1)$）。$T_i$ 为结构退化初始时间，服从对数正态分布（$T_i \sim LN(15,5)$）。

则首次期望维护时间 T_R：

$$T_R = \frac{\beta_0 - \beta_t}{\alpha} + T_i \tag{20}$$

式中，β_t 为桥梁结构目标可靠度。

本文中在进行结构检测后主要考虑三种维护决策：

① 不维护（No Maintenance，NM）；

② 预防性维护（Preventive Maintenance，PM）；

③ 实质性维护（Essential Maintenance，EM）。

（2）桥梁 LCC 分析方法

桥梁的生命周期成本分析，是要计算桥梁从设计方案、施工建造到运营维护，直至拆除的完整生命周期内的成本总和，并考虑资金的时间价值，以净现值（Net Present Value，NPV）作为成本的量化结果。本文根据对成本的生命周期划分，合并建设前期和施工的成本，形成 3 项成本的计算：建设施工成本、运营维护成本和拆除回收成本。分别计算如下：

① 建设施工成本，根据桥梁工程较为成熟的概预算内容，本文考虑建设期成本按公式（21）进行计算：

$$C_J = C_{J1} + C_{J2} + C_{J3} + C_{J4} \tag{21}$$

其中，C_J 为建设施工成本；C_{J1} 为建设安装工程费；C_{J2} 为广义的设备购置费；C_{J3} 为其他涉及费用；C_{J4} 为预备费。单位均为：万元。

② 运营维护成本：

$$C_o = C_{mhi} + C_{mpi} + C_{mti} + C_{mei} + C_{udi} + C_{uci} + C_{uti} + C_s \tag{22}$$

其中，C_{mhi} 为第 i 年桥梁管养成本；C_{mpi} 为第 i 年桥梁预防性维护成本；C_{mti} 为第 i 年桥梁实质性维护的成本；C_{mei} 为第 i 年桥梁实质性维护的成本；C_{udi} 为第 i 年通行受阻延误成本；C_{uci} 为第 i 年绕行交通成本；C_{uti} 为第 i 年运输损失成本；C_s 为运营阶段社会支付成本。

③ 拆除成本，当桥梁生命周期即将结束时，会根据桥梁实际情况选择全部或部分拆除，并根据桥梁使用状况进行回收再利用，此部分产生的资金投入，除人、财、物的核算，还应考虑固体废弃物、其他排放造成的影响成本。

3.2 案例分析

（1）工程概况

案例采用 1.4.2 节案例，桥梁信息见 1.4.2 节。

（2）指标维护策略优化过程

本文对案例桥梁进行以最大允许失效概率、生命周期环境影响与生命周期成本作为指标的维护策略优化。首先确定桥梁使用生命为 $T_L = 100$ 年，采用帕累托优化确定结构维护策略关键参数：

给定参数：β_0，T_i，α，β_t，T_{PD}；

确定参数：T_{XI}，T_Z；

以达到优化目标：

$$\begin{cases} P_f \leqslant P_{f,\text{target}}; \\ \text{minmize LCC} \qquad 0 \leqslant t \leqslant T \\ \text{minmize LCA}(E_I \text{ 值}) \end{cases} \tag{23}$$

本文采用基于遗传算法（Genetic Algorithms，GA）的 Pareto 多目标优化算法，从生成的 157 种维护策略中得到 16 种最优化策略，表 12 给出了这 16 种最优策略的相关参数，以及没有预防性维护策略的相关参数。

基于 LCA、LCC 与结构时变可靠度的最优维护策略　　　　　表 12

编号	PF_{TOTAL} ($\times 10^{-4}$)	LCC (万元)	LCA	预防性维护次数	实质性维护次数	预防性维护起始时间	预防性维护周期	实质性维护时间
NO PM	2.69	6830.98	3990.4	0	5	—	—	(29, 43, 57, 71, 85)
1	2.59	6661.93	2929.4	9	2	15	10	(43, 72)
2	1.47	6699.29	3208.2	7	3	16	12	(38, 62, 86)
3	1.67	6695.57	3208.2	7	3	17	12	(38, 62, 86)
4	1.98	6691.98	3208.2	7	3	18	12	(38, 62, 86)
5	2.48	6688.53	3208.2	7	3	19	12	(38, 62, 86)
6	1.3	6723.75	3208.2	7	3	21	12	(33, 57, 81)
7	1.4	6720.69	3208.2	7	3	22	12	(33, 57, 81)
8	2.19	6676.06	3262.4	8	3	15	11	(43, 67, 91)
9	1.33	6703.17	3262.4	8	3	15	12	(38, 62, 86)
10	1.79	6694.14	3262.4	8	3	16	11	(43, 67, 91)
11	1.81	6691.12	3262.4	8	3	20	11	(38, 62, 86)
12	2.12	6687.59	3262.4	8	3	22	10	(38, 62, 91)
13	2.41	6670.29	3316.6	9	3	16	10	(43, 72, 96)
14	2.12	6673.1	3316.6	9	3	17	10	(43, 67, 96)
15	1.68	6691.19	3316.6	9	3	18	10	(38, 67, 96)
16	2.04	6687.2	3316.6	9	3	19	10	(38, 67, 96)

由表 12 中策略 1 与没有进行预防性维护的情况对比，可以看出，由于桥梁的运营维护阶段时间较长，边际效应导致 LCC 最高的无预防性维护策略（LCC＝6830.98 万元），只比优化策略中的最优策略 1（LCC＝6661.93 万元）高出 169.05 万元，可见分析的量级与桥梁造价相比不算大，但这恰恰是考虑了成本的时间价值，并且具有实际意义。

由表 12 策略 2～7 可以看出，案例桥梁运营期进行预防性维护 7 次，实质性维护均为 3 次，预防性维护周期均为 12 年。图 15 给出了 LCC 与累积失效概率，维护起始时间之间的关系，可以看出，随着维护成本的降低，累积失效概率的也随之增大，当然，这些优

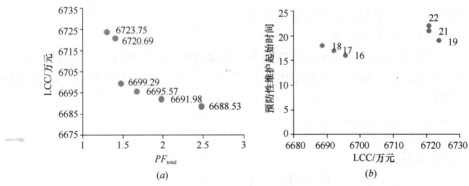

图 15　全生命周期进行 7 次预防性维护时的成本与累积失效概率、维护起始时间
(a) 失效概率与维护成本；(b) 维护成本与维护起始时间

化策略都是符合初始设定要求的,并且是最优化的解集之一。在考虑LCA值不变的情况下,可以根据对成本和安全的考量做出选择。

由表12中策略8~12可以看出,案例桥梁运营期进行预防性维护8次,实质性维护均为3次,预防性维护的周期集中在10~12年。图16给出了LCC与累积失效概率、维护起始时间的关系,可以看出,随着维护成本的降低,累积失效概率的也随之增大这一趋势始终保持,而成本和维护起始时间并没有太大的相关性,可以在此范围内根据实际情况进行选择。

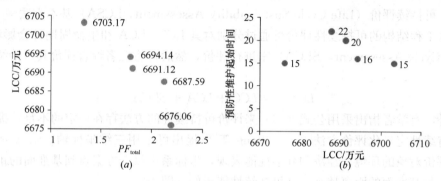

图16 全生命周期进行8次预防性维护时的成本与累积失效概率、维护起始时间
（a）失效概率与维护成本；（b）维护成本与维护起始时间

策略5和策略13的LCC与累积失效概率比较接近,但由于策略5预防性维护时间开始在先,使得进行首次实质性维护的时间较策略13推迟了5年,对控制成本的不确定性和环境影响的不确定性产生了积极作用。

由表12策略13~16可以看出,案例桥梁运营期进行预防性维护9次,实质性维护3次,预防性维护周期均为10年。与前两种优化策略集相比较,本种生命周期成本较低,实质性维护发生的时间推迟,但是环境影响较大。图17给出了LCC与累积失效概率、维护起始时间的关系,随着维护成本的降低,累积失效概率的也随之增大,而成本和维护起始时间并没有太大的相关性,选择时需要结合实际情况做出进一步权衡。

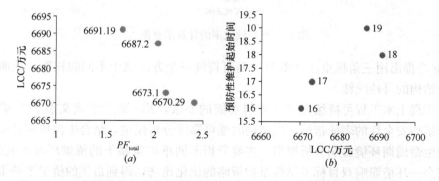

图17 全生命周期进行9次预防性维护时成本与累积失效概率、维护起始时间
（a）失效概率与维护成本；（b）维修成本与维护起始时间

这16种优化策略中,预防性维护的起始时间均匀分布在桥梁运营的第15年到第22年,维护周期集中分布在10、11与12年,这不仅与工程中的实际情况相符,验证了优化

模型的合理性,更给出了案例桥基于全生命周期成本和环境影响的维护优化策略的基本参数。

4. 基于可靠度的土木工程结构可持续性评价

4.1 基本方法

(1) 可持续评价(Life Cycle Sustainability Assessment,LCSA)基本方法

土木工程结构的可持续性评价可通过分别对其 LCC、LCA 和生命周期社会影响(Social Life Cycle Assessment,SLCA)来进行评价,然后再将三者综合评价其可持续性 LCSA,即,

$$LCSA = LCC + LCA + SLCA \tag{24}$$

此外,有学者指出采用公式(24)来评价可持续性的方法存在一定的不足,提出了不同的可持续性定量化评价方法[36-38]。Wang 等[37]提出可采用三维坐标轴,x、y 和 z 轴分别代表评价对象的环境、经济和社会性能表现,坐标系中每个方案点到基准面的距离为可持续值,与基准面的距离越远,其可持续性能越好(图18)。

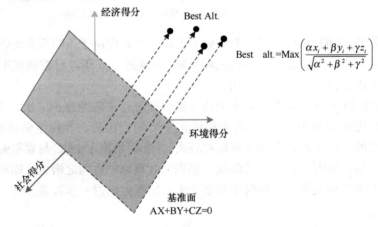

图18 可持续效率的计算示意图

Dong[38]提出用三角权重法(图19)对可持续三个方面赋予不同的权重,从而来评价土木工程结构的可持续性。

为了实现土木工程可持续发展技术与功能的要求,Xie 等[39]和武文杰等[40]采用失效概率作为结构安全性的表征指标,以结构时变可靠度分析模型,结合生命周期环境评价建立了桥梁生命周期环境影响分析模型。主要分析不同维护策略下的桥梁环境影响的变化,进行了安全—环境影响双目标下结构维护策略的优化比选,得到最优的桥梁维护策略。这些研究综合考虑了安全性和环境性能,还需要将安全性与经济、环境和社会性几个维度综合,开展全面的土木工程可持续性量化评价。

(2) 社会影响评价(SLCA)的基本方法

SLCA(Social Life Cycle Assessment)是一种社会影响评价的工具,该工具旨在评

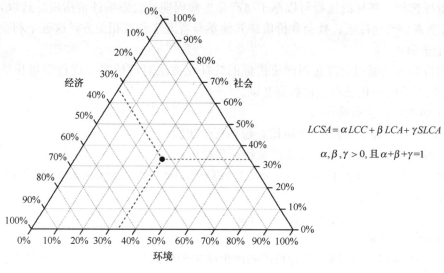

图 19 三角权重法

价产品的社会及社会经济方面的影响,包括产品全生命周期中潜在的正面及负面影响[41]。这里所说的全生命周期包含了产品原材料的开采与加工、制造、物流运输、使用与再利用、维护保养及最终的废弃。S-LCA 是环境全生命周期影响评价(E-LCA)的另一种类型,侧重对社会影响方面的评估。

SLCA 与其他社会影响评价方法相比,最大的不同点主要在于其对产品全生命周期的考察。SLCA 所关注的社会方面的影响主要包括在产品全生命周期中直接对利益相关方产生的正面或负面影响,当研究规模扩大时,有时也需要考虑间接的影响。这些影响与企业的行为、社会经济的进程、社会资本的积累情况相关。目前,关于 SLCA 在土木工程及其相关产品中已经有若干应用,例如,Fan 等基于联合国环境署(UNEP)和国际环境毒理学和化学学会(SETAC)出版的产品《产品生命周期社会影响评价指南》[41](以下简称为《指南》),对绿色住区的社会影响进行了社会影响评价[42]。张雨雄[43]将 S-LCA 的方法应用到装配式建筑当中,建立了装配式建筑的社会影响评价指标体系。高铸成[44]对该指标体系进行进一步本土化研究,并对其中的不确定性和参数敏感性进行了量化分析。

同 E-LCA 相同,开展建筑的 SLCA 也需要合理的方法和科学的过程,主要包括下述步骤:

1)建筑的 S-LCA 目的和范围的确定

① SLCA 目的:在于量化研究建筑生命周期的潜在社会影响,研究各项建筑技术措施对于建筑社会影响的作用。

② SLCA 的功能单元:建筑的社会影响,并不是其各个部件单元社会影响的线性总和,而是作为建筑这一个完整的整体对社会所产生的影响,因此考虑整个建筑作为一个功能单元。

③ SLCA 的系统边界:①时间边界:研究范围内的生命周期阶段;②物理边界:资源、材料、预制构件和建设阶段等实际阶段。

2)建筑的 SLCA 生命周期清单分析

① 指标选择：指标的选择可以基于《产品生命周期社会影响评价指南》选取，包括：工人、消费者、当地社区、社会和价值链其他参与者五个利益相关方，这五个利益相关方涉及 31 个子分类。

② 指标输入：通过合理选择评价指标的特征化方法对评价指标进行定量化处理，从而为标准化以及归一化处理提供数据基础。

3) 生命周期社会影响评价

影响评价是 S-LCA 的第三个阶段，包括三个步骤：

① 选择影响分类和特征化方法以及特征化模型；

② 将清单数据关联到特定的（Social Life Cycle Impact Assessment，SLCIA）子分类和影响分类中（分类）；

③ 确定/计算子分类指标的结果（特征化）。

4) 生命周期解释

生命周期解释是以得出结论为目的的评价结果分析过程。ISO 14044 中定义了三个主要步骤：

① 重要问题的识别；

② 研究的评估（包括对整体性和一致性的考量）；

③ 结论、建议和报告。

4.2 SLCA 相关标准[45]

（1）ISO 生命周期社会影响评价的发展

《指南》以 ISO 14040：2006《生命周期评价——原则与框架》和 ISO 14044：2006《生命周期评价——要求与指南》为骨架，确定了 SLCA 的四个评价步骤，即研究目标和范围、清单分析、影响评价及结果解释，并对每一个步骤都有详细的规定和描述，使得在此之后的实例研究有了充分的参考标准。SLCA 不仅可以单独使用，也可以与 LCA 结合使用，从而充分补充了产品生命周期的社会方面的影响。

在清单分析环节中，《指南》提出了双重社会影响的清单分类，一种是从利益相关方出发，另一种是从影响类型出发。这也为 SLCA 研究提供了数据库发展和软件设计的基础。作为第一份 SLCA 研究领域国际组织发布的指导性文件，《指南》的面世意味着SL-CA 在今后的研究中有导则可依循。

2010 年，UNEP/SETAC 提出了 SLCA 方法论手册，对产品 SLCA 基于利益相关方的影响分类进行了详细阐述[46]。利益相关方主要包括工人、当地社区、社会、消费者、价值链参与者等五方面，影响分类涉及薪资公平、技术发展、公平竞争等 31 个子分类。该方法论手册对每个子分类都进行了严谨的定义，列举了与该项分类相关的国际公约与协定，提出了全人类共同目标和建议性的规定。同时，该文件建议了每个分类应采取何种类型的数据（定量、半定量或定性）进行评价，详细列示了每个分类所需数据的参考数据来源，并给出了不同情况下进行评价可以采用的清单指标。这份文件的提出，对 SLCA 领域的研究在理论上做了进一步的完善，也为 SLCA 实际应用提供了一致有效且可靠灵活的帮助，对 S-LCA 的发展有着重要的意义。但是《指南》只是提出了对产品进行 SLCA 分析的框架，而建筑有自己的特殊性，不能单纯看成产品。

（2）欧盟建筑生命周期社会影响评价的发展

2014 年，CET/TC 350 发布了 EN16309（Sustainability of construction works-Assessment of social performance of buildings-Calculation methodology）[47]，该标准是在原有欧洲建筑可持续规范基础上进行的一个优化，并在考虑建筑功能和技术特性的基础上提供了建筑性能评价的特殊方法和要求。图 20 列出了欧盟可持续规范的评价框架。该评价体系是针对于定量化指标所表现出的建筑特性和影响进行评价，主要从以下几个方面进行评价：①可进入性；②适用性；③健康和舒适；④对于周边的影响；⑤维修；⑥安全和安保。但 EN16309 没有提出评估方法，在测量性能结果的水平、等级或基准线上没有进行规定，它的建立主要是基于客户需要、建筑管理、国家标准和建筑评估认证体系。不过在以下几个方面 EN16309 给出了要求：①评价对象的描述；②适用于建筑水平的系统边界；③指标列表和应用这些指标的步骤；④在报告和讨论中的结果展示；⑤应用于这个标准所需要的数据；⑥核查。

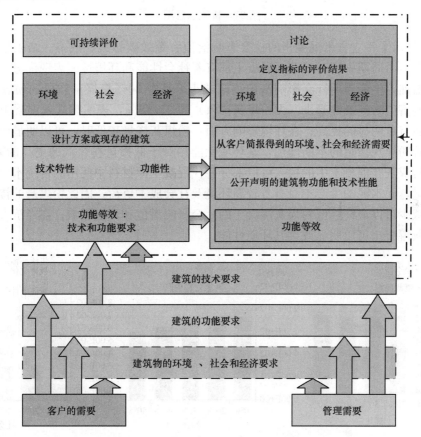

图 20　欧洲标准中建筑的可持续评价概述[47]

4.3　案例分析

本文将技术指标（可靠度）综合考虑到可持续性评价体系中，即通过衡量单位可靠度的环境影响、经济性能和社会影响来评价工程结构的可持续性，并以采用 5 组不同粉煤灰掺量[48]（0%，20%，30%，40%，50%）的 C50 混凝土设计的 9×20m 预应力混凝土简支梁桥为例进行研究,见图 21。桥梁上部结构根据《公路桥梁通用图》进行设计[49]。对于下部结构,采用双柱墩和钻孔桩基础。桩墩和桩基的直径分别为 0.9m 和 1.0m[50]。

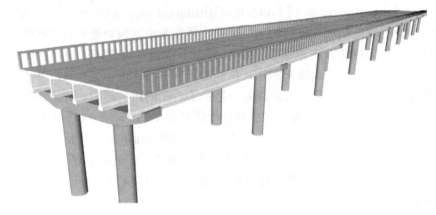

图 21 桥梁示意图

图 22 显示了每立方米混凝土的社会影响潜值、等效碳排放、成本。结果显示,粉煤灰混凝土在生命周期中相对于普通混凝土都有着社会性能和环境性能表现优势。随着粉煤灰掺量的增加,生命周期社会影响潜值逐渐减小,其中 30% 粉煤灰掺量的粉煤灰混凝土的社会影响潜值为基准混凝土(无粉煤灰掺入)的 93.48%,表明粉煤灰混凝土在改善社会效益层面有着明显的提高;在环境影响方面,粉煤灰掺量较低时,可以明显改变其环境性能,随着粉煤灰掺量的不断增加,其环境性能相对于低掺量粉煤灰的提高的并不明显。这是由于粉煤灰在混凝土中的掺入可以部分替代混凝土原材料中使用的传统水泥,减少水泥的使用量,而水泥的生产本身有着较高环境及社会成本负荷,因此粉煤灰的掺入使得水泥用量的减少可以最终降低粉煤灰混凝土的社会影响潜值和环境影响;在经济方面,低掺量的粉煤灰混凝土可以明显改善其环境性能。

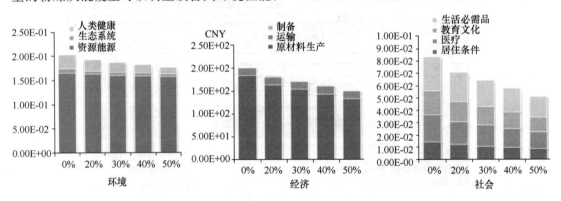

图 22 不同粉煤灰掺量下单位混凝土(m^3)生命周期环境-社会成本对比

由上述研究可知,随着混凝土中粉煤灰掺量的增加,单位粉煤灰混凝土的社会和环境影响会降低;但是粉煤灰掺量的增加会导致混凝土的实际强度的降低,从而导致构件的可靠度和使用年限相应降低。因此有必要对钢筋混凝土梁的生命周期可靠度和使用年限进行计算,从而评价整个生命周期内的粉煤灰混凝土梁可持续性。在本文中不考虑对构件的加固维修，即当构件达到其目标可靠度时（本文假定目标可靠度为4.2），即可以认为构件失效。因此，构件的使用寿命 T 可以用首次期望维护时间 T_R 来计算。不同粉煤灰掺量下钢筋混凝土梁的使用寿命见表13。

$$T = E(T_R) = (\beta_0 - \beta_t)E\left(\frac{1}{\alpha}\right) + E(T_i) \tag{25}$$

不同粉煤灰掺量下钢筋混凝土梁的初始可靠度及使用寿命　　　　表13

粉煤灰混凝土配合比编号及掺量	编号1 基准混凝土	编号2 掺量20%	编号3 掺量30%	编号4 掺量40%	编号4 掺量50%
初始可靠度	4.5130	4.3148	4.3449	4.2523	4.0296
目标可靠度	4.2	4.2	4.2	4.2	4.2
使用年限	30.38	28.64	27.21	25.45	21.24

桥梁的社会、环境和经济影响的年度值可以通过使用年限和对应的总生命周期影响来获得。将不同粉煤灰掺量下的钢筋混凝土梁的年均社会，环境，经济影响进行标准化处理，得到社会、环境和经济影响的得分，并进一步计算出可持续值，见图23。

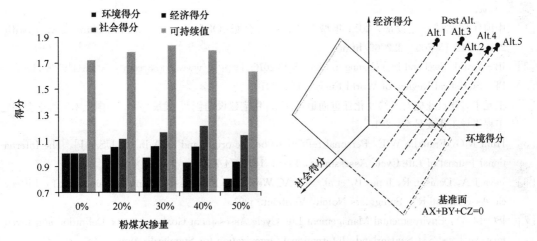

图23　不同粉煤灰掺量下环境—经济—社会得分和最终可持续值的比较

从图23（左）可以看出随着粉煤灰掺量的增加，该桥梁的环境得分随之下降，而经济得分随之增加，但经济得分随粉煤灰掺量的变化不明显，而社会评分随着粉煤灰掺量的变化呈现反V形，即呈现先增加后降低的趋势。随着粉煤灰掺量的增加，可持续值呈先增加后降低趋势。从图23（右）中可以直观看出方案3，即粉煤灰掺量为30%时，最终可持续性能表现最好。因此，在实际结构中应谨慎使用大掺量粉煤灰混凝土，并应积极探索粉煤灰的最佳掺量。

5. 结语与展望

土木工程相关的建筑和基础设施在建设和运营过程中消耗了大量的能源和资源，并产生很大的环境污染，是人类可持续问题重要的源头。本文针对土木工程结构在整个生命周期内的各个阶段，建立了综合考虑技术性能指标（可靠度）、经济指标（LCC）、环境指标

（LCA）和社会指标（S-LCA）的可持续评价指标体系，并通过不同的案例对相关问题进行详细说明。希望这些工作能为世界和中国重要性越来越凸显的土木工程可持续性评估，提供方法支持和实践示范

然而，土木工程的可持续评价仅刚刚开始，还很不完善，仍有很多问题需要进一步研究：①目前全球尚未建立完全针对土木工程的清单数据库，中国的清单数据库更是缺乏，故应加强基础数据的研究，为工程结构可持续评价提供基础支撑；②土木工程结构具有与普通产品完全不同的特点，应思考可持续评价方法如何更好地结合工程实际的特征并将其包含在评价过程中；③土木工程结构的性能具有很强的时变性和空间相关性，同时，经济、环境和社会指标也会随时间和空间的变化而变化，在未来的研究中，应构建时空相关的动态可持续评价体系，使得决策更加科学合理；④在可持续评价过程中包含了诸多的假设，故确定性可持续评价结果的准确性有待提高，在未来的研究中，应考虑不确定性和变异性的影响，构建工程结构基于不确定性的可持续评价方法，并积极探索将随机方法应用到可持续评价中的可能性。

参考文献

[1] 中国住房和城乡建设部，2014年城乡建设统计公报[OL]. http://www.mohurd.gov.cn/wjfb/201507/t20150703_222769.html.

[2] IBIS World. Steel and Iron Casting in China 2010-2016. http://www.doc88.com/p-5025844167284.html.

[3] BP. Statistical Review of World Energy[R]. 2010.

[4] 王元丰. 建筑业应对气候变化任重而道远[C]//住宅建设的创新/发展（六）. 中国土木工程学会住宅工程指导工作委员会，2010.

[5] Hunt R G, Franklin W E. Personal reflections on the origin and LCA in the USA[J]. The International Journal of Life Cycle Assessment，1996，1(1)：4-7.

[6] Fava J A, Denison R, Jones B, et al. SETAC Workshop Report: a Technical Framework for Life-cycle Assessment[R]. Smugglers Notch, Vermont. 1991.

[7] ISO 14041: Environmental Management-Life Cycle Assessment-Goal and Scope Definition and Inventory Analysis[S]. Switzerland: International Organization for Standardization, 1998.

[8] ISO 14042: Environmental Management-Life Cycle Assessment-Life Cycle Impact Assessment[S]. Switzerland: International Organization for Standardization, 2000.

[9] ISO 14043: Environmental Management-Life Cycle Assessment-Life Cycle Interpretation[S]. Switzerland: International Organization for Standardization, 2000.

[10] UNEP. Life Cycle Assessment: What It Is and How to Do It[R]. Paris, France: 1996.

[11] Haes H, Jolliet O, Norris G, et al. UNEP/SETAC life cycle initiative: background, aims and scope[J]. The International Journal of Life Cycle Assessment，2002，7(4)：192-195.

[12] Udo De Haes H Ed. Towards a Methodology for Life Cycle Impact Assessment[M]. Society of Environmental Toxicology and Chemistry, 1996.

[13] ISO 14044: Environmental Management, Life Cycle Assessment, Requirements and Guidelines[S]. Switzerland: International Organization for Standardization, 2006.

[14] PE International. GaBi Paper Clip Tutorial, Handbook for Life Cycle Assessment (LCA) using the GaBi Sotware[M]. 2011.

[15] Fava J A. Will the next 10 years be as productive in advancing life cycle approaches as the last 15

Years? [J]. The International Journal of Life Cycle Assessment, 2006, 11(1): 6-8.

[16] 许方强. 基于可持续发展的桥梁性能评估研究[D]. 北京: 北京交通大学, 2009.

[17] 顾道金. 建筑环境负荷的生命周期评价[D]. 北京: 清华大学, 2006.

[18] 庞博. 基于 LCA 的预应力钢筋混凝土连续刚构桥的环境影响分析[D]. 北京: 北京交通大学, 2011.

[19] Kägi T, Dinkel F, Frischknecht R, et al. Session "Midpoint, endpoint or single score for decision-making?"-SETAC Europe 25th Annual Meeting, May 5th, 2015[J]. The International Journal of Life Cycle Assessment, 2016, 21(1): 129-132.

[20] 王蕊. 建筑物化阶段 CO_2 和 PM2.5 排放及其不确定性分析[D]. 北京: 北京交通大学, 2018.

[21] JTG/T J22—2008. 公路桥梁加固设计规范[S]. 北京: 人民交通出版社, 2008.

[22] 沈蒲生. 混凝土结构设计原理(第三版)[M]. 北京: 高等教育出版社, 2007.

[23] Zhang Y R, Wu W J, Wang Y F. Bridge life cycle assessment with data uncertainty[J]. The International Journal of Life Cycle Assessment, 2016, 21(4): 569-576.

[24] Maurice B, Frischknecht R, Coelho-Schwirtz V, et al. Uncertainty analysis in life cycle inventory. Application to the production of electricity with French coal power plants [J]. Journal of Cleaner Production, 2000, 8(2): 95-108.

[25] Geisler G, Hellweg S, Hungerbühler K. Uncertainty analysis in life cycle assessment (LCA): Case study on plant-protection products and implications for decision making[J]. The International Journal of Life Cycle Assessment, 2005, 10(3): 184-192.

[26] Rypdal K, Flugsrud K. Sensitivity analysis as a tool for systematic reductions in greenhouse gas inventory uncertainties[J]. Environmental Science & Policy, 2001, 4(2-3): 117-135.

[27] Guo M, Murphy R J. LCA data quality: sensitivity and uncertainty analysis [J]. Science of the Total Environment, 2012, 435: 230-243.

[28] 解会兵. FRP 加固钢筋混凝土简支 T 梁桥抗弯可靠度研究[D]. 北京: 北京交通大学, 2016.

[29] 郁佳杰. 考虑 CFRP 抗弯加固方式的 RC 梁的可靠度及环境影响评价[D]. 北京: 北京交通大学, 2016.

[30] GB/T 50283—1999. 公路工程结构可靠度设计统一标准[S]. 北京: 中国计划出版社, 1999.

[31] JTG/T H21—2011, 公路桥梁承载能力检测评定规程[S]. 北京: 人民交通出版社, 2011.

[32] Pang B, Yang P, Wang Y, et al. Life cycle environmental impact assessment of a bridge with different strengthening schemes[J]. The International Journal of Life Cycle Assessment, 2015, 20(9): 1300-1311.

[33] 武文杰. 基于不确定性的钢筋混凝土桥梁量化可持续性评价[D]. 北京交通大学, 2013.

[34] Cornell C A. Bounds on the reliability of structural systems [J]Journal of the Structural Division, 1967, 93 (1): 171-200.

[35] Frangopol D M, Gharaibeh E S, Kong J S, et al. Optimal network-level bridge maintenance planning based on minimum expected cost[J]. Transportation Research Record: Journal of the Transportation Research Board, 2000, 1696(1): 26-33.

[36] 王元丰. 土木工程中国土木工程可持续发展的挑战与应对技术路径[C]//第 559 次香山科学会议报告文集, 北京, 2016.

[37] Wang J J, Wang Y F, Sun Y W, et al. Life cycle sustainability assessment of fly ash concrete structures[J]. Renewable & Sustainable Energy Reviews, 2017, 80: 1162-1174.

[38] Dong Y H. Life Cycle Sustainability Assessment Modeling of Building Construction[D]. The University of Hong Kong, 2014.

[39] Xie H B, Wu W J, Wang Y F. Life-time reliability based optimization of bridge maintenance strategy considering LCA and LCC [J]. Journal of Cleaner Production, 2018, 176: 36-45.

[40] 武文杰, 王元丰, 解会兵. 基于 LCA 和时变可靠度分析的桥梁维护策略优化[J]. 公路交通科技, 2013, 30(9): 94-100.

[41] Guidelines for Social Life Cycle Assessment of Products[M]. UNEP/Earthprint, 2010.

[42] Fan L, Pang B, Zhang Y, et al. Evaluation for social and humanity demand on green residential districts in China based on SLCA[J]. The International Journal of Life Cycle Assessment, 2018, 23(3): 640-650.

[43] 张雨雄. 工业化建筑生命周期社会影响评价[D]. 北京: 北京交通大学, 2016.

[44] 高铸成. 基于 S-LCA 的装配式建筑预制和施工阶段社会影响评价 [D]. 北京: 北京交通大学, 2018.

[45] 王元丰, 周硕文, 罗玮, 高铸成. 欧盟建筑社会影响评价标准及对我国的启示[C]//王清勤. 可持续建筑与城区标准化. 北京: 中国建筑工业出版社, 2017: 19-28.

[46] Catherine N. Introducing the UNEP/SETAC methodological sheets for subcategories of social LCA [J]. The International Journal of Life Cycle Assessment, 2011, 16: 682-690.

[47] BS EN 16309: 2014+A1: 2014, Sustainability of Construction Works-Assessment of Social Performance of Buildings-Calculation Methodology[S]. London: The British Standards Institution, 2014.

[48] 王帅民. 大掺量粉煤灰混凝土抗压性能试验研究[D]. 郑州: 郑州大学, 2011.

[49] 交通部专家委员会. 中华人民共和国交通行业-公路桥梁通用图[M]. 人民交通出版社, 2008.

[50] 王约发. 预应力简支桥梁结构可靠度分析及结构退化维修[D]. 南昌: 华东交通大学, 2009.

12 现代竹结构的进展

肖 岩[1,2]，单 波[3]，李 智[1]

(1. 南京工业大学土木工程学院，南京 211816；2. 美国南加州大学，洛杉矶 90089；
3. 湖南大学建筑安全与节能教育部重点实验室，长沙 410082)

摘 要：本文系统总结了现代工程竹（glubam）结构的进展，给出薄片和厚片两种不同胶合竹的构造及加工制备工艺，以及基本力学性能。基于现有木结构规范和材料性能试验方法，给出了胶合竹材强度特征值及设计值的计算方法。研究并建造了一系列格鲁斑竹结构示范工程。作者认为规范管理部门可以考虑将比较成熟的竹结构研究和应用成果纳入到现有的木结构规范体系，形成具有中国特色的现代竹木结构设计体系。

关键词：工程竹；胶合竹；glubam

Research and application progress of modern bamboo structures

Xiao Y[1,2], Shan B[3], Li Z[1]

(1. College of Civil Engineering, Nanjing Tech University, Nanjing 211816, China；
2. Department of Civil and Environmental Engineering, University of Southern California,
Los Angeles 90089, USA；
3. Hunan University MOE Key Laboratory of Building Safety and Energy Efficiency,
Changsha 410082, China)

Abstract: This paper reports the progress of modern engineered bamboo structures. The glued laminated bamboo or glubam is introduced in details including the configurations and manufacturing processes of two different types with thin-layer laminated bamboo and thick-layer laminated bamboo. Their basic mechanical properties, such as tension, compression, bending and shearing strengths were compared. According to the existing timber standards and material performance testing methods, strength characteristic values and design values of laminated bamboo material were proposed. Several demonstration modern bamboo structures were constructed with laminated bamboo based on mechanical characteristics of laminated bamboo. The need to update the current timber structures code is suggested to include bamboo structures.

Keywords: Engineered bamboo; laminated bamboo; glubam

1. 前言

可持续发展是当今世界土木工程的趋势，在此背景下，作为绿色材料的竹材越来越受到广泛关注。用竹材建造房屋和桥梁在竹产区均有着悠久历史[1]。现代竹结构可以定义为经过工程计算分析，采用现代加工及施工技术所建造的以竹材为主要材料的结构。按制作构件的材料现代竹结构又大致可以分为圆竹结构及工程竹结构。工程竹一般指的是通过胶合而重新组合的竹基材料。

竹材的最传统及最直接的应用是采用圆竹。国内外学者都曾对圆竹在建筑中的应用做了大量的研究。除测定竹材的物理力学性能[2]外，还进行了整竹杆件承载能力[3]，螺栓连接及竹屋架[4-5]的试验。竹材自重轻，便于架设多层建筑的建造和维修时的脚手架，所以竹脚手架在中国曾一度大量应用[6]。在南美洲等竹材产区，采用瓜竹制作的圆竹结构被比较广泛地用于建造桥梁和建筑[7]。国际标准化组织 International Organization for Standardization（ISO）于 2004 年颁布了针对圆竹结构的标准《Bamboo-Structural Design》(ISO 22156)[8]。哥伦比亚是第一个将圆竹结构纳入到规范的国家[9]。

采用胶合压制技术制成的胶合竹出现在 20 世纪后期，最初主要用于地板、汽车底板、包装箱等，后来由于国内木材缺乏，胶合竹板被广泛用于混凝土结构的模板。21 世纪初，国内外一些学者和工程师开始试图将胶合竹用于建筑及桥梁工程中。参照工程木（engineered timber）的称谓，这些以竹材作为基材的结构用胶合竹材也可以统称为工程竹（engineered bamboo）。作者的团队对工程竹做了 10 余年的研究和工程应用探索，研发了基于二次胶合压制的格鲁斑（Glubam）胶合竹[10]。

2. 格鲁斑胶合竹的制作工艺

由于竹子的自然属性，其单个竹竿的出材率远远小于单个树干，所以需要将竹材进行合理的集成制造满足结构设计需要的体积及体型的构件。胶合竹的制造工艺主要包括加工胶合竹板（plybamboo）的热压工艺和后续根据设计需要加工构件的冷压工艺。胶合竹结构的制作过程可以比拟为钢结构的加工，加工胶合竹板类似热轧钢板，形成尺寸比较规则便于运输的板材，而后续的冷压加工类似在钢构件加工厂将板材进行切割和连接（钢结构为焊接，竹结构为冷压胶合）制成各种满足设计要求的结构构件。

格鲁斑胶合竹的原材料一般为毛竹，而根据加工竹板所采用的竹片厚度又可以分为薄片胶合和厚片胶合。

薄片胶合也称为竹帘胶合板，首先将毛竹劈成竹条并剖成厚度约 2mm、宽度约 20mm 的竹篾，然后经过编帘、干燥、施胶、再干燥、组坯，最后在 150℃高温和 5MPa 的压力下热压约 20 分钟，制成厚度约为 30mm 的胶合竹板，其尺寸一般为长度 2440mm、宽度 1220mm。结构用薄片胶合竹板的竹纤维配置一般为沿纵向配置 80%，沿横向配置 20%，即配置为 4:1 [10]。

厚片胶合也称为竹集成材。首先将圆竹加工成宽度约 20mm、厚度约 5～8mm 的竹条，然后在 150℃的高温和 5MPa 的压力下热压约 20 分钟制成竹板，压制过程中一般还需要在板的侧向施加约 1.5MPa 的压力。比起薄片胶合，厚片胶合对于竹条的精度及压制过程的精度要求较高，得材率相对低，因此造价相对较高。此外，热压工艺本身耗能较高，且压制的厚度有限，对时间控制要求较严。图 1 给出两种胶合竹板的对比。

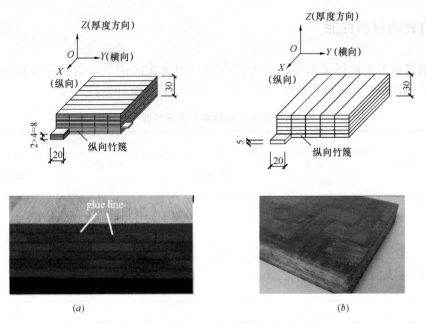

图 1　两种胶合竹板的构造
(a) 薄片胶合竹板；(b) 厚片胶合竹板

格鲁斑胶合竹构件一般都由多层胶合竹板，在常温下施胶加压叠合，制成构件。对于尺寸较大的构件，在冷压的同时需要对长度有限的板材进行接长，一次性整体胶合成型。接长一般可以采用齿接，并分层错开。将切割并开齿后的单板材按叠铺要求编号，然后按由下至上的原则分层在单板的指接头和叠合面涂刷粘结剂，再将叠铺好的板材移入特制的冷压机上，然后施加压力，待粘结剂固化达到要求后即可进行下一步加工。图 2 为笔者课题组研制的组合式多功能冷压机，冷压机由标准节段构成，因此冷压机的加工长度可根据加工构件的要求进行改变，且可以双向施加压力，压力通过同步液压系统加载。从冷压机上卸下构件胚体，进行必要的外观加工，如切边、刨光等处理，对于厚度差异较大的构件，可进行砂光定厚。根据设计，可能还需要对构件进行最后的加工，如钻孔或安装连接件等。

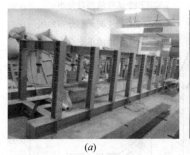

图 2　冷压工艺
(a) 组合式多功能冷压机；(b) 压制中的胶合竹梁

3. 胶合竹材的材料性能

课题组目前为止进行了大量的有关胶合竹材的材性试验[11]，获得了各种强度数据。表1给出了薄片及厚片胶合竹的拉、压、弯、剪平均值及统计参数。

两种胶合竹板材料的力学性能　　　　表1

强度参数	薄片胶合竹		厚片胶合竹	
	平均值（MPa）	变异系数	平均值（MPa）	变异系数
抗拉强度	$f_{t,ox}=83.0$	19.3%	$f_{t,ox}=106.7$	19.0%
抗压强度	$f_{c,oz}=51.0$	5.1%	$f_{c,ox}=71.2$	6.3%
	$f_{c,oy}=26.0$	11.5%		
弯曲强度	$f_{m,oz}=99.0$	12.1%	$f_{m,oz}=112.6$	13.5%
剪切强度	$\tau_{xy}=14.7$	13.5%	$\tau_{yx}=11.4$	34.0%
	$\tau_{zx}=4.6$	30.2%	$\tau_{zx}=4.5$	19.1%

注：$f_{t,ox}$ 表示沿 X 坐标轴方向的抗拉强度；$f_{c,ox}$，$f_{c,oy}$，$f_{c,oz}$ 分别表示沿 X，Y 及 Z 坐标轴方向的抗压强度；$f_{m,oz}$ 表示沿 Z 坐标轴方向的抗弯强度；τ_{xy} 表示剪切面平行于 Y 坐标轴，且与 X 坐标轴垂直的抗剪强度；τ_{yx} 表示剪切面平行于 Y 坐标轴，且与 X 坐标轴垂直的抗剪强度；τ_{zx} 表示剪切面平行于 X 坐标轴，且与 Z 坐标轴垂直的抗剪强度。

需要指出的是目前有关胶合竹材的材性实验并无成熟的标准，笔者在试验时遵循的是国内外木结构材料的试验方法和规程。除了出于现实的考虑外，本身也和目前的研发定位有关，因为格鲁斑胶合竹是作为格鲁斑胶合木的替代品研发的。作为天然材料的木材变异较大，其材性试验往往需要很大的试件量。而胶合竹材是经过一定程序加工而成的人工材料，变异性相对减小。图3给出了有关胶合竹剪切强度的试件量和统计值的对比，62个试件及任意30个试件的统计参数差别甚微[12]。因此，在今后建立有关胶合竹的材性试验标准时可以适当合理减少试件数目。

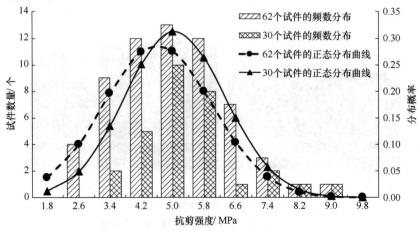

图3　不同试件数的概率分布与正态性检验

在上述材性试验数据的基础上，按照中国规范《工程结构可靠性设计统一标准》GB 50153—2008[13]，强度标准值 f_k：

$$f_k = \mu_f - 1.645\sigma_f \tag{1}$$

式中：μ_f 为清材小试件强度试验结果的均值；σ_f 为试验结果的标准差；f_k 为清材小试件强度的标准值（概率分布的 0.05 分位值）。

设计值 f_d，以《木结构设计手册》[14]为例：

$$f_d = K_P K_A K_Q f_k / \gamma_R \tag{2}$$

式中：f_d 为胶合竹强度设计值；K_P 为方程准确性系数；K_A 为尺寸误差影响系数；K_Q 为材料强度折减系数；γ_R 为抗力分项系数，顺纹受拉取值 1.95，顺纹受弯取值 1.60，顺纹受压取值 1.45，顺纹抗剪取值 1.50。

胶合竹强度设计值 f_d 在试验数据与概率计算所获得的标准值 f_k 的基础上，还应考虑以下几个方面的因素：（1）该设计值 f_d 在规定作用组合的效应 S_d 作用下，代入相应的抵抗力方程 $R_d = R(f_d, \alpha_d)$ 后，$R-S>0$ 的概率在 $\phi\{\beta\}$ 以上，其中 β 为目标可靠度，α_d 为几何参数设计值，一般取为实测值。这个目标，主要通过抗力分项系数 γ_R 来实现。（2）考虑建筑材料的实际使用环境与获得其清材小试件强度的标准值 f_k 的实验室环境的差异而引起的强度折减。对竹木材料而言，需要考虑的因素主要是材料的天然缺陷，相对含水率[15]（实验室相对湿度为 65±3%，温度为 20±2℃），加载时间与尺寸差异。而这些影响因素，通过系数 $K_{Qi}(i=1,2,3,4)$ 给予考虑。（3）尺寸误差影响系数以 K_A 表示。值得注意的是，此处的尺寸误差与（2）所述的尺寸差异是不同的概念，前者描述的是实验室小试件与足尺试件之间的尺寸效应，而后者指的是足尺试件本身的构造误差。（4）除此以外，抗力方程 R_d 与以上各个系数自身的不确定性，还通过方程精确性影响系数 K_P 给予考虑。

考虑到胶合竹材本质上是一种重组胶合材料，与胶合木类似，故按照《胶合木结构设计规范》GB/T 50708—2012[16]的相关规定（附录 D 根据构件足尺试验确定胶合木强度等级，条文 D.0.2）与《木结构设计规范》GB 50005—2003[17]，目前可偏于保守的按 TC17A 对格鲁斑胶合竹材进行强度设计值的取值。后续在针对工程胶合竹材使用环境及时间对其材料设计值影响的研究完备后，可进一步给出胶合竹材更为科学和节约材料的设计值。

4. 胶合竹结构构件研究成果

作者团队针对胶合竹各种构件及其连接件进行了一系列研究，包括梁，柱及轻型剪力墙等。

4.1 胶合竹柱

关于胶合竹柱承载力的研究表明，胶合竹柱的承载力高于《木结构设计规范》GB 50005—2003[17]的计算值，与美国 LRFD 规范计算值接近，如图 4 所示。

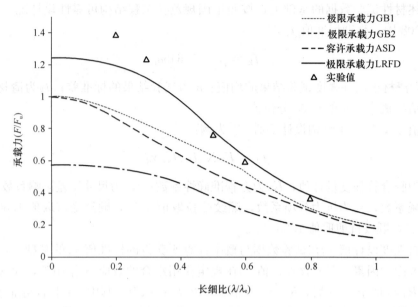

图 4 胶合竹柱实验结果与规范计算式对比

4.2 胶合竹梁

和现代木结构一样，胶合竹的一个主要结构构件是结构梁构件。由于胶合竹板材的生产工艺、运输等限制，其长度收到一定的规格限制，而结构大梁的尺寸一般比原材料的竹胶合板长，所以在二次胶合时需要接长。作者团队设计的竹梁采用如图 5 所示的齿接方式，并将每层的齿接错位以保证每个有接长的截面只有一层竹板有齿接。实验结果表明，齿接对竹梁的初始刚度没有影响，而只影响竹梁的极限强度。极限强度可以通过对受拉区减少了一层竹板的梁截面进行近似计算[18]。

图 5 竹梁的齿接

4.3 轻型框架剪力墙

竹结构轻型框架剪力墙墙体合理借鉴了北美轻型木结构墙体的构造，是轻型竹木结构的主要抗侧力构件，主要由四部分组成：墙体骨架，覆面板，面板—骨架连接件以及墙角抗倾覆锚固构件（hold-down）。墙体骨架由顶梁板、底梁板和墙骨柱组成，材料可以采用木规格材或胶合竹材，其中墙骨柱间距一般为 400mm，顶（底）梁板与骨柱之间采用钉连接。覆面板采用胶合竹板。面板—骨架连接件是影响轻型墙体性能的重要因素，一般采用钉连接，其中钉直径和钉间距还需满足剪力墙的要求。墙角抗倾覆锚固构件可以传递剪力并限制墙体在侧向力作用下的翻转。图 6 给出了此类墙体的主要结构组成及其构造方式。

作者课题组先后进行了竹胶合板覆面轻型木结构框架剪力墙[19]和 glubam 全竹轻型竹结构剪力墙[20]的低周往复加载实验研究。结果表明，轻型竹结构剪力墙的抗震性能与轻

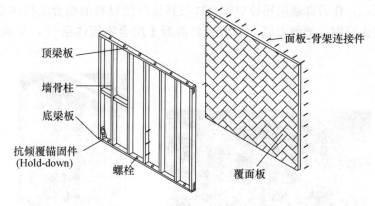

图 6　竹木轻型框架墙体构造详图

型木结构剪力墙类似其性能非常接近,具有推广使用的可行性,特别是轻型木结构剪力墙的覆面板应当可以用竹胶板代替。

5. 胶合竹结构体系

笔者课题组目前研究了一系列格鲁斑竹结构的结构体系,包括装配式轻型竹结构框架房屋[21],装配式竹结构板房[22],框架房屋和厂房,竹结构桥梁[23,25],桁架[25,26]等。图 7 给出了一些示范的案例照片。

图 7　胶合竹结构体系

(a) 装配式轻型竹结构框架房屋;(b) 装配式竹结构板房;(c) 框架房屋和厂房;
(d) 竹结构桥梁;(e) 竹结构三角桁架

如图 8 所示，作者课题组还针对胶合竹与其他传统材料的组合结构体系进行了研究，包括胶合竹覆面轻钢骨架剪力墙[27]，胶合竹混凝土组合梁板体系[28]，及钢竹组合屋架及网架[25,29,30]。

图 8 胶合竹组合结构体系
（a）钢拉杆竹屋架试验；（b）轻钢竹板剪力墙试验；（c）钢、竹梁混凝土组合梁板；（d）钢竹组合网架

6. 胶合竹结构的设计规程进展

为了使竹材能够更为广泛地应用于现代建筑，有必要建立相关技术规程。关于胶合竹装配式板房，笔者的研究团队提出了相关设计和施工建议[22]。

作者认为，一些竹结构构件的设计可以纳入到现有的木结构设计体系中，这样可以反映中国的竹木结构特色。现阶段，采用全竹或部分采用竹材的轻型剪力墙体系应当已经比较成熟，因此建议将有关轻型木骨架竹覆面板剪力墙及轻型竹结构剪力墙的设计条款纳入进木结构设计规范。

目前国内外研究一般认为墙体覆面板与骨架之间的钉连接件的承载力确定了墙体的抗侧向能力，故剪力墙的抗剪承载力设计值计算公式可表述为：

$$V = \sum f_\mathrm{d} l \tag{3}$$

$$f_\mathrm{d} = f_\mathrm{vd} k_1 k_2 k_3 \tag{4}$$

其中：f_vd 为采用结构板材作为覆面板的剪力墙的抗剪强度设计值（kN/m），其中采用木基材料的覆面板与骨架的设计值参见表 2；f_d 为剪力墙抗剪承载力设计值；l 为平行于荷载方向的剪力墙墙肢长度（m）；k_1 为木基结构板材含水率调整系数[14]；k_2 为骨架构件材料树种的调整系数[14]；k_3为强度调整系数，仅用于无横撑水平铺板的剪力墙[14]。

采用竹基材料作为覆面板或骨架材，在获知其单个面板-骨架钉连接件承载力设计值的基础上，亦可按上述公式（3）、（4）计算。同时研究表明，按既有木基材料的相关数据进行设计，是偏于保守的。

剪力墙的抗剪强度设计值 f_{vd}（kN/m）　　　　表 2

最小名义厚度（mm）	最小钉入深度（mm）	普通钉直径（mm）	面板直接铺于骨架构件			
			面板边缘钉间距（mm）			
			150	100	75	50
7	31	2.8	3.2	4.8	6.2	8.0
9	31	2.8	3.5	5.4	7.0	9.1
9	35	3.1	3.9	5.7	7.3	9.5
11	35	3.1	4.3	6.2	8.0	10.5
12	35	3.1	4.7	6.8	8.7	11.4
12	38	3.7	5.5	8.2	10.7	13.7
15	38	3.7	6.0	9.1	11.9	15.6

7. 结语及展望

现代竹结构已经有了较为丰富的研究积累，具备了形成规范化和工业化推广的条件。本文总结了格鲁斑glubam胶合竹的制作工艺及其材料性质，重点介绍了其主要结构构件的研究结果，结构体系运用，并对胶合竹的设计规程进展进行了介绍。本文作者认为：格鲁斑glubam胶合竹是具备工业化前景的竹基建材，具备替代欧美工业化木基建材glulam的潜力。今后业界应在前期工程实践与科学研究基础上，形成设计规程指导工程应用和推广。

参考文献

[1] 肖岩，单波. 现代竹结构[M]. 北京：中国建筑工业出版社，2013.
[2] 工程材料教研组. 北京市用毛竹性质研究[J]. 清华大学学报（自然科学版）. 1955，(6)：139-170.
[3] YU W K, CHUNG K F, CHAN S L. Column buckling of structural bamboo [J]. Engineering Structures. 2003，25(6)：755-768.
[4] 陈肇元. 有关竹屋架的几个问题[J]. 清华大学学报（自然科学版），1958，4(2)：269-286.
[5] 黄熊. 屋顶竹结构[M]. 北京：建筑工程出版社，1959.
[6] CHUNG K F, YU W K. Mechanical properties of structural bamboo for bamboo scaffoldings [J]. Engineering Structures，2002，24(4)：429-442.
[7] ARCHILA-SANTOS H F, ANSELL M P, WALKER P. Low carbon construction using guadua bamboo in Colombia [J]. Key Engineering Materials，2012，517：127-134.
[8] ISO 22156 Bamboo-Structural design[S]. Switzerland：International Organization for Standardization，2004.
[9] NSR-10 Reglamento colombiano de construccion sismo resistente: titulo g-estructuras de maderay es-

tructuras de guadua[S]. Columbia, 2010.

[10] XIAO Y, YANG R Z, SHAN B. Production, environmental impact and mechanical properties of glubam[J]. Construction and Building Materials, 2013, 44(3): 765-773.

[11] ASTM Standard D143-2009 Standard test methods for small clear specimens of timber [S]. West Conshohocken: ASTM International, 2009.

[12] XIAO Y, WU Y, LI J, et al. An experimental study on shear strength of glubam[J]. Construction & Building Materials, 2017, 150: 490-500.

[13] GB 50153—2008 工程结构可靠性设计统一标准[S]. 北京：中国建筑工业出版社，2009.

[14] 龙卫国. 木结构设计手册[M]. 北京：中国建筑工业出版社，2005.

[15] GB/T 1928—2009 木材物理力学实验方法总则 [S]. 北京：中国标准出版社，2009.

[16] GB/T 50708—2012 胶合木结构技术规范[S]. 北京：中国建筑工业出版社，2012.

[17] GB 50005—2003 木结构设计规范[S]. 北京：中国建筑工业出版社，2005.

[18] Xiao Y, Zhou Q, Shan B (2010) Design and construction of modern bamboo bridges. Journal of Bridge Engineering 15(5): 533-541.

[19] Yan Xiao, Zhi Li and Rui Wang. Lateral Loading Behaviors of Lightweight Wood-Frame Shear Walls with Ply-Bamboo Sheathing Panels [J]. ASCE Journal of Structural Engineering, 2015, 141(3): B4014004.

[20] Rui Wang, Yan Xiao and Zhi Li. Lateral Loading Performance of Lightweight Glubam Shear Walls [J]. ASCE Journal of Structural Engineering, 2017: 04017020.

[21] 肖岩，陈国，单波，等. 竹结构轻型框架房屋的研究与应用[J]. 建筑结构学报，2010，31(6): 195-203.

[22] 肖岩，佘立永，单波，等. 装配式竹结构房屋的设计与研究[J]. 工业建筑，2009，39(1): 56-59.

[23] XIAO Y, ZHOU Q, SHAN B. Design and Construction of Modern Bamboo Bridges[J]. Journal of Bridge Engineering, 2010, 15(5): 533-541.

[24] 肖岩，李磊，杨瑞珍，等. 胶合竹梁桥蠕变及承载力试验研究[J]. 建筑结构，2013，43(18): 86-91.

[25] 谢桥军，肖岩. 大跨度胶合竹结构屋架受力性能研究[J]. 建筑结构学报，2016，37(4): 47-53.

[26] XIAO Y, CHEN G, FENG L. Experimental studies on roof trusses made of glubam[J]. Materials and Structures, 2014, 47(11): 1879-1890.

[27] GAO W C, XIAO Y. Seismic behavior of cold-formed steel frame shear walls sheathed with ply-bamboo panels[J]. Journal of Constructional Steel Research, 2017, 132: 217-229.

[28] 单波，梁龙辉，肖岩，等. 胶合竹-混凝土复合式凹槽连接性能的试验研究[J]. 工业建筑，2015，45(4): 18-25.

[29] 杨秋旺，肖岩，吴越，等. 新型钢-胶合竹组合网架的试验研究[J]. 南京工业大学学报(自然科学版). 2016, 38(5): 7-12.

[30] Xiao Y. and Wu Y. Steel and glubam hybrid space truss, Engineering Structures, 2018, accepted.

13 组合混凝土结构基本概念和原理

肖建庄,张青天,丁 陶

(同济大学土木工程学院)

摘　要:从对混凝土结构发展历程的梳理中,提出了组合混凝土结构的概念,即在不同的层次上(材料、构件、结构)组合不同种类的混凝土,对混凝土材料以及结构进行精细化和优化设计,从而使其性能提升,更好满足安全性、适用性和耐久性,实现可持续性。在这个思路下,完成了符合组合混凝土结构概念的案例分析,并从中指出了其关键的科学问题,剖析了组合混凝土结构设计原理。最后对组合混凝土结构的发展趋势,包括界面设计、施工方式与可持续性评价进行了分析与展望。

关键词:组合混凝土结构;界面性能;优化设计;3D打印;可持续性

Fundamental concept and principle of composite concrete structures

Xiao Jianzhuang, Zhang Qingtian, Ding Tao

(School of Civil Engineering, Torgji University)

Abstract: The present paper introduces a fundamental concept for a novel structure through the development process of concrete, i. e., the composite concrete structures. Taking the mechanical properties of various kinds of concretes into consideration, concrete in structures could be combined in different levels (material, component and structure) according to this concept. Through optimization design, the safety, serviceability and durability of concrete structures could be significantly improved, and the sustainability of concrete structures could be achieved consequently. Based on the proposed concept, case studies were carried out, the design principle and existing problems were analyzed. Finally, the future development of composite concrete structures, including interface design method, new construction technique and the sustainability evaluation, are suggested and analyzed.

Keywords: Composite concrete structures; Interfacial properties; Optimization design; 3D-printing; Sustainability

1. 引言

　　混凝土作为建筑业的主要结构材料,其用量在近20年快速增加,2016年中国商品混

说明:本文的主要内容,已经在《同济大学学报》2018年46卷第二期上发表。

凝土用量已达17.9亿立方米，消耗量跃居世界第一[1]。同时，混凝土的生产会排放大量CO_2，且能耗总量大[2]。

纵观混凝土的发展历程，自波特兰水泥发明以来，混凝土经历了多个阶段，普通混凝土通常由水、水泥、粗骨料和细骨料组成，直至外加剂的加入，诞生了高强混凝土、高性能混凝土、超高性能混凝土，最后随着性能更优材料的加入，产生了纤维增强型水泥基材料（Engineered Cementitious Composites，ECC）、活性粉末混凝土等。混凝土自出生以来即带有混合的概念。此外，以往在设计钢筋混凝土结构时，人们更偏向于通过设计配筋来增强混凝土结构的性能，关注的重点往往是钢筋这个层面，从而钢筋的强度与各方面的性能不断得到提升。随着环境恶化与资源短缺的情况越来越严重，混凝土结构的可持续性越来越受到关注，而决定其可持续性的重要因素即为混凝土材料自身，因此需要把注意力转移到混凝土材料本身的发展上，以应对建筑业下一个100年的发展。

目前，再生混凝土材料与结构技术作为混凝土结构可持续发展的途径之一，已有较为成熟的研究[3]，经过合理设计和科学施工，再生混凝土可以作为结构混凝土用于实际工程中。需要指出的是，无论是普通混凝土还是再生混凝土结构，目前的结构设计通常仅关注其受力状态，功能主要通过外围护、外保护加以保障。整个构件与结构通常只采用单一的混凝土材料，常常造成不必要的浪费。

然而，纵观自然界生物进化过程，为适应自然界的变化，各生物不断优化和完善自身的组织结构与性能，其特点突出表现在选用合适的组合与复合形式，最大限度地减少材料消耗，来满足自身的功能需求。随着一系列新型水泥基材料，如海水海砂混凝土、纤维混凝土、轻质混凝土、橡胶混凝土、ECC以及新型施工方式（3D打印、可拆装施工等）的出现，混凝土工程完全可以结合不同水泥基材料的特点，设计出优化的混凝土构件与结构。基于此，提出了"组合混凝土结构"（Composite Concrete Structures）的新概念。

2. 组合混凝土结构概念

2.1 混凝土与其他材料的组合

将混凝土和其他材料，如钢材和纤维增强复合材料（Fiber Reinforced Polymer，FRP），组合在一起的结构形式已得到深入研究[4-5]，如图1所示，钢-混凝土组合结构、FRP-混凝土组合结构以及FRP-混凝土-钢组合结构形式已在工程中得以应用。这些组合

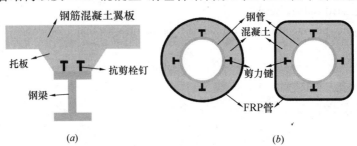

图1 混凝土与其他材料的组合
(a) 钢-混凝土组合形式；(b) FRP-混凝土-钢组合形式

将混凝土材料和其他材料各自的优势充分发挥了出来。但是，这种组合形式引发的界面问题十分突出，如钢-混凝土组合梁，为了防止纵向剪切破坏，需要增加抗剪栓钉，往往增加了设计困难和施工成本。

2.2 组合混凝土结构的雏形

过去在一般的混凝土结构中，已有组合混凝土的一些雏形，如图 2 所示，在框架结构的不同高度处，柱子采用不同强度等级的混凝土；楼盖结构中，在梁、板、柱、墙和节点处，采用不同强度等级的混凝土，这些均为不同强度上的组合形式，是组合混凝土的雏形。

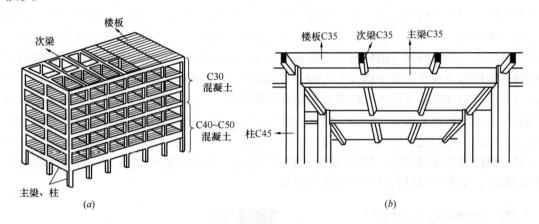

图 2 组合混凝土结构的雏形
(a) 框架结构；(b) 楼盖结构

但总的来说，采用这样的组合时，混凝土强度等级往往不会差别很大，如 C30~C50，除了抗压强度有所差别外，弹性模量和抗拉强度差别较小。在其结构设计以及计算分析上，也往往差别不大，如本构关系模型的选择、分析模型参数的选取等方面，没有加以区分。这样的高低强混凝土界面处理采取一些简单措施即可，如调整混凝土配合比设计，控制浇筑顺序，布置构造钢筋等。

2.3 组合混凝土结构的发展

近年来，随着对混凝土材料研究的不断深入，诞生了一系列由特殊材料组成的、具有特殊功能的水泥基材料，包括海水海砂混凝土、再生混凝土、纤维混凝土、轻质混凝土、橡胶混凝土、ECC 等。其中，随着纤维混凝土与 ECC 这类具有高延性的水泥基材料的出现，逐渐克服了传统普通混凝土受拉性能较差的缺点，这使混凝土自身即可通过这些材料的组合满足各种受力要求。综合考虑这些特种水泥基材料的特点，根据性能需求的不同，将其在材料、构件以及结构等层次进行组合，可以实现材料和结构的最优化配置，实现"组合混凝土结构"。相对于混凝土与其他材料的组合形式，不同混凝土间的组合更具有相容性，经过适当处理，不同混凝土间的界面可以达到良好的整体性[6]。

混凝土至少可从三个层次上进行组合，分别为：

(1) 材料层次：组合不同材料制备混凝土，包括骨料（天然骨料、再生骨料、轻骨料、金属骨料）以及功能性材料以及外加剂等；

（2）构件层次：可以在构件截面和轴向上进行组合。在构件截面上组合不同强度或功能混凝土，形成梯度或者分层，组合类型包括强度（高强混凝土、低强混凝土）、种类（再生混凝土、海水海砂混凝土）、组分（纤维、橡胶）、功能（防水、抗火、隔音）等；在构件轴向上组合不同的混凝土，组合方式包括按受力区域分段组合（梁塑性铰区以及中间段）、按约束情况分段组合（变形转动要求不同的区域）等；

（3）结构层次：组合不同混凝土构件，将最合适的构件放在最需要（受力上和功能上）的地方。

凡是符合上述组合特点的结构均可称为"组合混凝土结构"。"组合混凝土结构"的概念与当前"组合结构"以及"混合结构"有一定的区别，如图3所示，"组合结构"是指由组合结构构件组成的结构，以及由组合结构构件与钢构件、钢筋混凝土构件等组成的结构，其中组合结构构件是由钢材或其他非水泥基材料与钢筋混凝土组合能整体受力的结构构件。这种组合是将混凝土以及与混凝土性能相差较大材料（如钢材、FRP材料）进行组合，而"组合混凝土结构"的组合都是水泥基材料。"混合结构"是相对于单一结构如钢筋混凝土结构、木结构、钢结构而言的概念，是指由多种结构形式组合而成的共同承受水平和竖向作用的建筑结构，其中各结构可构成独立的受力系统，例如钢框架与钢筋混凝土核心筒组成的框筒结构。"组合混凝土结构"与之区别同样在于其组合材料的性能相容性，同时，各水泥基材料共同形成受力体系。

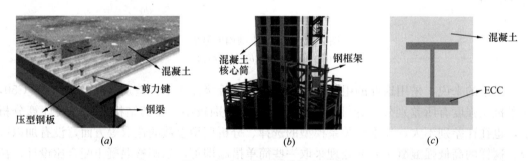

图3 组合混凝土结构概念区分
(a) 组合结构；(b) 混合结构；(c) 组合混凝土结构

3. 组合混凝土结构原理

目前，国内外尚未有针对"组合混凝土结构"的概念与研究，但有一些新颖的构件与结构形式，符合本文提出的组合混凝土结构的概念，现对其进行案例分析。

3.1 组合混凝土材料

在制备混凝土材料时，将不同的骨料进行组合，可使混凝土在使用功能、力学性能上得以提升。例如，为降低大坝混凝土因碱硅酸反应而产生的膨胀，刘文潮等[7]进行了设计优化，采用组合骨料混凝土（砂岩粗骨料＋大理岩细骨料）比全砂岩骨料混凝土具有更良好的变形性能、热学性能以及抗渗性能。严雄风等[8]选择天然沸石、钢渣和浮石作为原材料形成组合骨料,制备出脱氮除磷效果良好的植生混凝土。不仅在功能上,力学性能上也会有所改善。李坛等[9]提出了大粒径再生骨料(最大粒径为80mm)与一般骨料的组合骨料混凝土。破碎效率得到了显著提升,同时可使骨料总表面积减小,从而减少水泥的用量。如图4(a)所示,通过试验,表明可使组合骨料混凝土的抗压强度相比废旧混凝土明显提高。吴波等[10]也提出了更大尺寸的大尺度废旧混凝土块体[图4(b),特征尺寸约50~400mm]与新混凝土的组合物,而使用FRP约束可以显著提升其性能[11]。军事和水利上在浇筑混凝土过程中,也会埋入大量的大块石、毛石、片石等大粒径的石块,即块石混凝土、刚玉块石混凝土、毛石混凝土、抛石混凝土等,来进行大体积混凝土的浇筑。

近年来,将纤维组合在混凝土中的研究越来越多,随着高强高弹模纤维(聚乙烯纤维、PE纤维,等)的出现,混凝土的性能和功能得到改变。传统的钢纤维混凝土的拉伸延性为0.5%~1.0%;经过特殊设计的聚乙烯醇纤维增强水泥基复合材料(PVA-ECC)的抗拉强度约为3~7MPa,拉伸极限应变约为2%~4%。研究表明[12-13],新型的聚乙烯纤维增强高延性水泥基复合材料(PE-ECC)的抗压强度介于30~150MPa之间,抗拉强度介于5~20MPa之间,是普通混凝土的3~10倍,平均拉伸应变达8%,最大拉伸应变达到12%以上,接近建筑钢材水平,极大地增强混凝土的性能。

此外,针对混凝土表面质量问题,为提高混凝土的防护寿命,产生了外渗表面强化与修复材料的混凝土组合形式[14]。如图4(c)所示,利用混凝土具有可渗透性的特点,将活性物质渗入内部并与内部组分发生复杂的化学反应生成新的物质,自适应地与混凝土结合在一起,从而强化表面以及阻止外界的有害物质进入混凝土。这种组合方式,对新建混凝土结构的加强以及旧混凝土建筑的修补与加固都适用。

综上所述,在材料层次上,组合不同水泥基材料可从力学性能和功能上,实现优化结构的目的,形成组合混凝土结构,同时也可组合功能性材料,对结构起到强化加固以及耐久维护的作用。

  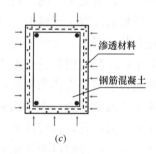

(a) (b) (c)

图 4 组合混凝土材料
(a) 大粒径再生骨料组合骨料混凝土[9];(b) 大尺度废旧混凝土块体组合骨料混凝土[10]
(c) 混凝土外渗功能材料原理图

3.2 组合混凝土构件

根据梁的受力特点,一般采用的是在截面层次上下组合的形式,作为受压和受拉区域转换的交界面,梁的中性轴刚好可以作为组合混凝土构件间界面。设计时叠合面需满足一定的抗剪承载力要求,叠合梁的受力行为已有较为系统的研究[15]。结合再生混凝土特点,肖建庄等[16]提出了再生混凝土叠合梁,如图 5 (a)、(b) 所示,完成了 C 形叠合梁与 U 形叠合梁的抗弯与抗剪试验。预制梁段由于受力与耐久性考虑,采用较低的再生粗骨料取代率。抗弯试验中,预制段为普通混凝土,后浇段为再生粗骨料取代率为 100% 的再生混凝土;抗剪试验中,预制段取代率为 70%,后浇段为 100%。结果表明,再生混凝土叠合梁截面的形状与梁的力学性能无明显相关,叠合面未发生对承载力和变形不利的破坏,连接完好。ECC 与普通混凝土梁的组合形式试验表明[17],与 ECC 组合可使叠合梁设计方式更加灵活,甚至可以做到无筋形式;还可以结合新型的 3D 打印施工方式,将 ECC 外壳进行打印,再浇筑构件内部混凝土。相比于一般混凝土,ECC 能够更好地与增强筋材协同变形,使得构件整体受力效率更高。同时 ECC 组合梁能在荷载较大时依然保持裂缝宽度在较小的一段范围内,可有效保护梁内部钢筋或 FRP 筋材不受外部有害介质侵蚀,提高梁的耐久性,其具体构造如图 5 (c) 所示,构件外壳受拉区域采用 ECC 预制构件,内部采用普通钢筋混凝土。另外,在构件层次上,由于梁端塑性铰区域受力情况较梁中段复杂,可以将塑性铰区域与梁中段交界面作为组合界面,形成分段梁[18]。

图 5 组合混凝土梁
(a) C 形梁预制段[16];(b) U 形梁预制段[16];(c) ECC-RC 组合梁截面[17]

组合混凝土柱的一般构造为,首先浇筑柱的口型外壳或内芯,再浇筑柱的其余部分。这样根据混凝土材料的不同,即可形成组合混凝土柱。一般柱的柱芯位置受力较小,因此柱在截面层次上通常采用内外的组合方式。外部形成口字型,采用力学性能相对较好的混凝土,并配置钢筋进行加强,可以在工厂预制保证其质量,内部可浇筑相对较差的材料并利用口字型外壳作为模板,同时外壳也会对内部混凝土产生一定的约束作用,提高内部混凝土的力学性能。由于约束作用,内部和外部混凝土间的界面性能也得到加强。结合柱截面的受力特点,肖建庄等[19]完成了半预制再生混凝土柱的抗震试验与分析,如图 6 (a)、(b) 所示,柱的口型外壳采用普通混凝土并配置钢筋,柱芯采用再生混凝土。试验结果表明,组合柱的形式可以改善再生混凝土的力学性能,起到施工改性的作用。如图 6 (c) 所示,结合 ECC 材料的组合混凝土柱试验表明[20],试件在反复荷载的作用下,出现许多细密裂缝,平行裂缝带相互交叉,表现出了明显的多缝开裂特征;在破坏时,未出现保护层剥落,构件完整性良好;ECC 材料对构件耗能能力有显著提升,组合柱各方面参数指标均呈现出良好的抗震性能。

此外,一些组合构件应用了"梯度材料"的基本概念[21],其设计原理是根据构件的使用要求,通过连续改变材料的组成和结构,使材料成为性能和功能均呈连续平稳变化的一种非均质复合材料,从而保证其物理、力学、化学甚至生物特性的连续变化,以适应不

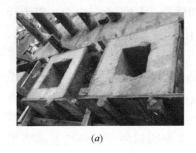  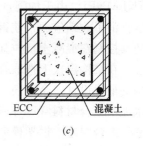

(a) (b) (c)

图 6　组合混凝土柱

(a) 外部预制口型柱[19]；(b) 内部预制口型柱[19]；(c) ECC 组合混凝土柱[20]

同环境特殊功能的要求。将梯度材料的这种原理应用在混凝土构件中，可形成另一种组合混凝土构件的设计思路。Shen[22]提出了一种水泥基功能梯度材料，如图 7（a）所示，采用纤维含量作为梯度参数，试验结果显示，按梯度分布的各层材料之间具有良好的黏结性能，梯度分布可以很好地减小界面的应力与开裂，且分层数量越多，效果越明显，这一结果也被其他学者证实[23]。肖建庄等[24]利用再生粗骨料取代率这一指标，在混凝土板截面上将取代率按一定梯度分布，开展了再生混凝土梯度板的试验研究。如图 7（b）所示，再生混凝土梯度板在厚度方向上分为了三层，再生粗骨料取代率从上至下分别为 50%、100%、0%。试验结果显示，不同混凝土层间没有发生滑移，仍符合平截面假定。一般再生混凝土的弹性模量随着再生粗骨料取代率的增加而降低，当受压区浇筑的混凝土弹性模量较高时，板的刚度会有所提升。另外，梯度分布是解决组合混凝土构件中各混凝土间界面问题的有效方法，为组合混凝土构件的设计提供了新的路径。

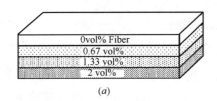 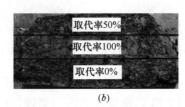

(a) (b)

图 7　组合混凝土板

(a) 纤维梯度板[22]；(b) 再生骨料梯度板[24]

3.3　组合混凝土结构

面对超长混凝土结构收缩徐变所致开裂问题，采用掺加膨胀剂的补偿收缩混凝土，用膨胀加强带取代后浇带，不仅能够解决超长混凝土结构的收缩裂缝问题，实现结构的连续浇筑，而且具有施工方便，周期短，结构整体性较好等优点。图 8（a）为设置膨胀加强带后，混凝土结构内部应力变化情况[25]。未设置膨胀加强带的普通混凝土结构，在温度收缩作用下，其应力分布曲线为 ABCDE，应力由两端向中部逐渐增大，在 B、D 两点处，达到极限状态，裂缝产生；当超长结构整体采用小掺量膨胀剂的补偿收缩混凝土时，其收缩应力得到一定程度降低，应力分布曲线变为 FGHIJ，在 G、I 两点处，应力达到极限状态，裂缝产生；当整体采用小掺量补偿收缩混凝土，并在合理部位 G、I 处采用大掺量补偿收缩混凝土设置膨胀加强带时，结构内部应力状况得到有效改善，应力分布曲线变为

FKLMNPJ。由于加强带处加大了膨胀剂掺量，化学膨胀能转化为构件预压应力，很大程度抵消了收缩应力，使得应力曲线从L、N两点处重新增长，大大降低了结构最大应力，使其控制在混凝土抗拉强度范围内。

在结构层次上选取合适的预制构件，并通过各类节点将其沿高度方向组合，可达到结构的最优设计。余江滔等[26]进行了组合混凝土框架的尝试，并采用自主研发的超高延性水泥基复合材料（Ultra-high ductility cementitious composites，UHDCC）完成了无筋建造。组合混凝土框架模型如图8（b）所示。利用UHDCC优异的拉伸延性和耗能能力提升关键部位的抗震性能，在框架制作中采用UHDCC浇筑框架的①区域（包括底层柱脚和一层框架节点区域），即地震易损部位。为验证无筋建造和组合混凝土结构的理念，①区域没有配置纵向和横向钢筋。图8（b）中的②区域（包括梁、板和柱）为普通钢筋混凝土构件，是组合混凝土框架的预制部分。框架③部分为UHDCC构件与普通钢筋混凝土构件的连接节点，通过高流动性聚合物砂浆实现二次浇筑和拼装。试验结果说明，在结构层面上适当地组合不同性能的梁、板、柱、节点等构件可获得良好的结构性能，证明了组合混凝土结构的整体安全性。

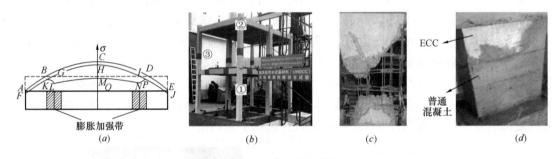

图8 组合混凝土结构

（a）膨胀加强带补偿收缩[25]；（b）UHDCC组合混凝土框架[26]；（c）ECC加固前[27]；（d）ECC加固后[27]

此外，在修复加固中，采用ECC进行组合，也可得到良好的效果。张远森等的试验显示[27]，以ECC修复后的剪力墙较之原试件［图8（c）和（d）］，剪力墙的承载能力基本得到恢复，在保证承载能力的前提下，延性得到提高，剪力墙的破坏模式，由脆性破坏转化为延性破坏，依靠ECC与钢筋良好的变形协调性，提高了角部钢筋的利用率。然而，由于ECC与混凝土的黏结界面的存在，试件首先从界面处出现裂缝并不断形成主裂缝，并延伸至墙脚，最终导致试件破坏。ECC与混凝土的黏结界面为构件的薄弱部位，需要采取相应的措施。

3.4 组合混凝土结构中的科学问题

上述符合组合混凝土概念的材料与结构，在一定程度上证实了组合混凝土的可行性与优越性。而正如前文所述，目前一系列新型水泥基材料种类繁多，并不断有新型的材料产生，如何进行组合是其基本问题。本文提出的一种组合混凝土设计思路如图9所示。首先，根据所需的工程背景，提出相应的性能需求，按照这些需求选择相应的组合材料，并对其进行性能测试，经过不断优选与优化，最终满足相应的性能需求。材料选择完毕后，即进行组合结构的设计，最终对其进行可持续性评价。在这个思路下，组合混凝土结构设

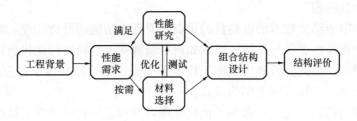

图 9 组合混凝土结构设计框架

计中主要存在以下科学问题:

(1) 界面问题

由于组合混凝土结构是各种水泥基材料的组合,这些材料间的界面是设计的关键问题。过去在混凝土结构设计中,新老混凝土界面仅作为构造处理,而在组合混凝土结构中,由于需要保证构件的整体性,其设计的重要性颇为显著。在设计时就要考虑不同混凝土间的界面结构、形成过程以及不同混凝土间界面的处理方式。

组合混凝土结构中主要存在着混凝土-混凝土的界面,这与传统的钢－混凝土界面在性质上存在着较大的不同。通常认为,新老混凝土结合面黏结力的主要来源包括范德华力、骨料咬合力、化学作用力和表面张力。有黏性的水泥浆渗透到老混凝土的表面空隙中,硬化后的新老混凝土相互交错黏合,包括 $Ca(OH)_2$、Aft、C-S-H 在内的水化产物在老混凝土的孔隙或裂缝中生长,从而形成一定的黏结强度。从宏观的角度看,结合面黏结力主要受混凝土表面粗糙度的影响,也可以通过涂刷界面剂来提高界面的化学作用。同时,由于混凝土结构本身配筋的存在,穿过叠合面的箍筋会因为销栓作用而提供额外的界面抗力。传统钢-混凝土组合结构叠合面的抗剪性能十分薄弱,在结构受剪时,通常通过增加剪力键来提供结合面的抗剪强度[28]。因此,从某种角度上说,在实际应用时,组合混凝土结构在界面抗剪性能上,比钢-混凝土结构有明显优势。而其具体的设计方式与方法则有待提出新的设计体系。

(2) 施工问题

传统的施工方式中,已经包含有组合的概念,当采用现场浇筑施工方式时,混凝土构件通常全采用一种混凝土,由于混凝土凝结时间以及构件与结构的尺寸原因,产生了分段浇筑的形式,这就产生了新老混凝土的组合,当结构高度很高时,在底部浇筑强度较高的混凝土,上部随着楼层升高逐渐降低混凝土的强度等级,这就产生了结构上高强度与低强度混凝土的组合。随着预制装配式施工方式的发展,半预制的构件形式开始出现,这类构件在施工中可以减少施工脚手架的搭设,属于构件上的组合形式,全预制构件则可以看做是混凝土结构上的组合,可以运用不同的混凝土预制构件以达到结构上的优化,同时,除少量连接处的后浇混凝土外,施工现场的工作仅为吊装和拼接,可以极大地加快施工进度。

而如果混凝土的组合形式较为复杂,尤其是采用梯度分布时,传统的支模板、绑钢筋、浇筑混凝土的施工步骤将会显得十分困难,并且难以实现。在施工过程中,不同混凝土间的界面也会根据施工方式的不同而产生变化。因此,面对这种新型的组合混凝土构件与结构,需要研发新型的施工技术来实现。

（3）可持续性问题

发展新型混凝土结构技术的最终目的是提升混凝土结构的可持续性。组合混凝土结构是对混凝土的一次新的思考与总结，思考如何对新混凝土材料进行优化设计，如何对旧混凝土材料加固增强，如何有效利用新施工技术于混凝土结构的建设中，总结出力学性能、使用功能最优化的可持续混凝土结构形式。

虽然在材料与结构优化上，做到了优化处理，节省了材料或增强了结构性能。而考虑结构的生命周期，在拆除后阶段，由于使用不同混凝土的组合，其材料成分将会变得极为复杂。目前的再生混凝土技术，针对这类复杂水泥基材料的再利用尚无成套的解决方式，例如普通混凝土与纤维混凝土的组合，如何对其进行有效分离、分选与资源化处理仍然是个难题。

4. 组合混凝土结构建造

4.1 组合混凝土结构设计

（1）界面设计

在组合混凝土结构的界面设计中，建议也采用 $S \leqslant R$ 的计算理论以达到定量的设计。可以借鉴新老混凝土界面的研究成果[29]，提高组合混凝土结构的界面性能。一般来讲，界面粗糙度越大，新旧混凝土黏结强度越高，但是也有试验表明过大的粗糙度并不能获得较高的黏结剪切强度。因此，需要定量评估黏结粗糙度[30]，可采用膨浆界面剂和聚合物界面剂等改善粗糙度[31]。采用特殊构造的钢筋（或复合材料筋）连接也可以有效地加强不同混凝土间的作用。不同混凝土间界面的受力示意图如图10所示，影响界面的外部受力作用 S 特征参数包括，界面的压应力、界面的剪应力以及界面的形状等，其作用可以通过计算得来；界面本身的抗力作用 R 的影响因素包括，界面两侧混凝土的基本性质，不同混凝土黏结时水化机理，界面粗糙度，界面剂的利用情况以及界面处的配筋情况等，其作用与施工方法和材料性质相关，工程设计时的抗力量化也还存在一定困难。从目前的研究成果来看，其基础理论研究还有待深入，需进行后续深入的试验与分析，以明确组合混凝土结构界面受力和劣化机理。

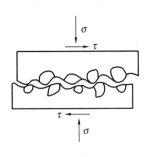

图10 混凝土-混凝土界面分析示意图

（2）计算方法

普通混凝土结构在截面上、构件上可看作均匀单一的材料，因此可以采用一般结构力学、材料力学以及基本的混凝土设计原理进行设计。而组合混凝土结构由于在各层次（材料、构件、结构）上进行了组合，材料的参数均有较为显著的不同，需要重新建立本构关系，并对计算方法进行改进。可以运用弹塑性力学的理论，通过有限元等方法进行计算设计。基于柔度法且沿单元长度积分的弹塑性纤维单元[32]，可以模拟截面特性复杂、参数沿长度方向连续变化的构件，可看作组合混凝土结构设计的一种方法，不同混凝土间界面的模拟也是计算分析的关键问题。目前各常用的设计软件，如 PKPM 等，在参数输入上设置较为单一,材料的组合无法表达和体现。因此为推广组合混凝土结构的设计,如何将此计算部分合理简化并准确应用也是后续研究的方向。

(3) 设计理念

组合混凝土结构的设计理念是一种精细化、一体化的设计过程,需要考虑力学性能的设计与使用功能结合,做到结构-功能一体化设计;在设计时,同时需要考虑施工的过程,相对于一般的混凝土结构,组合混凝土结构的施工过程对组合构件、结构的力学性能有更大的影响,尤其对不同混凝土间的界面形成有着直接影响,因而设计时也需要做到设计-施工一体化设计。因此,组合混凝土结构设计需要建立一套新的设计体系,可以结合BIM (Building Information Modle)[33]等新型技术,符合可持续性设计的新要求。

4.2 组合混凝土结构施工

预制装配式的施工方式,为组合混凝土结构的施工提供了便利,预制混凝土构件间的连接形式、构造细节以及施工方式可直接应用在组合混凝土结构中,相对于钢-混凝土以及FRP-混凝土组合构件间复杂的节点形式,其施工更为简便。同时,随着新型施工技术发展,组合混凝土结构的形式将会更加优化。近年来诞生了一系列新型的施工技术,包括有3D打印施工技术、模块化工业建造技术和可拆装施工技术等。

3D打印技术,是一种增材建造技术[34],它通过将材料逐层叠加的方式完成实体部件的制造[如图11 (a) 所示]。与传统的去除材料加工技术(减材制造)不同,3D打印没有剪裁过程,因此不会产生边角料,从而使原材料的利用率增加。目前运用较多的方法为挤压法以及铺层法,前者是利用机械喷嘴将"油墨"材料挤压喷出,循环往复成型,后者则是将"油墨"材料一层一层的堆叠成型。随着超高延性水泥基材料的出现,无筋建造的3D打印成为了可能。3D打印这种新型的施工方式可以有效地解决组合混凝土结构施工难的问题,运用不同的"油墨"材料,即不同种类的混凝土,可以很容易的形成组合混凝土结构在材料以及构件层次上的组合。

对于模块化工业建造技术[35][如图11 (b) 所示],模块化的概念来源于电子硬件工程,通过对电路板进行模块化分隔设计,然后进行组装,可以完全控制数以万计的电子元件单元。建筑结构模块化过程是指对建筑物的空间功能进行的类型划分,以建筑物为主体研究对象,将内部构成元素进行统一分解,然后组合的过程,利用模块(子结构)与模块之间的组合方式,按照设计要求以及建筑学原理而搭建结构的过程。模块化设计注重建筑空间使用的标准化、建筑资源配置的统一化以及建筑内部管线的管理专业化。混凝土结构模块以及其施工已有一定的工程尝试,其组装的可以不是单纯的结构性构件,还包括各种装饰层、保温层、防水层等功能性器件,甚至包括一些简单的设备以及管线等,结合工业装配化,今后将会越来越多地运用这种模块化的思想。

另外,可拆装技术[36-37]是将可拆装设计(Design for Deconstruction,DfD)的概念引入混凝土的设计中,如图11 (c) 所示,这样在混凝土构件需要拆除时,可以将可拆装构件再次运用到新建建筑中,从而赋予混凝土构件的二次生命。这种结构设计时就考虑到拆卸和再安装过程,由于可拆装技术拆卸时效率高,构件的利用率高,因此是对预制装配式施工方式的进一步提升。利用这种可拆装结构思想,可以形成组合混凝土结构在结构层次上的组合,并能提高混凝土结构的可持续性。

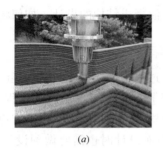  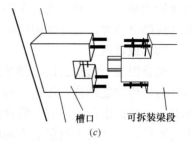

图 11　新型施工方式
(*a*) 3D 打印施工技术[34]；(*b*) 模块化工业建造技术；(*c*) 可拆装施工技术[37]

4.3　组合混凝土结构可持续性评价

对于混凝土结构的可持续性定义，肖建庄认为[38]：混凝土结构可持续性是指在从混凝土材料开采与运输、混凝土结构设计与建造、混凝土结构使用与维护一直到混凝土结构拆除与资源化的生命周期内，混凝土结构在满足安全性和适用性的前提下，具备资源消耗最小、环境影响最低、经济和社会因素相协调的总能力。

组合混凝土结构在其材料设计、构件设计、结构设计以及施工上，均是对混凝土的优化处理，其概念完全符合混凝土结构可持续性的要求，可持续性评价主要包括以下几点：

（1）组合混凝土结构的结构性能评价。根据前述的一些组合混凝土构件的原理及案例分析，经过合理的设计以及施工，可以达到现行混凝土结构的标准，并能得到优化的效果。

（2）组合混凝土结构的生态环境评价。组合混凝土结构，在材料使用方面，可以根据需求广泛地应用再生材料技术（再生混凝土、再生纤维等）以及绿色环保的材料（地聚合物、海水、海砂等）；在施工方面，预制装配式施工等新型的施工工艺，对环境负荷小，结合可拆装技术，还能使构件循环使用。

（3）组合混凝土结构的经济效能评价。组合混凝土结构在材料方面，可以更容易地做到就地取材，减小运费；在施工方面，相对于一般混凝土结构可以缩短工期，减少模板、脚手架等措施费，表现出良好的经济效益。

综合上述结构、环境、经济上的评价，组合混凝土结构相对于一般混凝土结构，具有更好的可持续性，其具体分析还需结合结构生命周期评价方法，如 LCA（Life Cycle Assessment）、LCC（Life Cycle Cost）等[39-40]，对混凝土结构的可持续性进行更为具体的评价。

5. 结语

（1）组合混凝土结构的基本思想，来源于混凝土科学与工程快速发展，可以通过在不同的层次上（材料、构件、结构）组合不同种类的混凝土加以实现，是精细化设计思想发展的必然结果，也是混凝土材料与结构发展的一次新飞跃。

（2）利用组合混凝土结构的概念和原理，对新混凝土材料与结构进行优化设计，对旧混凝土材料与结构加固补强，对施工技术进行创新和变革，从而使混凝土结构性能提升，满足安全性、适用性、耐久性，实现可持续性。

（3）不同类型混凝土结合界面的形成机制、损伤机理与受力过程规律以及组合混凝土结构界面的设计与施工方法，将是今后研究的重点。

（4）梯度化以及仿生设计等新颖的方法，也可为组合混凝土结构的设计开拓思路。在再生混凝土材料与结构创新发展的基础上，组合混凝土结构将会成为实现混凝土结构可持续发展的又一重要途径。

参考文献

[1] 国家统计局. 中国统计年鉴—2016[M]. 北京：中国统计出版社，2016.

[2] Magee L, Scerri A, James P, et al. Reframing social sustainability reporting: towards an engaged approach[J]. Environment Development & Sustainability, 2013, 15(1): 225-243.

[3] 肖建庄. 再生混凝土[M]. 北京：中国建筑工业出版社，2008.

[4] Teng J G, Yu T, Wong Y L, et al. Hybrid FRP-concrete-steel tubular columns: Concept and behavior[J]. Construction & Building Materials, 2007, 21(4): 846-854.

[5] 曾岚，李丽娟，陈光明，等. GFRP-再生混凝土-钢管组合柱轴压力学性能试验研究[J]. 土木工程学报，2014，(S2)：21-27.

[6] 王标，王凯，王真. 新老混凝土黏结性能的提高途径[J]. 建筑技术，2010，(01)：26-27.

[7] 刘文潮，蔡华龙. 浅析不同骨料组合对大坝混凝土性能的影响[J]. 人民长江，2009，(18)：88-90+98.

[8] 严雄风，刘迎云，虢清伟，等. 脱氮除磷植生混凝土组合骨料优选研究[J]. 混凝土，2016，(01)：93-95+102.

[9] Li T, Xiao J, Zhu C, et al. Experimental study on mechanical behaviors of concrete with large-size recycled coarse aggregate[J]. Construction & Building Materials, 2016, 120: 321-328.

[10] Wu B, Zhang S, Yang Y. Compressive behaviors of cubes and cylinders made of normal-strength demolished concrete blocks and high-strength fresh concrete[J]. Construction & Building Materials, 2015, 78: 342-353.

[11] Teng J G, Zhao J L, Yu T, et al. Behavior of FRP-Confined Compound Concrete Containing Recycled Concrete Lumps[J]. Journal of Composites for Construction, 2016, 20(1): 04015038.

[12] 王雨，丁高升，张桂媛. 渗透型混凝土保护剂的应用[J]. 新型建筑材料，2008，(03)：69-72.

[13] 薛伟辰，杨云俊. 混凝土叠合梁受力性能与设计方法研究进展[J]. 混凝土与水泥制品，2008，(01)：44-48.

[14] 林建辉，余江滔，Victor C Li. 超高韧度水泥基复合材料经亚高温处理后的性能[J]. 硅酸盐学报，2015，43(5)：604-609.

[15] Yu K Q, Wang Y C, Yu J T, et al. A strain-hardening cementitious composites with the tensile capacity up to 8%[J]. Construction and Building Materials, 2017, 137: 410-419.

[16] Xiao J Z, Pham T L, Wang P J, et al. Behaviors of semi-precast beam made of recycled aggregate concrete[J]. Structural Design of Tall & Special Buildings, 2013, 23(9): 692-712.

[17] Hung C C, Chen Y S. Innovative ECC jacketing for retrofitting shear-deficient RC members[J]. Construction & Building Materials, 2016, 111: 408-418.

[18] 肖建庄，林壮斌. 一种再生混凝土分段梁及其施工方法[P]. 中国专利：CN104032892A，2014-09-10.

[19] Xiao J Z, Huang X, Shen L. Seismic behavior of semi-precast column with recycled aggregate concrete[J]. Construction & Building Materials, 2012, 35(10): 988-1001.

[20] Wu Chang, Pan Zuanfeng, Shaoping Meng. Seismic behavior of steel reinforced ECC columns under constant axial loading and reversed cyclic lateral loading. Materials and Structures. 2017, 50(1): 1-15.

[21] Koizumi, M. FGM activities in Japan[J]. Composites Part B: Engineering, 1997, 28(1), 1-4.

[22] Shen B, Hubler M, Paulino G H, et al. Functionally-graded fiber-reinforced cement composite: Processing, microstructure, and properties[J]. Cement & Concrete Composites, 2008, 30(8): 663-673.

[23] Yang J J, Hai R, Dong Y L, et al. Effects of the Component and Fiber Gradient Distributions on the Strength of Cement-based Composite Materials[J]. Journal of Wuhan University of Technology-Mater. Sci. Ed. 2003, 18(2): 61-64.

[24] Xiao J Z, Sun C, Jiang X H. Flexural behavior of gradient slabs with recycled aggregate concrete[J]. Structural Concrete, 2015, 16(2): 249-261.

[25] 黄威, 潘钻峰, 沈雷鸣, 等. 膨胀加强带在超长混凝土结构中的应用[J]. 建筑施工, 2016, (09): 1222-1224.

[26] 余江滔, 詹凯利, 俞可权, 等. 超高延性混凝土的无筋框架振动台试验超高延性混凝土的无筋框架振动台试验研究报告. 上海: 同济大学, 2017.

[27] 张远淼, 余江滔, 陆洲导, 等. ECC修复震损剪力墙抗震性能试验研究[J]. 工程力学, 2015, 32(1): 72-80.

[28] 史晓宇, 陈世鸣, 裘子豪. 组合板剪切-黏结机理及承载能力试验[J]. 同济大学学报(自然科学版), 2012, (05): 666-672.

[29] 谢慧才, 李庚英, 熊光晶. 新老混凝土界面黏结力形成机理[J]. 硅酸盐通报, 2003, (03): 7-10+18.

[30] 韩菊红, 袁群, 张雷顺. 新老混凝土黏结面粗糙度处理实用方法探讨[J]. 工业建筑, 2001, (02): 1-3.

[31] 潘东芳, 乔运峰, 夏春, 等. 新老混凝土界面处理材料的试验研究[J]. 混凝土, 2006, (09): 60-61+64.

[32] 聂利英, 李建中, 范立础. 弹塑性纤维梁柱单元及其单元参数分析[J]. 工程力学, 2004, (03): 15-20.

[33] 刘照球, 李云贵, 吕西林, 等. 基于BIM建筑结构设计模型集成框架应用开发[J]. 同济大学学报(自然科学版), 2010, (07): 948-953.

[34] 马义和. 3D打印建筑技术与案例[M]. 上海: 上海科学技术出版社, 2016.

[35] 俞宝达, 俞宝明. 从模块化建筑到模块化施工[J]. 浙江建筑, 2013, 30(5): 56-59.

[36] Crowther P. Design for Disassembly-Themes and Principles[J]. BDP Environment Design Guide, 2005.

[37] 肖建庄, 丁陶, 张青天. 一种可拆装的混凝土构件施工方法[P]. 上海: CN105155773A, 2015-12-16.

[38] 肖建庄. 可持续混凝土结构导论[M]. 北京: 科学出版社, 2017.

[39] 顾道金, 朱颖心, 谷立静. 中国建筑环境影响的生命周期评价[J]. 清华大学学报(自然科学版), 2006, (12): 1953-1956.

[40] 崔云华, 陈柏昆. 基于生命周期成本分析的多层宿舍经济评价[J]. 青海大学学报(自然科学版), 2016, (04): 104-108.

14 铝合金结构在我国的应用和发展

张其林

(同济大学土木工程学院)

摘 要：本文介绍了铝合金结构在国内人行桥、门架、空间网格结构等方面的应用，总结了国内近年来在空间网格结构板式节点方面的研究情况，简要介绍了国家标准《铝合金结构技术标准》的重编情况，最后提出了铝合金结构在我国进一步发展需要进行的相关研究工作。

关键词：铝合金结构；应用；发展

Applications and developments of aluminium structures in China

Qilin Zhang

(School of Civil Engineering, Tongji University)

Abstract: In this paper the applications of aluminum structures in footbridges, portal frames and spatial grid structures in China are introduced, the researches in recent years on working behaviors of the TEMCOR joint of aluminium spatial grid structures are reviewed. The reediting of the national code "Technique Specification for Aluminium Structures" is briefly introduced. Finally the relative researches work is suggested for the further development of the aluminium structures.

Keywords: Aluminium Structures; Applications; Developments

1. 前言

　　铝合金材料的自重和弹性模量均为钢材的三分之一左右，强度接近钢材，具有良好的耐腐蚀性能。因为这些特点，铝合金材料在建筑和结构工程中获得了广泛的应用。建筑幕墙和大型金属屋面系统已经大量应用了铝合金型材和板材。除此之外，铝合金材料在人行桥梁、门式刚架、空间网格结构中已获得较多的应用。近年来，随着产能的增加，已经开展了铝合金材料在其他建筑和结构工程中的应用研究。

　　我国首部国家标准《铝合金结构设计规范》于2007年颁布并正式实施，《铝合金结构工程施工质量验收规范》于2010年颁布并正式实施。国家标准《铝合金结构设计规范》主要参考了欧洲铝合金结构设计标准，在大量的针对轴心受力、受弯、压弯和拉弯等基本构件，以及焊接、螺栓连接等连接件及连接节点实验研究和数值分析的基础上编制完成了。但限于当时条件，无论是设计规范还是施工质量验收规范均没有针对不同的结构体系进行设计和施工验收方面的规定。

本文介绍了铝合金结构在我国的工程应用和实践，对应用最广的铝合金空间网格结构板式节点的研究作了回顾，对铝合金结构在我国进一步应用发展应作的工作提出了建议。

2. 铝合金工程结构的应用和实践

2.1 人行桥梁

铝合金材料最早应用于人行桥的工程案例是1933年美国匹兹堡斯菲尔德区的跨河桥（图1），其桥面板采用了铝合金材料。1946年，美国纽约建成了全球首座全铝人行桥。1946～1963年，美国共建成了9座铝合金人行桥，其中尚有8座目前仍在使用。

图1 美国匹兹堡斯菲尔德区的跨河桥　　图2 加拿大魁北克省阿尔维达铝拱桥

1949年，加拿大魁北克省阿尔维达建成了跨越塞右奈河的铆接铝拱桥（图2），跨径88.4m，桥桥墩高约15m，2个车行道。所用的铝合金为2014-T6，总重163t。1949～1985年间，英国建造了约35座铝合金结构桥梁；1950～1970年间，德国建造了约20座铝合金结构桥梁。

图3 杭州庆春路人行桥

我国于2006年建成的杭州市庆春路人行天桥是首座铝合金人行桥（图3），材料进口于德国，牌号AlMgSi1（F31）EN AW 6082-T6，由德国设计和安装，主跨36.81m，11t。2007年建成的天津市蚌埠海河桥桥面板采用了6005-T5铝合金型材。标准面板长5m、宽0.25m，由27mm厚的上部翼缘板和3mm厚且底部带有加强的马蹄形构造的肋板组成，桥长192m。2007年建成的上海徐家汇人行天桥由国内自主研发和生产装配（图4），采用6061-T6，自重15t，载重量50t，长度33m，宽6m，拱形矢高2.7m。2008年建成的北京西单商业区人行桥（图5），主桥长84m，主跨38.1m，宽8m，两侧四个方向共8部上行电动扶梯，采用AW6082-T6铝合金。2013年建成的杭州新解百"口"字形人行桥，总长217.55m，宽4.8m，

图 4　上海徐家汇人行桥　　　　　　图 5　北京西单人行桥

重 226t，可满足 4000 余人同时通行，采用 6082-T6 型材。该桥分为 4 个主桁架、4 个过渡桁架，组成一座闭合环形人行天桥，其中 A 桁架长 53m，B 桁架长 44m，C 桁架长 44m，D 桁架 42m。2015 年，南宁市建成了 2 座人行桥，据当地报道称，至 2018 年计划建成 20 座铝合金人行桥。2018 年，北京启动了将已建成投入使用多年的东单南北钢天桥更换为铝合金天桥的工程。

2.2 门式刚架

铝合金门式刚架最早作为一种来自德国的产品在国内的临时建筑（展厅、仓库等）中开始得到了应有。近年来，铝合金门式刚架已逐渐拓展应用到永久建筑中。

2015 年建成的上海浦东世博源洛克公园体育馆（图 6）跨度 25m，边高 8m，弧形屋面，屋面墙面围护材料为 PVC 膜，面积 25m×36m。主体梁柱铝用量约 10.2t。同年建成的山西省实验中学篮球馆（图 7）跨度 35m，6.5m 边高，局部二层，屋面围护材料为 PVC，墙面为彩钢板系统，面积 35m×90m。主体梁柱铝用量约 30t。2016 年建成的合肥安徽省进口商品直销中心展厅（图 8）跨度 30m，6.8m 边高，屋面围护为充气膜，墙面为玻璃幕墙，面积 30m×80m。主体梁柱铝用量约 21t。同年建成的东莞马士基临时仓库（图 9）跨度 30m，6.8m 边高，屋面围护为 PVC，墙面为彩钢板系统，面积 30m×90m。主体梁柱铝用量约 20t。2016 年上海克虏伯不锈钢展厅（图 10）跨度 20m，5.4m 边高，屋面围护为充气膜，墙面为模块玻璃墙，面积 20m×35m。主体梁柱铝用量约 6t。2017 年建成的中农临时仓库（图 11）跨度 40m，5.4m 边高，屋面围护为 PVC 膜，墙面为彩钢夹芯板，面积 40m×106m。主体梁柱铝用量约 35t。

图 6　上海浦东世博源洛克公园体育馆　　　　　　图 7　山西省实验中学篮球馆

图8　合肥安徽省进口商品直销中心展厅　　　图9　东莞马士基临时仓库

图10　上海克房伯不锈钢展厅　　　图11　中农临时仓库

铝合金门式刚架的连接和构造已经实现了标准化、产品化和装配化（图12）。梁柱节点和屋脊节点一般采用内插芯材、外覆加强板的构造方式，柱脚采用销轴铰接连接。支撑构件钢拉杆，围护面材选用PVC膜材、彩钢板等轻质材料。

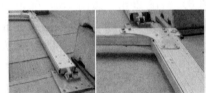   

图12　铝合金的连接节点和支撑

2.3　空间网格

自1996年起，铝合金网格结构开始在国内得到了应有。最早的铝网格结构是采用TEMCOR板式节点的成套美国技术、产品和系统。1996～2010年，除了美国系统外，国内相关单位也尝试采用了螺栓球网架、插接毂式节点、铸铝节点、螺栓连接桁架等体系（图13）。表1列举了1996～2010年国内建成的部分铝合金空间网格结构工程。

　盘式节点　　　螺栓球节点　　　铸铝节点　　　螺栓连接节点　　　毂式节点
图13　早期采用的铝合金网格结构节点形式

1996～2010年国内建成的铝合金空间网格体系   表1

序号	项目名称	竣工日期	概况
1	天津市平津战役纪念馆	1996年	单层球面网壳，底平面直径45.6m，矢高33.83m，最大球面直径48.945m，节点采用弧形板连接
2	上海国际体操中心	1997年	单层扁球面网壳，其平面直径68m，矢高11.9m，球面曲率半径55.37m，节点采用弧形板连接
3	上海浦东游泳馆	1997年	双层圆柱面正放四角锥网壳，平面尺寸（54.0～58.0)m×72.0m（一边带圆弧形），网壳厚2.4m，曲率半径100m
4	上海马戏城	1999年	单层球面网壳，直径50.6m，矢高28m
5	北京城建集团北苑宾馆游泳池屋面铝合金网架	1999年	双层网架，跨度13.6m，长度33.2m，网架厚度1m，螺栓球节点
6	北京航天实验研究中心零磁试验室铝合金网架	2000年	双层网架，长度30m，跨度22m，高度12m，展开面积约1900m^2，螺栓球节点
7	无锡汽车销售中心铝网架	2000年	周边支承双层网架，最大跨度19.0m，展开面积410m^2，网架厚度1m，螺栓球节点
8	北京香青园游泳馆屋盖	2000年	双层筒壳，跨度16m，长度30.6m，厚度0.9m，展开面积833m^2，螺栓球节点
9	上海科技馆	2001年	单层椭球面网壳，长轴长67m，短轴长51m，矢高42.2m
10	上海植物园展览温室	2001年	双层网壳、网架高度2m，平面尺寸81m×66m，主屋面桁架最大跨度24m
11	湖北清江游泳馆屋顶	2001年	双层单曲率弧形网架，长38m，宽23.226m，展开面积930m^2，螺栓球节点
12	航天509所铝合金网架	2002年	双层网架，螺栓球节点，展开面积约460m^2
13	湖南长沙招商服务中心	2005年	单层球面网壳，球面直径42m，矢高23m
14	上海悦达广场云顶状铝合金网壳	拟建	云状单层网壳，由3个球面和2个马鞍面相交得到，最大跨度约42m，投影面积约4000m^2

2010年以来，铝合金网格结构在我国开始进入快速发展的轨道，建筑外形由单一规则穹顶发展成为自由曲面形式，结构跨度以及构件截面也随之要求而不断增大，但结构体系和节点形式却发展统一为单层网格和板式节点。但相当数量的具有较大难度的工程实践及其相关的设计研究积累，已经填补了铝合金网格结构的设计、制作和安装空白，领先于世界。

表2为2010年以来国内建成的铝合金空间网格结构主要工程案例。图14～图18给出了部分工程图片。

表 2

2010 年以来国内主要铝合金网格结构工程

序号	项目名称	结构形式	尺寸	材料牌号	主要构件截面和尺寸	节点形式	施工技术	工期	竣工年份	设计单位
1	中国现代五项赛事中心游泳击剑馆	单层三向相交网壳	最大跨度 90m，矢高 8.5m	6061-T6	H型截面：H450×200×8×10	板式节点	满堂脚手架	3个月	2010	西南院
2	上海辰山植物园	单层自由曲面三向网格网壳	尺寸分布为：203m×33m×20.5m；128m×100m×17m；110m×34m×14m	6061-T6	H型截面：H300×200×8×10	板式节点	满堂脚手架	4个月	2011	上海院
3	重庆国博中心	单层自由曲面三向网格网壳	25 万平方米	6061-T6	H型截面：H300×150×8×10 H300×150×6×10	板式节点	脚手架		2013	北京院
4	苏州大阳山温室展览馆	单层球面网壳	A馆直径 98.2m，B馆直径 79.5m	6061-T6	H型截面：H350×140×8×10	板式节点	满堂脚手架	2个月	2014	苏州中铁院
5	南京牛首山佛顶宫小穹顶	单层三向网格椭球面网壳	小穹顶长约 150m，宽约 100m，矢高约 40m	6061-T6	H型截面：H450×180×8×12	板式节点	满堂脚手架	2个月	2015	华东院
6	南京牛首山佛顶宫大穹顶	单层三向网格自由曲面网壳	大穹顶长约 250m，宽约 112m	6061-T6	H型截面：H550×220×9.5×11；箱型截面：□550×290×12×22	板式节点	分块吊装、曲面滑移	3个月	2015	华东院
7	南昌保税区主卡口	单层自由曲面壳体	投影为直径 60m 半圆，高 11m	6061-T6	H型截面：H400×180×8×11	板式节点	满堂脚手架	1个月	2016	北京院
8	北京新机场航站楼核心区铝结构	单层椭球面网壳	8 个长 58m，宽 28m	6082-T6	H型截面：H250×150×6×10	板式节点	高空钢平台散装	2个月	2017	北京院
9	海南海花岛奇珍馆	单层自由曲面壳体	117m×38.8m	6061-T6	H型截面：H450×200×8×12	板式节点	满堂脚手架	2个月	2018	同济院
10	成都郫县体育馆	单层落地网壳	长 192m，宽 65.5m	6061-T6	H型截面：H460×200×8×12	板式节点	满堂脚手架	3个月	2018	西南院
11	上海南部综合体拉斐尔云廊	单层自由曲面壳体	长 730m，宽 130m	6061-T6	H型截面：H550×250×9×16	板式节点	分块吊装、整体提升	5个月	在建	华东院

图 14　现代五项游泳击剑馆

图 15　辰山植物园

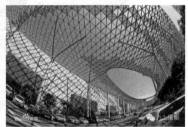

图 16　重庆国博中心

图 17　南京牛首山佛顶宫大小穹顶

图 18　上海南部综合体—拉斐尔云廊

3. 空间网格结构板式节点性能的研究

铝合金空间网格结构近年来的应用发展很快，相关的研究工作也很多。铝合金在高温下力学性能退化显著，可焊性差，故机械连接是铝合金空间网壳结构的主要连接形式。其中板式节点（图19，又称盘式节点、TEMCOR节点、圆盘盖板节点、锚栓盘节点等）是通过不锈钢螺栓或锁紧螺栓（Lockbolt）将上下两块圆形节点板和Ⅰ型杆件翼缘紧密连接而成的，由节点板传递杆件内力。其中，锁紧螺栓一般采用美国HUCK螺栓（图20），这类螺栓有一定的预紧力，预紧力达不到高强螺栓的水平，但工作性能又不同于普通螺栓。

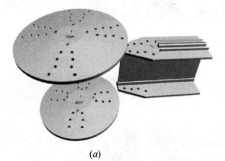 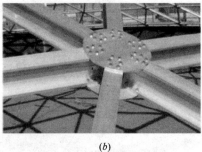

(a)　　　　　　　　　　　(b)

图 19　铝合金空间网格的板式节点
(a) 节点板和构件；(b) 结构体系中的节点

图 20　HUCK 自紧螺栓

因其连接的简易性，板式节点已成为当今铝合金空间网格结构中的主要节点形式，但因其构造的特殊性又使其成为整个结构设计中的关键环节。

板式节点主要适用于构件主要受压的铝合金穹顶结构，其设计依据是《美国铝合金设计手册》[1,2]。但目前在国内这类节点已经广泛应用于构件受弯剪轴力共同作用的自由曲面铝合金网格结构。对这类节点的破坏模式、承载性能、抗弯刚度等的研究虽然已有很多，但目前尚未形成可直接应用的研究成果。2014年，郭小农等[3]设计了4个铝合金板式节点试件试验，研究板式节点在面外纯弯矩作用下以及弯剪联合作用下的破坏模式，提出了纯弯矩作用下节点盘发生块状拉剪破坏时的抗弯承载力公式，并考虑了杆件撬力下节点盘局部弯曲对节点抗弯承载力的折减系数 k_1。接着郭小农等[4,5]以加载模式、抗剪键种类、节点盘厚度以自变量进行了14个节点试样的试验，分析弯矩转角曲线知该节点为典型的半刚性节点，节点盘厚度增加和设置抗剪键都会显著提高节点初始弯曲刚度。对于3杆加载时，加载杆和非加载杆间存在交互现象。2015年郭小农[6]针对14个节点试样试验的其中两种的破坏模式：节点板块状拉剪破坏、节点盘中心区屈曲，提出了理论公式。节点的刚度对铝合金板式节点网壳稳定性能的影响不容忽视。熊哲等[7]为此提出了考虑铝合金板式节点半刚性的杆件单元力学模型，在弯矩-转角实验的基础上，通过数值分析研究发现铝合金板式节点的半刚性特性显著降低了网壳结构的承载力。2014年，王元清[8]对设计了3个足尺的两端夹支的双肢节点试样，在节点盘中心竖向加载，试验自变量为工字型杆件尺寸和节点盘厚度，破坏模式为受拉翼缘最外排螺孔处脆性断裂，分析弯矩转角曲线知节点为半刚性节点，且刚度较大，但延性较低。2017年，柳晓晨、王元清[9]设计了5个双肢节点试样，试验变量、加载方式同2014年的试验，破坏模式除了工字型杆件翼缘的拉断、还发生了节点盘块状撕裂，螺栓群剪断。对于前两种破坏模式提出了类似郭小农[6]中的计算公式；对于螺栓群剪断破坏，假定破坏时塑性重分布，各个螺栓承担的荷载相同，并考虑连接长度较长导致的螺栓应力不均匀，提出螺栓群剪断的计算公式。2014年，韦申[10]也进行了两端夹支的双肢节点试样，在节点盘中心竖向加载的试验，破坏模式为受拉翼缘最外排螺孔处脆性断裂。根据试验的弯矩-转角曲线，提出简化的双线性弯矩-转角曲线，并可用于网壳整体稳定分析。2018年，施明哲[11]根据节点区纯弯，弯剪联合两种加载工况，和抗剪键有无两个控制变量，设计了6个两端夹支的双肢节点试样的试验。纯弯工况下，有无抗剪键均发生受拉翼缘沿螺孔拉断；弯剪联合工况下，无抗剪键，受拉螺栓螺帽拉脱，有较弱的抗剪键，发生腹板鼓屈，有较强的抗剪键时，发生腹板沿着螺孔处断裂。抗剪连接件在弯剪联合工况下的作用更明显，极大提高初始剪切刚度，极大提高受拉翼缘拉断承载力。抗剪连接件和非加载方向的短梁对于节点剪切刚度的贡献占到80%，而节点盘面变形占20%。

现有研究大都围绕节点在纯弯、弯剪联合作用下的工作性能。但对各种受力情况下的自紧螺栓连接本身的工作性能、破坏模式和机理、设计方法等的研究不多。另一方面，随着铝合金网格结构工程的跨度增大、复杂程度增加，高度500左右的大截面铝合金型材的应用日渐增多，对大截面型材板式节点在各种内力及其组合作用下的工作性能研究刚刚开始。

4. 铝合金结构规范重编工作

2017年9月4日召开了国家规范《铝合金结构技术标准》的启动会，《铝合金结构技术标准》是原《铝合金结构设计规范》和《铝合金结构施工质量验收规范》的合并修订规范，计划2019年10月完成报批稿。根据铝合金结构的应有和发展情况，重编的《铝合金结构技术标准》除了整合设计和施工验收两方面的内容外，还将增加关于铝合金新材料、新连接、新节点和不同结构体系等部分的内容，增加关于抗风抗震设计方面的内容。《铝合金结构技术标准》的主要章节和修改内容见表3所示。

《铝合金结构技术标准》的主要章节和修改内容   表3

1	总则	8	拉弯和压弯构件的计算 ——增加钢铝组合、异形型材
2	术语和符号	9	面板的计算 ——增加异形面板
3	基本规定 ——增加抗震设计、抗风设计	10	连接和节点的计算 ——增加自攻螺丝、拉铆钉等，增加节点域刚度计算等
4	材料 ——增加6082、7系	11	构造要求
5	构件有效截面和有效厚度 ——增加大截面	12	结构设计 12.1 一般规定 12.2 增加网格结构的设计 12.3 增加刚架和框架结构的设计 12.4 增加塔架结构的设计 12.5 增加人行桥梁的设计
6	受弯构件的计算 ——增加钢铝组合、异形型材	13	施工质量验收 13.1 原材料及成品进场 13.2 零部件和组装工程 13.3 连接工程 13.4 安装工程 13.5 分部（子分部）工程竣工验收
7	轴心受力构件的计算 ——增加钢铝组合、异形型材		附录

5. 结论和展望

随着铝合金材料行业的发展和进步，国家经济和社会发展的需求，铝合金结构在我国还有很大的发展和应用空间。以下方面的研究有助于促进铝合金结构的健康发展和推广应用：

（1）高性能铝合金材料的力学性能研究；

（2）铝合金结构体系的拓展研究，铝合金结构在通讯塔架、低层框架结构中的应用拓展金额研究；

（3）新型铝合金连接件和新型铝合金结构连接节点的研究和应用；

（4）铝合金结构的抗风和抗震设计方法研究；

（5）铝合金结构性能化抗火设计研究。

参考文献

[1] Aluminum Association. Aluminum design manual [S]. 8th ed. Washington D C：Aluminum Association. 2005

[2] Aluminum Association. Aluminum design manual [S]. 8th ed. Washington D C：Aluminum Association. 2015

[3] 郭小农，邱丽秋，罗永峰，等．铝合金板式节点受弯承载力试验研究[J]．湖南大学学报（自科版），2014，41(4)：47-53.

[4] 郭小农，熊哲，罗永峰，等．铝合金板式节点承载性能试验研究[J]．同济大学学报（自然科学版），2014，42(7)：1024-1030.

[5] Guo XN, Xiong Z, Luo YF, et al. Experimental investigation on the semi-rigid behaviour of aluminium alloy gusset joints[J]. Thin-Walled Structures，2015，87：30-40.

[6] Guo XN, Xiong Z, Luo YF, et al. Block tearing and local buckling of aluminum alloy gusset joint plates[J]. Ksce Journal of Civil Engineering，2016，20(2)：820-831.

[7] 熊哲，郭小农，罗永峰，等．节点刚度对铝合金板式节点网壳稳定性能的影响[J]．天津大学学报（自然科学与工程技术版），2015(s1)：39-45.

[8] 王元清，柳晓晨，石永久，等．铝合金网壳结构盘式节点受力性能试验[J]．沈阳建筑大学学报（自然科学版），2014(5)：769-777.

[9] 柳晓晨，王元清，石永久，等．铝合金网壳结构盘式节点承载力计算方法研究[J]．建筑结构，2017(12)：68-73.

[10] 韦申，杨联萍，张其林，等．铝合金单层网壳螺栓连接节点试验研究[J]．建筑钢结构进展，2014，16(4)：46-50.

[11] Shi M, Xiang P, Wu M. Experimental investigation on bending and shear performance of two-way aluminum alloy gusset joints[J]. Thin-Walled Structures，2017，122：124-136.

[12] GB/T 228.1—2010, "Metallic Materials Tensile Testing part 1：Method of test at room temperature, General administration of quality supervision,"*Inspection and quarantine of the People's Republic of China*, 2011.

[13] Deng H, Chen W, Bai G, et. al. "Experimental study on shearing behavior of lockbolted lap connection for aluminum alloy plates," *Journal of Building Structures*, vol. 37, no. 1, pp. 143-149, 2016.

15 Mechanics of Biological Materials: A Civil Engineering Approach

Lihai Zhang

Department of Infrastructure Engineering

The University of Melbourne, VIC 3010, Australia

Tel: +61 3 83447179; Email: lihzhang@unimelb.edu.au

Abstract: Recent progress in the development of bioinspired materials and structures have created a unique interdisciplinary combination of engineering and life sciences through an engineering-mechanics-based focus. Civil engineers always have the great potential of contributing to the understanding of the mechanics of biological materials and structures. First, engineers learn from nature how to solve complex structural problems and to transfer the basic ideas of these solutions to man-made structures (e.g. new engineering bearing materials and self-healing concrete). Second, the theories developed in civil engineering discipline to describe can characterize the natural materials (e.g. soil and rock) and manmade materials (e.g. steel and concrete) can be used to describe the fundamental mechanical and biological behavior of biological tissues (e.g. articular cartilage and bone). This paper presents our recent research works on modelling biological tissues using a civil engineering approach (i.e. theory of porous media). The outcomes for this research could potentially contribute to the development of bioinspired materials and structures.

Keywords: Bioinspired materials and structures, multidisciplinary research, theory of porous media

1. Introduction

Articular cartilage is the smooth, glistening white tissue that covers the surface of the diarthrodial joints. The function of articular cartilage in joints is to increase the area of load distribution and to provide a wear resistant bearing [17]. The composition and structural properties of cartilage allow it to achieve and maintain proper biomechanical function over much of a human lifespan [12].

In synovial joints the contacting porous surfaces are articular cartilage. These surfaces are bathed and infiltrated in synovial fluid. The viscosity of synovial fluid depends strongly on the presence of the molecules hyaluronan and lubricin [11]. Nature has successfully combined both a soft porous bearing (i.e. around 3mm thick cartilage) and hydrodynamic lubrication (i.e. synovial fluid) to achieve ultralow friction, even at relatively high contact force (up to 5 times of body weight experienced during stair climbing [28]). However much remains to be discovered about the mechanisms involved. Most now accept that lubrication of synovial joints occurs in a so-called 'mixed-mode'. Mixed-mode lubrication means there is a combination of 'hydrodynamic lubrication' and 'boundary lubrication'[8]. It is the details of how the relative importance of hydrodynamic lubrication and boundary lubrication are influenced by consolidation of cartilage at a rough interface that is still poorly understood, but the subject of this study. A primary difficulty is that two competing theories have been presented: 'weeping lubrication' and 'boosted lubrication'. Weeping lubrication refers to fluid that 'weeps' from the porous, deformable cartilage tissue into the 'gaps' at the contact interface [18], whereas in 'boosted lubrication' the fluid flow is in the opposite direction, i.e. from the contact interface into the cartilage [3]. For example, cartilage's famous low friction coefficient (of around 0.005-less than ice sliding on ice) is actually temporary[19]. Under steady-state conditions, following repeated loading, the friction coefficient ranges between 0.05 and 0.1 (i.e. 10 to 20 times larger). The low initial friction coefficient is due to hydrodynamic lubrication dominating in the early stages of mixed-mode lubrication, but friction increases as the contribution of boundary lubrication increases relative to that of hydrodynamic lubrication [5]. How quickly boundary lubrication assumes greater significance will depend on what is happening at the contact interface. 'Weeping lubrication' is driven by increasing asperity contacts. Pore fluid enters the cartilage-cartilage contact from the cartilage only so long as the cartilage is consolidating. However a 'boosted lubrication' posits that fluid trapped in the contact interface is squeezed from the contact interface into the cartilage tissue [3]. The flow into the cartilage filters and concentrates larger molecules (hyaluronan and lubricin) [11] at the cartilage surface, as size exclusion prevents them moving into the cartilage tissue with the fluid. Instead, the larger molecules remain on the surface and aid boundary lubrication. Further, previous studies have shown varying level of surface roughness contributes to the lubrication of articular cartilage [27]. Understanding the mechanisms leading to low friction in biological joints is expected to lead to biologically-inspired innovation in engineering design. With some very promising early innovations [9; 16], understanding the mechanics of basic biological joint lubrication will broaden and deepen engineering innovation further.

2. Method

As shown in Figure 1, the two opposing articular cartilage tissues (i.e. soft porous bearings) can be treated as fluid saturated porous media with a solid phase (i.e. cartilage extracellular matrix) and a fluid phase (i.e. interstitial fluid). The movement of fluid at contact interface will be modelled by assuming different boundary conditions (i.e. 'weeping lu-

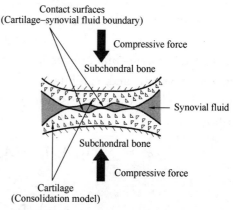

Figure1　Schematic of synovial joints

brication' versus 'boosted lubrication'). A key obstacle toward understanding lubrication of synovial joints has been the correct formulation of the boundary conditions describing the interaction between opposing cartilage surfaces and synovial fluid. As shown in Figure 1, the complexity of these boundary conditions is in great part due to the roughness of the articular surface [1], which results in a time dependent stress boundary condition due to incomplete separation or closure of two contact surfaces. Specifically, we have published a non-linear large deformation poroelastic model of cartilage consolidation under physiological relevant loading [20, 31].

The time-dependent position of the particle in the current Eulerian configuration is given by

$$x(X^{\alpha},t) = X^{\alpha} + u^{\alpha}(X^{\alpha},t) \tag{1}$$

where X^{α} is the position of the α-constituent in material coordinates and u^{α} is the α-constituent displacement, i.e. solid phase ($\alpha=s$) or fluid phase ($\alpha=f$). The sum of volume fraction of solid phase (ϕ^s) and volume fraction of fluid phase (ϕ^f) must satisfy $\phi^s + \phi^f = 1$. For incompressible solid and fluid phases, we obtained,

$$\nabla \cdot [\phi^f(v^f - v^s)] + \nabla \cdot v^s = 0 \tag{2}$$

where v^s and v^f are the velocity of solid and fluid phase respectively, and $\nabla \cdot$ denotes the divergence operator. Ignoring the body and inertial forces, the momentum equation of the solid phase is given by,

$$\nabla \cdot \sigma^{\alpha} + \pi^{\alpha} = 0 \tag{3}$$

where σ^{α} is the Cauchy stress tensor for the α-constituent and π^{α} is the momentum exchange between the phases. By defining all fluid quantities with respect to the solid phase, we obtain

$$\dot{W} + k \cdot \nabla_0 p = 0 \tag{4}$$

where ∇_0 is the material gradient operator on Ω_0^s, and $\dot{W}$ the Lagrangian fluid velocity relative to the solid phase, that is the material time derivative of Lagrangian relative fluid displacement W which is defined as

$$\dot{W} = J^s F^{s-1} \cdot [\phi^f(v^f - v^s)] \tag{5}$$

In addition, the Lagrangian permeability tensor is given by

$$k = J^s F^{s^{-1}} \cdot \kappa \cdot F^{s^{-t}} \tag{6}$$

Recent experimental studies have shown that cartilage responds differently in tension and compression, and exhibits a so-called "tension-compression nonlinearity" behaviour[2, 12, 26]. Thus, the elastic stress σ_E^s resulting from solid phase deformation can be described by,

$$\sigma_E^s = \sum_{a=1}^{3} \left\{ \lambda_1[A_a:E] \text{tr}(A_aE)A_a + \sum_{\substack{b=1\\b \neq a}}^{3} \lambda_2 \text{tr}(A_aE)A_b \right\} + 2\mu E \tag{7}$$

$$\lambda_1[A_a:E] = \begin{cases} \lambda_{-1}, & A_a:E < 0 \\ \lambda_{+1}, & A_a:E > 0 \end{cases} \tag{8}$$

where E is the strain tensor, $A_a = a_a \otimes a_a$ the texture tensor corresponding to each of the three preferred material directions defined by unit vector a_a ($a_a \cdot a_a = 1$, no sum over a), and the term $[A_a : E]$ represents the component of normal strain along the preferred direction a_a. The volumetric strain dependent permeability can be described as [10; 15],

$$\kappa = \kappa_0 \left(\frac{J^s - 1 + \phi^f}{\phi^f} \right) \exp[M(III - 1)/2] \qquad (9)$$

where κ_0 is the zero-strain permeability and exponent $M = 4$. The value of $III = (J^s)^2$ and $J^s = \det F^s$ is the volume change of solid phase, and must be a positive value to prevent self-interpenetration of each constituent in the continuum [15].

3. Results and Discussions

The research presents an integrated state-of-the-art computational model for describing the interaction between consolidation and lubrication of a synovial joint. Ourporo-nonlinear-elastic model [25; 29; 30] can accurately predict hydrostatic pressures, strain profiles and the degree of consolidation throughout cartilage subject to dynamic loading.

Figure 2 shows friction coefficient obtained from their experimental result and degree of consolidation (for normal cartilage) from our computational result. It can be seen that both degree of consolidation and friction coefficient increase with respect to loading time with a similar rate, confirming that the frictional force in articular cartilages increases with the load carried by solid phase.

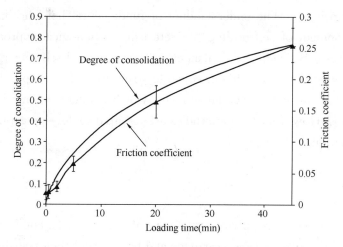

Figure 2 Time evolution of normal cartilage consolidation and friction coefficient of cartilage on cartilage contact [20].

By providing a fundamental understanding of the lubrication behaviour of synovial joint resulting from the exudation and imbibition of interstitial fluid within two soft biological porous bearings (e. g. cartilage), the research outcomes of this research could potentially contribute to the investigation of complex materials and systems in basic science.

This understanding is essential for the development of bio-inspired smart materials, which have a variety of practical applications in the fields of daily life, industry, and agriculture (e. g. cartilage-inspired lubrication system).

Greene et al (2014) incorporated the key features of cartilage into the development of a cartilage-inspired lubrication system which could potentially be implemented in civil engineering practice (e. g. bridge bearing) [9]. The system consists of a network of interconnected cellulose fibers which has a similar morphology of the superficial zone of the cartilage. The poroelastic deformable solid network has a relatively low permeability which allows the generation of high fluid pressure. In addition, the solid porous network contains a high concentration of immobilized, highly charged macromolecules which give a large osmotic pressure gradient that enhances the mechanical stiffness of the network. Finally, boundary lubricant molecule is provided to protect against wear.

Our developed non-linear large deformation poroelastic model of cartilage consolidation has also been implemented in recent years to describe the bone fracture healing process [21-24; 29; 32-35] which could potentially contribute to the development of self-healing materials applied in civil engineering practice (e. g. self-healing concrete). The most common form of bone healing after fracture consists of a total of five phases, i. e. inflammation, granulation tissue formation, cartilage callus formation, lamellar bone deposition and remodeling to original bone contour [6]. Particularly, at the early of healing, mechano-regulation plays an important role during bone healing [4; 7; 14; 35]. For example, the cellular response is triggered based on the mechanical stimuli such as octahedral shear strain and interstitial fluid velocity [13]. The understanding of fundamental bone healing process could lead to the development of self-healing concrete using a biological approach. Inspired by bone healing process, Sangadji and Schlangen (2013) proposed self-repair mechanism with the aim of prolonging the service life of concrete structures. The proposed control scheme involves crack detection sensors and an activator which triggers the injection of healing agent to seal cracks. However, further studies are required to develop application criteria in the practical situation.

4. Conclusions

There is great interest inapplying the theory of porous media and continuum mechanics to model biological tissues (e. g. cartilage and bone), and then transfer the fundamental understanding of the biological process into the development of bio-inspired engineering materials.

However, researchers are still facing huge challenge. For example, the success of a bioinspired load bearing materials (e. g. bridge bearing) depends on the incorporation of the key features of the articular cartilage system, i. e. a deformable porous material with a relatively low permeability for fluid flow, a high concentration of highly charged macromolecules and a suitable boundary lubricant molecule. Furthermore, the development of bio-inspired self-healing of concrete represents a wide interdisciplinary area involving the process of immobilization the bacteria in the concrete and activating them once water penetrates fresh cracks. So far, the effects of the bacteria on controlling of the maximum crack width and length under different environmental conditions (*e. g.* temperature and humility) still remain as open questions. Nevertheless, the attraction of biomimetrics is that it provides sustainable solutions for civil engineers using lessons from the natural world.

References

[1] Ateshian GA. A theoretical formulation for boundary friction in articular cartilage. J Biomed Eng 119: 81-86, 1997.

[2] Burasc PM, TW Obitz, SR Eisenberg, D Stamenovic. Confined and unconfined stress relaxation of cartilage: appropriateness of a transversely isotropic analysis. Journal of Biomechanics 32: 1125-1130, 1999.

[3] Charnley J. The lubrication of animal joints in relation to surgical reconstruction by arthroplasty. Annals of the rheumatic diseases 19: 10-19, 1960.

[4] Claes LE, CA Heigele, C Neidlinger-Wilke, D Kaspar, W Seidl, KJ Margevicius, P Augat. Effects of mechanical factors on the fracture healing process. Clin Orthop Relat Res 355: S132-S147, 1998.

[5] Coles JM, DP Chang, S Zauscher. Molecular mechanisms of aqueous boundary lubrication by mucinous glycoproteins. Current Opinion in Colloid & Interface Science 15: 406-416, 2010.

[6] Doblare M. Modelling bone tissue fracture and healing: a review *1. Eng Fract Mech 71: 1809-1840, 2004.

[7] Epari D, G Duda, M Thompson. Mechanobiology of bone healing and regeneration: in vivo models. Proceedings of the Institution of Mechanical Engineers, Part H: Journal of Engineering in Medicine 224: 1543-1553, 2010.

[8] Gleghorn JP, LJ Bonassar. Lubrication mode analysis of articular cartilage using Stribeck surfaces. J Biomech 41: 1910-1918, 2008.

[9] Greene GW, A Olszewska, M Osterberg, H Zhu, R Horn. A cartilage-inspired lubrication system. Soft Matter 10: 374-382, 2014.

[10] Holmes MH, VC Mow. The Nonlinear Characteristics of Soft Gels And Hydrated Connective Tissues In Ultrafiltration. J Biomechanics 23: 1145-1156, 1990.

[11] Jay GD, JR Torres, ML Warman, MC Laderer, KS Breuer. The role of lubricin in the mechanical behavior of synovial fluid. Procceedings of the National Academy of Sciences of the United States of America 104: 6194-6199, 2007.

[12] Krishnan R, FE Seonghun Park, GA Ateshian. Inhomogeneous cartilage properties enhance superficial interstital fluid support and frictional properties, but do not provide a homogenous state of stress. Journal of Biomechanical Engineering 125: 569-577, 2003.

[13] Lacroix D, PJ Prendergast. A mechano-regulation model for tissue differentiation during fracture healing: analysis of gap size and loading. Journal of Biomechanics 35: 1163-1171, 2002.

[14] Lacroix D, PJ Prendergast, G Li, D Marsh. Biomechanical model to simulate tissue differentiation and bone regeneration: Application to fracture healing. Med Biol Eng Comput 40: 14-21, 2002.

[15] Levenston ME, EH Frank, AJ Grodzinsky. Variationally derived 3-field finite element formulations for quasistatic poroelastic analysis of hydrated biological tissues. Computer Methods in Applied Me-

chanics and Engineering 156: 231-246, 1998.

[16] Ma S, M Scaraggi, D Wang, X Wang, Y Liang, W Liu, D Dini, F Zhou. Nanoporous substrate-infiltrated hydrogels: a bioinspired regenerable surface for high load bearing and tunable friction. Advanced Functional Materials 25: 7366-7374, 2015.

[17] Maroudas A, P Bullough. Permeability of articular cartilage. Nature 219: 1260-1261, 1968.

[18] McCutchen CW. Joint lubrication. Bulletin of the Hospital for Joint Diseases Orthopaedic Institute 43: 118-129, 1983.

[19] Merkher Y, IE S. Sivan, A Maroudas, G Halperin, A Yosef. A rational human joint friction test using a human cartilage-on-cartilage arrangement. Tribology Letters 22: 29-36, 2006.

[20] Miramini S, DW Smith, L ZHANG, BS Gardiner. The spatio-temporal mechanical environment of healthy and injured human cartilage during sustained activity and its role in cartilage damage. Journal of the Mechanical Behavior of Biomedical Materials 74: 1-10, 2017.

[21] Miramini S, L Zhang, M Richardson, P Mendis. Computational simulation of mechanical microenvironment of early stage of bone healing under locking compression plate with dynamic locking screws. Applied Mechanics and Materials 53:281-286, 2014.

[22] Miramini S, L Zhang, M Richardson, P Mendis. The role of locking plate stiffness in bone fracture healing stabilized by far cortical locking technique. International Journal of Computational Methods 15: 1850024-1850021-1850015, 2018.

[23] Miramini S, L Zhang, M Richardson, P Mendis, P Ebeling. Influence of fracture geometry on bone healing under locking plate fixations: A comparison between oblique and transverse tibial fractures. Medical Engineering and Physics 38: 1100-1108, 2016.

[24] Miramini S, L Zhang, M Richardson, M Pirpiris, P Mendis, K Oloyede, G Edwards. Computational simulation of the early stage of bone healing under different Locking Compression Plate configurations. Computer Methods in Biomechanics and Biomedical Engineering 18: 900-913, 2015.

[25] Smith DW, BS Gardiner, JB Davidson, AJ Grodzinsky. Computational model for the analysis of cartilage and cartilage tissue constructs. J Tissue Eng Regen Med: 2013.

[26] Soltz MA, GA Ateshian. A conewise linear elasticity mixture model for the analysis of tension-compression nonlinearity in articular cartilage. Journal of Biomechanical Engineering 122: 576-586, 2000.

[27] Soltz MA, IM Basalo, GA Ateshian. Hydostatic pressurization and depletion of trapped lubricant pool during creep contact of a rippled indentor against a biphasic articular cartilage layer. Journal of Biomechanical Engineering 125: 585-593, 2003.

[28] Taylor WR, MO Heller, G Bergmann, GN Duda. Tibio-femoral loading during human gait and stair climbing. Journal of Orthopaedic Research 22: 625-632, 2004.

[29] Zhang L. 2015. Computational Modeling of Bone Fracture Healing by Using the Theory of Porous Media. In: Liu Z editor. Frontiers in Applied Mechanics: Imperial College Press

[30] Zhang L, BS Gardiner, DW Smith, P Pivonka, AJ Grodzinsky. A fully coupled poroelastic reactive-transport model of cartilage. Molecular & Cellular Biomechanics 5: 133-153, 2008.

[31] Zhang L, S Miramini, BS Gardiner, DW Smith, AJ Grodzinsky. Time evolution of deformation in a human cartilage under cyclic loading. Annals of Biomedical Engineering 43: 1166-1177, 2015.

[32] Zhang L, S Miramini, M Richardson, P Ebeling, D Little, Y Yang, Z Huang. Computational modelling of bone fracture healing under partial weight-bearing exercise. Medical Engineering and Physics 42: 65-72, 2017.

[33] Zhang L, S Miramini, M Richardson, P Mendis, P Ebeling. The role of impairment of mesenchymal stem cell function in osteoporotic bone fracture healing. Medical Engineering and Physics 40: 603-610, 2017.
[34] Zhang L, S Miramini, M Richardson, M Pirpiris, P Mendis, K Oloyede. The effects of flexible fixation on early stage bone fracture healing. International Journal of Aerospace and Lightweight Structures 3: 181-189, 2013.
[35] Zhang L, M Richardson, P Mendis. The role of chemical and mechanical stimuli in mediating bone fracture healing. Clinical and Experimental Pharmacology and Physiology 39: 706-710, 2012.

16 Crumb Rubberized Concrete (CRC) Application in Structural Engineering

Yan Zhuge[1], Xing Ma[1], Danda Li[1], Jianzhuang Xiao[2]

(1. School of Natural and Built Environments, University of South Australia, Adelaide, Australia
2. Department of Structural Engineering, Tongji University, Shanghai, China)

Abstract: Solid waste materials represent a challenge to societies across the globe as they require the use of significant energy and resources for disposal. One of the materials that generates a significant carbon foot print is rubber from vehicle tyres. An estimated around 50 million equivalent passenger units (EPUs) of end-of-life tyres (ELT) are generated in Australia every year, but only a small proportion are recycled (30%), and the rest are dumped in landfills or are unaccounted for. These waste tires pose serious environmental and health risks. Therefore, there is an urgent need to develop a market to extend the application of recycled tyre products and one solution is to add rubber particles into concrete as partial replacement for the fine aggregate.

In the past few years, a series of research projects have been carried out at the University of South Australia (UniSA) aimed at developing Crumb Rubberized Concrete (CRC) for structural engineering application. This paper will provide an overview of the structural behavior of CRC elements based on the investigations conducted at UniSA, which include the stress-strain behavior, reinforced CRC beams, reinforced CRC slabs, bond behavior of profiled steel reinforced CRC composite slabs and FRP confined CRC columns. The structural elements were tested under either static, cyclic or impact loads.

Keywords: crumb rubber concrete (CRC); overview; structural behavior; reinforced CRC slab; composite slab; impact resistance; static flexural test

1. Introduction

Disposing of a large amount of end-of-life tyres is a serious problem throughout the world (Fig. 1 (a)). An estimated one billion Equivalent Passenger Units (1 EPU equals a standard passenger car tyre) reach the end of their life each year around the world and the number is increasing steadily. In Australia it was estimated that 48 million EPU tyres (equivalent to 500k tons) reached their end of life in 2009-10, however only 30% of them were recycled or properly managed [1]. The remaining tyres went to landfill, stockpiles, illegal dumps or overseas. In US, 700 million tyres need disposing of every year [2]. Due to their large volume, waste tyres are undesirable materials at landfill sites and cause many serious environmental problems: sheltering rodents, trapping water, providing a breeding ground for mosquitoes, consuming air space in landfills, polluting surface and ground water and increasing fire accident risk (Fig. 1 (b)).

At the same time, the concrete industry in Australia is facing the challenge of decreased availability of natural material resources such as natural sand, particularly along the most populous east coast. This decreased availability and the need to reuse waste materials to reduce their environmental impact, have combined to encourage the development of crumb rubber concrete (CRC). CRC used in this research refers to concrete with the fine aggregate partially replaced by crumbed rubber from waste tyres. To date recycled rubber has been used in a range of applications such as road pavements, playground surfaces or backfill for civil engineering projects but not in structural engineering applications. In the 2013/2014 financial year, an estimated 40% of all premixed concrete supplied in Australia (approximately 9.6 million cubic metres), was used in the residential market [3]. The application of reinforced CRC in the residential construction industry has great potential to reduce the environmental impacts of both waste tyres and exploitation of natural material resources.

Fig. 1 End-of-life tyres: (a) dump site; (b) fire risk [1]

The use of recycled rubber began in pavement engineering as rubber modified asphalt binders in 1960~1970s [4]. The idea of CRC appeared in the 1990s when researchers introduced rubber particles into Portland cement concrete as partial replacement for the aggregate component [5, 6]. Compared with conventional concrete (CC), research to date has shown that CRC has a better acoustic insulation, higher toughness, better impact resistance, reduced shrinkage and cracking and improved damping properties. However, the previous studies also showed that CRC compressive strength usually decreases with increasing rubber replacement. The reduced compressive strength and lower elastic modulus has limited the application of CRC in structural engineering to date.

Recent research [7] has proven that higher strength rubberized concrete could be achieved through a range of measures such as rubber pre-treatment, introducing various additives such as silica fume, steel fibre and chemical admixtures into the concrete mixture, the use of high strength cement, optimal rubber content and good grading of combined sized rubber. High-strength rubberized concrete could have potential structural applications such as reinforced CRC beams [8, 9], columns [10] and beam-column joints

[11], composite columns [12-15], and composite slabs [16, 17, 26]. The research results verified that using rubberized concrete could effectively improve ductility of structural components. More recently, the authors of this paper [18] studied the stress strain formulation of CRC in terms of strength and rubber percentages and its applications in reinforced CRC slabs.

Since 2013, a series of research projects have been carried out at the University of South Australia (UniSA) to explore the structural applications of CRC. This paper aims to provide an overview of the structural behavior of CRC based on the investigations conducted at UniSA, which includes the stress-strain behavior, reinforced CRC beams, reinforced CRC slabs, bond behavior of profiled steel reinforced CRC composite slabs, FRP confined CRC columns and post-tensioned segmental CRC columns.

2. Mechanical properties and stress-strain behavior of CRC

2.1 Preliminary tests on mechanical properties of CRC

A series of tests have been conducted to measure the mechanical properties of CRC [19, 20]. Several factors which could potentially affect the concrete properties, including water to cement ratio, rubber content, and cement content were considered in the tests. General purpose cement with specifi gravity of 3.15, according to Australian Standard was used as the binder material. Both 20mm and 10mm dolomite aggregates were used as coarse aggregate with a ratio of 1 : 2.34. River sand with a maximum aggregate size of 5mm was used as fine aggregate. Crumb rubber from waste tyres was supplied by Chip Tyre Pty Ltd. and was cleaned, containing no steel belting or steel fiber. It consists of two different nominal sizes: 1.18mm and 2.36mm at a weight ratio of 1 : 1 (Fig. 2). Rubbers were used with NaOH treatment or without any pre-treatment. Sixteen mixes were investigated including conventional concrete (CC), rubberized concrete (CRC) with rubber percentages varying from 0-20% by sand volume, SF as a partial replacement of cement weight by 0%, 5%, 10% and 15% were used in some mixes. Figure 3 shows the particle size distribution of the sand and rubber mix. The specific gravity of stone, sand and rubber were 2.73, 2.6 and 0.85 respectively. The concrete was mixed based on Australian Standard AS 1012.2 [21]. The water cement ratio of the control mix and normal CRC mix was 0.45 to 0.5.

Apart from conventional compressive

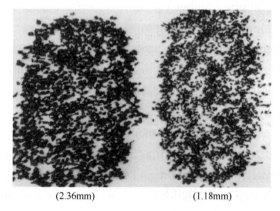

(2.36mm)   (1.18mm)

Fig. 2 Rubber Particles used at UniSA

and indirect tensile tests, impact resistance of CC and CRC was tested for specimens CRC series through a drop-weight test set up recommended by ACI Committee 544 [22], where 150mm×63.5mm disc specimens were fixed onto a flat base plate and a 63.5mm steel ball was placed on top of the specimen and positioned by a bracket. A 4.54kg standard, manually operated hammer with a 457mm drop was repeatedly released and dropped on the specimen until it broke apart. The number of blows at final breaking was recorded and assessed as the impact resistance ability of the specimen.

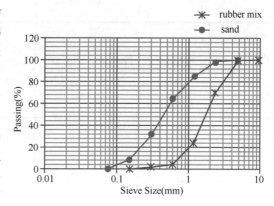

Fig. 3 Particle distribution of rubber mix and sand aggregate

The 7 and 28 days compressive and indict tensile strengths of CC and CRCs are shown in Table 1 [19, 20]. Modulus of elasticity and impact resistance are shown in Table 2 [19]. The test results indicated that strength and Young's modulus of CRC reduced with increasing rubber percentage rate. Pre-treatment of rubber particles for 0.5h in 10% NaOH solution was the best pre-treatment period in this study. However, the effect was not significant. Using SF as partial replacement of cement was not useful and showed some negative effects. The resuts also showed that impact resistance increased with rising rubber addition. An important result is that compressive strengths of CRC at 28 days were all higher than 32MPa, which are stronger than the generally used concrete C32 in residential structural applications.

7-day and 28-day compressive and indirect tensile strengths comparison　　Table 1

Specimen	Rubber (%)	Treatment method	SF (%)	7-day Average Strength (MPa)	28-day Average Strength (MPa)	28-day Strength reduction rate (%)	Indirect tensile Strength (MPa)
CC	0	—	—	38.8	48.4	—	3.6
CRC3	3	—	—	38.1	46.2	4.5	3.7
CRC6	6	—	—	37.2	44.4	8.3	3.5
CRC12	12	—	—	33.8	39.6	18.2	3.6
CRC18	18	—	—	30.6	37.0	23.6	2.8
M1	0	—	—	44.9	53.5	—	4.1
M2	20	NaOH (0.5h)	—	36.0	42.1	21.3	3.1
M7	20	—	—	31.2	35.9	32.9	2.7
M8	20	NaOH (1.0h)	—	33.5	38.6	27.9	3.1

continued

Specimen	Rubber (%)	Treatment method	SF (%)	7-day Average Strength (MPa)	28-day Average Strength (MPa)	28-day Strength reduction rate (%)	Indirect tensile Strength (MPa)
M9	20	NaOH (2.0h)	—	32.3	37.2	30.5	3.2
M10	0	—	5	44.4	55.4	—	4.3
M11	20	NaOH (0.5h)	5	32.3	39.7	28.3	3.4
M12	0	—	10	38.1	50.6	—	4.2
M13	20	NaOH (0.5h)	10	30.8	37.3	26.3	3.1
M14	0	—	15	37.7	47.7	—	3.9
M15	20	NaOH (0.5h)	15	29.5	36.8	22.9	3.1

Young's Modulus and Impact resistance comparison   Table 2

Specimen	Young's Modulus (MPa)	Impact resistance (Blow numbers pre-break)
CC	35702	10.3
CRC3	34002	13.3
CRC6	32262	14.2
CRC12	31832	17.1
CRC18	29556	16.5

2.2 Stress-strain behavior of CRC

In the literature, although there are a large number of investigations on CRC and its mechanical properties, there is a lack of comprehensive investigation and sufficient analysis of the effect of rubber particles on the compressive stress strain behaviour of CRC. For practical application, a semi-empirical full stress-strain model for CRC under uniaxial compression has been developed at UniSA and verified with available experimental data collected from literature [18].

The stress-strain curve for a typical specimen from each mix was chosen and plotted out for comparison, as shown in Fig. 4. These curves indicated that the shape of the stress-strain curve of CRC was similar to that of CC. It is obvious that the slope of the descending part of CC is higher than that of all CRC groups. The modified stress-strain model is shown in equations (1) to (8).

$$\sigma_{CRC} = f'_{c,CRC} \left\{ \frac{\rho_m \left(\frac{\varepsilon_{CRC}}{\varepsilon'_{CRC}} \right)}{\rho_m - 1 + \left(\frac{\varepsilon_{CRC}}{\varepsilon'_{CRC}} \right)^{\rho_m}} \right\} \quad (1)$$

$$\rho_m = \left[1.02 - 1.17 \left(\frac{E_p}{E_{CRC}} \right) \right]^{-0.74} \quad \text{if} \quad \varepsilon_{CRC} \leqslant \varepsilon'_{CRC} \quad (2)$$

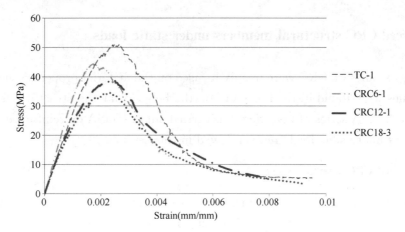

Fig. 4 Stress-strain curves of all concrete groups

$$\rho_m = \left[1.02 - 1.17\left(\frac{E_p}{E_{CRC}}\right)\right]^{-0.74} + (\alpha + \beta) \text{ if } \varepsilon_{CRC} > \varepsilon'_{CRC} \quad (3)$$

$$\alpha = (135.16 - 0.1744 f'_{c,CRC})^{-0.46} \quad (4)$$

$$\beta = 0.35 \exp\left(-\frac{9.11}{f'_{c,CRC}}\right) \quad (5)$$

$$E_p = f'_{c,CRC} / \varepsilon'_{CRC} \quad (6)$$

$$\varepsilon'_{CRC} = \left(\frac{f'_{c,CRC}}{E_{c,CRC}}\right)\left(\frac{v}{v-1}\right) \quad (7)$$

$$v = \frac{f'_{c,CRC}}{17} + 0.8 \quad (8)$$

Where σ_{CRC} is the stress of CRC, $f'_{c,CRC}$ is the compressive strength of CRC, ε_{CRC} is the CRC strain, ε'_{CRC} is the CRC strain at peak stress point, $E_{c,CRC}$ is the modulus of elasticity of CRC, E_p is the secant modulus of elasticity, ρ_m is the modified material parameter for both ascending and descending branch, α and β are coefficients of linear equation. There are two variables in this model, CRC compressive strength and modulus of elasticity.

The comparison between predicting model and experimental results is shown in Fig. 5.

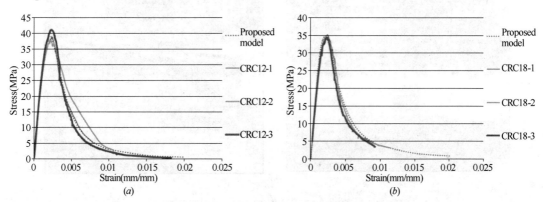

Fig. 5 Stress-strain relationship for CRC
(a) CRC12; (b) CRC18

3. Reinforced CRC structural members under static loads

Previous studies on CRC were mainly focused on the material properties and only very limited studies were found in the literature on the behavior at the structural level [8, 9, 14, 23, 26, 27]. A series of investigations carried out at UniSA on reinforced CRC structural members under staic load are overviewed in this section.

3.1 Reinforced CRC beams

Four reinforced CRC beam specimens were cast using the same mix design discussed in section 2 (CC, CRC6, 12 and 18). The rubber particles were treated with 10% NaOH [8]. The testing results indicated that CRC beams exhibited more deflection capacity compared with normal RC beams. Although the compressive strength reduced around 25%~30% when increasing rubber content up to 18%, the corresponding flexural ultimate moment capacity reduction of the reinfored CRC beams was only around 6%. Fig. 6 shows the failure mode of the specimens without and with 18% rubber content. Although the failure modes were generally similar, the width and size of the cracks were decreased for reinforced CRC beam (Beam 4). When concrete crushing occurred in the compression re-

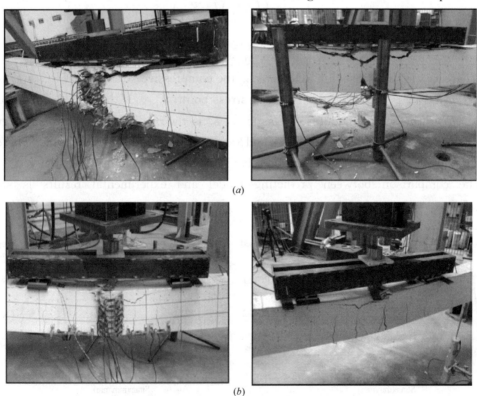

Fig. 6 Damage pattern of the beam specimens [8]
(a) Beam 1 (CC); (b) Beam 4 (CRC18)

gion, it was more explosive and and sudden for beam without any rubber (Beam 1). The testing results also indicated that CRC beams exhibited more deflection capacity compared with CC beams. An increase of 27.9% was observed in the deflection capacity of Beam 4 compared to Beam 1 [8]. Therefore, reinforced CRC beams could be used as flexural members.

3.2 Reinforced CRC slabs

To study the structural behavior of reinforced CRC slabs, four-point flexural bending tests were conducted experimentally for both CRC18 slabs and CC slabs [18]. The numerical simulation analysis for both tests was carried out through ABAQUS 3D modelling. The proposed stress-strain relationship of both CRC18 and CC (equation (1)-(8)) were used as input parameters in finite element modelling. $2200 \times 550 \times 100$mm sized one-way slabs for both CRC18 and CC were cast in plywood forms in two equal layers and compacted with an electronic vibrator. Following demoulding, the slabs were cured by being wrapped in polythene sheets and water-sprayed intermediately for 28 days before tests were conducted. Two slabs were cast for the CRC18 mix and one for the CC mix. N12 bars were used as the main longitudinal tension bars and N6 bars were adopted as lateral crack control bars in accordance with the Australian Standard AS3600 [24]. The concrete cover was kept as 25mm. The nominal yield strength of both bar types was 500MPa. Fig. 7 shows the schematic drawing of the test set-up. The clear span for all slabs was 2000mm and the overhang beyond each support at both ends was 100mm. The hydraulic loading system was connected to computer for data logging. A distributer beam was used to divide the load on the two loading points. A LVDT transducer was used to measure the maximum slab deflection at mid-span. Load was applied at a rate of 0.05kN/s. As the load was applied, the appearance and propagation of cracks on both sides of the slabs were recorded manually.

Fig. 7 Four-point flexural test setup

Fig. 8 gives the load-deflection relations for CC and CRC slabs. The test results show that CRC slab has a more ductile structural behavior during flexural loads. The load-carry-

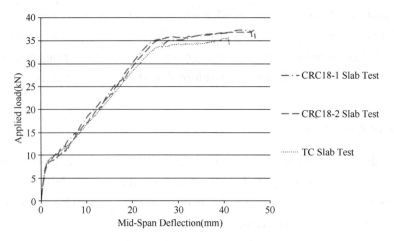

Fig. 8 Load-deflection relations for CC and CRC18 slabs

ing capacity of CRC slabs is comparable with CC slabs although the compressive strength of CRC was 30% lower than CC. Failure mode of CRC18 slab in ABAQUS simulation is shown in Fig. 9.

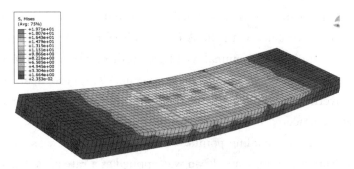

Fig. 9 Concrete stress distribution of CRC18 slab at ultimate load in ABAQUS simulation

3.3 FRP confined CRC columns

At UniSA, the behaviour of FRP-confined CRC columns has been investigated [14]. Five reinforced CRC columns with a diameter of 240mm and shear span of 1500mm were tested under axial compression load. The variablesin this study were the rubber content (0% and 20%) and the FRP confinement thickness (0, 2, and 4 layers). The role of using FRP-confinement was to overcome the CRC material deficiencies (compressive strength). The column geometry and confinement details are shown in Fig 10. The mix design was similar to that discussed in Section 2.1. The rubber particles were treated with 10% NaOH.

The testing results indicated that the structural behaviour of FRP-confined CRC columns and FRP-confined CC columns was comparable. The strength confinement effectiveness (SCE) of CC and CRC columns were 1.11 and 1.13, respectively. The drift confinement effectiveness (DCE) were 0.98 and 1.07 respectively (FRP-confined CRC column

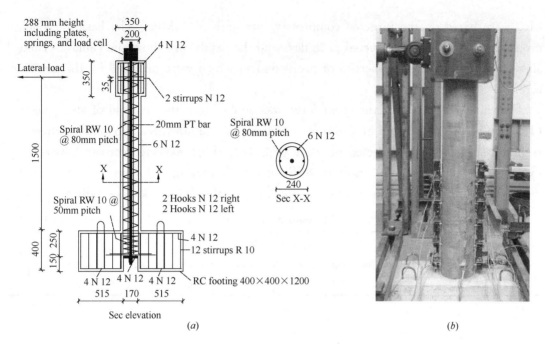

Fig. 10 Column geometry and testing setup [14]
(a) Column geometry and reinforcement details; (b) Testing set-up

was 9% higher than FRP-confined CC column). At a similar column axial capacity, CRC column with 4 FRP layers showed peak strength and ultimate drift 12.4% and 53.0% higher than CC column with 2 FRP layers. These results proved that using FRP-confined CRC in columns is a promising and effective alternative option. For example, as CRC DCE was 9% higher than that of the CC, FRP design thickness required for CRC may be decreased by 9%. This is attributed to the low modulus of elasticity and high Poisson's ratio of CRC which results in a relatively deformable concrete with lower crushing compared to the conventional concrete [14].

3.4 Bond behavior of profiled steel reinforced CRC composite slab

Composite slabs consist of in situ reinforced concrete topping and profiled steel sheeting in which the steel deck acts as permanent formwork during construction and as positive reinforcement for the slab during service. The composite action between profiled steel deck and concrete is affected by the compatibility of deformation properties between the two materials [25]. Therefore, it is expected that using of lighter in weight and more ductile CRC would benefit the bonding performance within the composite slabs. As the first step to investigate the CRC composite slab, the bond behaviour of profiled steel reinforced CRC composite slabs from small-scale push-off tests were carried out at UniSA [26].

The mix proportion of all concrete specimens used the same optimised mix design as shown in Table 3, the mix design was developed through a large number of trial mixes as part of the Australian Research Council funded project. The mixtures for both CC and

CRC were based on the targeted compressive strength of 25MPa. The basic mechanical properties testing were conducted at 28 days and the results are shown in Table 4. Table 4 also shows the material properties of profiled steel which were provided by the manufacturer.

For all series of tests, the shear force was applied into the centroid of steel sheeting (15.3mm above the bottom of sheeting as defined in manufacturer's booklet) where the resultant shear bond is expected to occur (Fig. 11). With expected contact between the loading plate and the top of sheeting soffit in the middle, mini rollers of 5mm diameter were positioned on soffit of sheeting to eliminate possible friction and swivelling of ram.

Optimised mix design — Table 3

	Aggregate (kg) 20mm	10mm	Concrete sand (kg)	Cement (kg)	Rubber (kg)	Water (kg)	WR (kg)	AE (kg)	Actual W/C
SG	2.77	2.65	2.61	3.15	0.97	1.00	1.075	1.002	
CC 25	539	458	885	285	0	203.5	0.56	0.3	0.71
CRC 25(st)	563	478	740	298	68.6	164.7	3.98	0.3	0.55

Material properties of specimens — Table 4

	Concrete properties			Profiled sheeting			
	f'_c (MPa)	E_c (MPa)	f_{cf} (MPa)	f_y (MPa)	f_u (MPa)	Mass (kg/m²)	Coverage (m²/t)
CC	29.05	2.06E+04	7.08	550	750	10.5	95.24
CRC	18.19	1.58E+04	5.45	550	750	10.5	95.24

f'_c = concrete compressive strength on testing day
E_c = Young's modulus
f_{cf} = modulus of rupture
f_y = minimum yielding stress of steel sheet
f_u = ultimate strength of steel sheet

The shear resistance-slip curvesare shown in Fig 12. It is worth mentioning that although the data seems scattered, the magnitude of ultimate shear resistance and the slip where the ultimate resistances occur, which are normally used in designing of the composite slabs, are surprisingly aligned. The difference between the unclamped CC and CRC specimens is only 1.2%, and the 2kN clamping force had raised this difference to 7.6%. The slip where the ultimate resistance was reached for unclamped CC and CRC specimens was the same. The clamping force had made only 4% discrepancy between the slip for CC and CRC specimens. The difference of ultimate resistance between clamped and unclamped specimens id less than 1kN (9.2% increase with clamping force), which indicates that the frictional bond between profiled sheeting and concrete, which is partially influenced by clamping force, does not play a crucial role in overall longitudinal shear resistance. The testing results further proved that although the compressive strength of CRC specimen was much lower than CC specimen, the ultimate resistances of these two specimens, which are normally used in designing of composite slabs, are almost the same with even slightly higher value for CRC. This indicates that the longitudinal shear resistance between steel profiled sheeting and concrete topping is not affected by the lower compressive strength of CRC concrete topping.

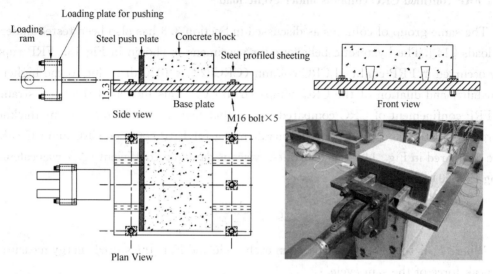

Fig. 11   Layout of test setup [26]

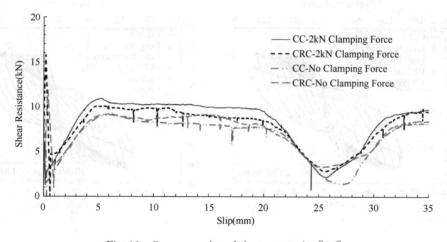

Fig. 12   Summary plot of the test results [26]

4. Reinforced CRC structural members under cyclic and impact loads

Although tests at the material level have shown that CRC has generally higher viscous damping compared to CC, investigations on structural behaviour of CRC under cyclic or dynamic load are rare and have shown contradictory results [27]. While some results indicated that the ductility and energy dissipation of CRC specimens improved significantly compared to CC specimens [11, 23]; other investigations indicated that the crumb rubber did not significantly affect the damping, and sometimes reduced the energy dissipation and viscous damping [14, 28]. The investigations carried out at UniSA on reinforced CRC structural members under cyclic and impact loads are overviewed in this section.

4.1 FRP confined CRC columns under cyclic load

The same group of columns as discussed in Section 3.3 has also been tested under cyclic loads [14]. The hysteretic behavour for specimens is shown in Fig 13. FRP rupture only occurred in FRP-confined CRC column (CRCF2) at 10% drift due to the higher deformability and dilation of the CRC compared to CC. This increased the hoop strains in the FRP-confinement of CRC compared to CC that had the same confinement thickness. Energy dissipation and equivalent viscous damping for FRP-confined CRC and CC column were compared in Fig. 14. The equivalent viscous damping coefficient (ζ_{eq}) was calculated using Eq. (9) [29].

$$\zeta_{eq} = \frac{1}{4\pi} \frac{E_d}{E_s} \tag{9}$$

Where E_d is the energy dissipated in each cycle and E_s is the stored energy measured at the peak force of the sam cycle.

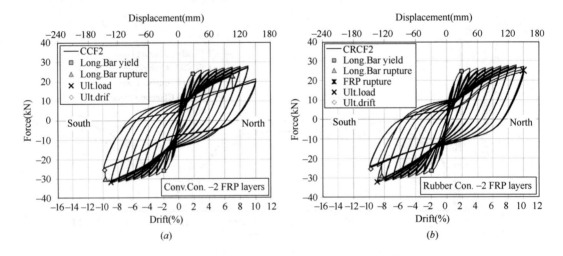

Fig. 13 Hysteretic behaviour for specimens [14]
(a) CCF2; (b) CRCF2

As shown in Fig 14, at the same lateral confinement, CRC column dissipated slightly less energy and with less ζ_{eq} than CC column along the whole test by an average value of 4.8% for both cases. Using FRP-confinement for CC was more efficient than for the CRC in terms of improving the energy dissipation and there is no significant difference in the behaviour of the CRC and CC in terms of damping properties. However, more research is required in this area.

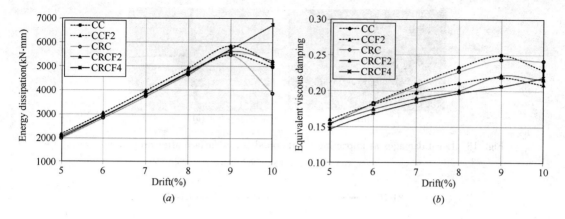

Fig. 14　Energy dissipation and viscous damping for specimens [14]
(a) Energy dissipation beyond 5% drift; (b) Equivalent viscous damping beyond 5% drift

4.2　Reinforced CRC slab under impact loads

Similar to the static flexural tests as described in section 3.2, CC and CRC slabs were also tested under the impact loads as shown in Fig. 15. A free-falling weight was dropped and impacted the top surface centre of the slabs through the use of a steel ball. This process was repeated for 10 drops during the impact testing on CRC and CC slabs with 6.64kg drop weight and 700mm drop height. As shown in Fig. 16, testing results indicated that local damage around the impact point in CRC slab was more severe than in CC slab. The main reason might be the reduced strength in CRC slabs. Strain-time history for steel reinforcement and top concrete at the mid-span for CC and CRC slabs was shown in Fig. 17. As it can be seen, the maximum stress responses in reinforcement (R1) and top concrete (C1) of CRC slab are much smaller than in CC slab. It can be concluded that CRC has obvious advantages for reinforced concrete slab applications.

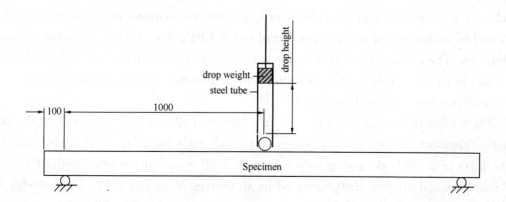

Fig. 15　Slab impact test set-up

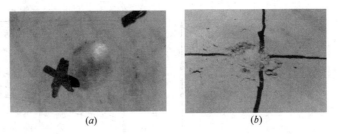

Fig. 16 Local damages at impacting points on slab top surface after drop-weight tests
(a) CC Slab; (b) CRC Slab

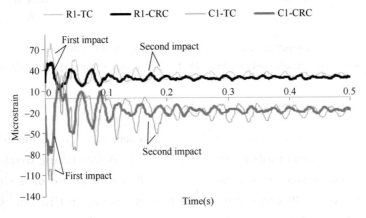

Fig. 17 Maximum reinforcement (R1) and concrete strain (C1) in CC and CRC slabs

4.3 Precast post-tensioned segmental FRP-confined CRC columns under seismic load

Eight circular segmental concrete columns with a diameter of 150mm and shear span of 1,425mm were manufactured, posttensioned, and tested under incremental displacement increasing reverse cyclic load [27]. Each column specimen consisted of an assemblage of four segments having a diameter of 150mm and a height of 300mm. All the top three segments were constructed out of CRC having a rubber content of 18% replacement of sand volume. The bottommost segment in four column specimens was constructed out of CC, and in the other four it was constructed out of CRC. The column geometry is shown in Fig. 18. The posttensioned (PT) bars were anchored at the top of the column loading head and in a recess in the footing of each column. The mix design and procedures are the same as those described in Section 3.3.

The testing results indicated that using a layer of FRP-confinement wrap on the bottommost segments had a significant effect on the column's behavior, peak loads, and ultimate drifts (Fig. 19). By adding only a layer of FRP wrap, at the same initial PT load, the peak load and ultimate drift increased by an average of 32 and 69%, respectively, for CC column and by an average of 44 and 89%, respectively, for CRC. The higher improvements in the peak load and ultimate drift capacity due to confinement when using CRC compared with CC were attributed to the low modulus of elasticity and high Poisson's ratio of CRC, which results in a relatively deformable concrete with lower crushing compared to CC. However, at the same initial PT force and confinement level, the peak load of CRC specimens was on average 10% lower than that of the CC specimens.

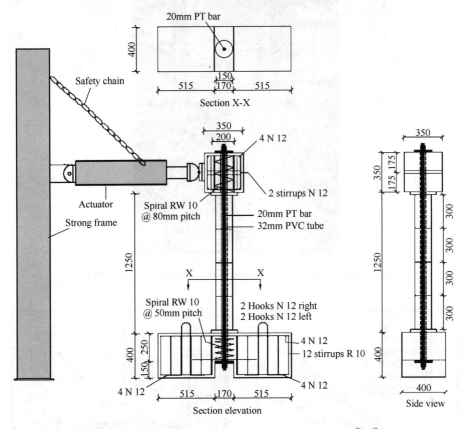

Fig. 18　Post-tensioned segmental column geometry [27]

As shown in Fig. 20 (a), using 18% rubber content increased the energy dissipation in specimens CRF50 and CRF100 by 45 and 17% respectively, compared to that of CC specimens (CF50 and CF100), which is contradictory to the results discussed in Section 4.1 of FRP confined columns. Regarding the equivalent viscous damping coefficient (ζ_{eq}), as shown in Fig. 20 (b), there is no significant difference in the behavior of the CRC and the CC in terms of the hysteretic damping of the segmental column system.

5. Conclusions

In the past few years, a series of research projects have been carried out at the University of South Australia (UniSA) aimed at developing Crumb Rubberized Concrete (CRC) for structural engineering application. This paper presented an overview of the structural behavior of CRC members based on the investigations conducted at UniSA,

Fig. 19 Large displacement capacity of the segmental column [27]

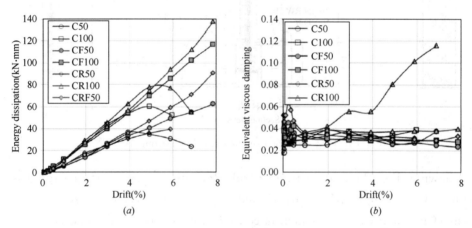

Fig. 20 Energy dissipation and viscous damping for columns [27]
(a) Energy dissipation; (b) Equivalent viscous damping

which include the stress-strain behavior, reinforced CRC beams, reinforced CRC slabs, bond behavior of profiled steel reinforced CRC composite slabs, FRP confined CRC columns and posttentioned CRC segmental columns. The structural elements were tested under either static, cyclic or impact loads. The findings and conclusions are summarised as follows:

• The strength and Young's modulus of CRC reduced with increasing rubber percentage rate. Pre-treatment of rubber particles for 0.5h in 10% NaOH solution was the best pre-treatment period in this study. However, the effect was not significant. The maximum reduction was around 30% when rubber content replacement up to 20% without any pre-treatment.

- An important result is that compressive strengths of CRC at 28 days were all higher than 32MPa, which are stronger than the generally used concrete C32 in residential structural applications. Therefore, CRC could be used in residential construction.

- A semi-empirical stress-strain model for CRC under uniaxial compression has been developed and verified with available experimental data collected from literature. The shape of the CRC curve was similar to that of CC. It is obvious that the slope of the descending part of CC is higher than that of all CRC groups.

- Although the compressive strength reduced around 25-30% when increasing the rubber content up to 20%, the corresponding flexural ultimate moment capacity reduction of the reinfored CRC beams was only around 6% and there was no reduction for reinforced or composite CRC slabs. Furthermore, the test results show that the CRC beams/slabs have a more ductile structural behavior under flexural loads compared to CC specimens.

- Using FRP-confined CRC in columns is a promising and effective alternative option. As CRC DCE was 9% higher than that of the CC, FRP design thickness required for CRC may be decreased by 9%. This is attributed to the low modulus of elasticity and high Poisson's ratio of CRC which results in a relatively deformable concrete with lower crushing compared to the conventional concrete.

- Investigations on structural behaviour of CRC under cyclic or dynamic load are rare and have shown contradictory results. While some results indicated that the ductility and energy dissipation of CRC specimens improved significatly compared to CC specimens; other investigations indicated that the crumb rubber did not significantly affect the damping, and sometimes reduced the energy dissipation and viscous damping. Further research is urgently required in this area.

References

[1] NEPC (2012) "Study into domestic and international fate of end-of-life tires-final report", Australian National Environment Protection Council Report.

[2] Tyremil(2016) Accessed online 6th June 2016 at http://www.tyremil.com/main/index.php/the-tyre-problem.

[3] CCAA(2013) Cement Concrete & Aggregates Australia Internal Report. Accessed online at http://www.concrete.net.au/industry/overview.php.

[4] Huff, B. J. (1977) "Rubber in asphalt pavements-method for utilizing all of our rubber waste", Elastomerics, vol. 109, 27-37.

[5] Eldin, N. N. and Senouci, A. B. (1994) "Measurement and prediction of the strength of rubberized concrete", Cement and Concrete Composites, vol. 16, no. 4, 287-298.

[6] Topçu, I. B. (1995) "The properties of rubberized concretes", Cement and Concrete Research, vol. 25, no. 2, 304-310.

[7] Li, D, Mills, J, Benn, T, Ma, X, Gravina, R & Zhuge, Y. (2016) 'Review of the performance of

high-strength rubberized concrete and its potential structural applications', Advances in civil engineering materials, 5 (1), 149-166.

[8] Hassanli, R., Youssf, O., and Mills, J. E. (2017) 'Experimental investigations of reinforced rubberized concrete structural members', Journal of Building Engineering, 10, 149-165.

[9] Mendis, ASM., Al-Deen, S., Ashraf, M (2017) 'Effect of rubber particles on the flexural behaviour of reinforced crumbed rubber concrete beams', Construction and Building Materials, 154, 644-657.

[10] Youssf, O., ElGawady, M. A., Mills, J E (2015) 'Experimental Investigation of Crumb Rubber Concrete Columns under Seismic Loading', Structures, 3, 13-27.

[11] Ganesan, N., Raj, B., and Shashikala, A (2013) 'Behavior of self-consolidating rubberized concrete beam-column joints', ACI Materials Journal, 110, 697-704.

[12] Duarte, APC., Silva, BA., Silvestre, N., de Brito, J (2016) 'Finite element modelling of short steel tubes filled with rubberized concrete', Composite Structures, 150, pp. 28-40.

[13] Duarte, APC., Silvestre, N., Brito, J., Júlio, E. and Silvestre, JD (2018) 'On the sustainability of rubberized concrete filled square steel tubular columns', Journal of Cleaner Production, 170, 510-521.

[14] Youssf, O., ElGawady, M. A., Mills, JE (2016) 'Static cyclic behaviour of FRP-confined crumb rubber concrete columns', Engineering Structures, 113, 371-387.

[15] Silva, A., Jiang, Y., Castro, JM., Silvestre, N., Monteiro, R (2017) 'Monotonic and cyclic flexural behaviour of square/rectangular rubberized concrete-filled steel tubes', Journal of Constructional Steel Research, 139, 385-396.

[16] Mohammed, B S(2010) 'Structural behavior and m-k value of composite slab utilizing concrete containing crumb rubber', Construction and Building Materials, 24, 1214-1221.

[17] Holmes, N., Dunne, K., and O'Donnell, J (2014) 'Longitudinal shear resistance of composite slabs containing crumb rubber in concrete toppings', Construction and Building Materials, 55, 365-378.

[18] Li, D, Zhuge, Y, Gravina, R & Mills, JE (2018) 'Compressive stress strain behavior of crumb rubber concrete (CRC) and application in reinforced CRC slab', Construction and building materials, 166, 745-759.

[19] Li, D, Mills, J, Benn, B & Ma, X (2014), 'Abrasion and impact resistance investigation of crumbed rubber concrete (CRC)', in ST Smith (ed), Australasian conference on the mechanics of structures and materials, Southern Cross University, pp. 1-6.

[20] Youssf, O., Mills, J. E. and Hassanli, R. (2016), "Assessment of the mechanical performance of crumb rubber concrete", Construction and Building Materials, 125, 175-183.

[21] Standards Association of Australia 1994, AS 1012.2—1994, *Methods of testing concrete -Preparing concrete mixes in the laboratory*, Standards Australia, Sydney.

[22] BADR, A. & ASHOUR, A. F. 2005. Modified ACI Drop-Weight Impact Test for Concrete. *ACI Materials Journal*, 102, 249-255.

[23] Son, K. S., Hajirasouliha, I. and Pilakoutas, K. (2011) 'Strength and deformability of waste tyre rubber-filled reinforced concrete columns', Construction and Building Materials, 25, 218-226.

[24] Standards Australia (2009)"Concrete Structures", AS 3600, Sydney, Standards Association of Australia.

[25] Cifuentes, H. & Medina, F. (2013) "Experimental study on shear bond behavior of composite slabs

according to Eurocode 4", Journal of Constructional Steel Research, Vol. 82, 99-110.

[26] Yi, O, Zhuge, Y., Ma, X., Gravina, R. and Mills, J. (2018)"mall-scale Testing on Bond Behaviour of Profiled Steel Reinforced CRC Composite Slabs", FIB Congress 2018.

[27] Hassanli, R., Youss O. and Mills, J. E. (2017)" Seismic performance of precast posttensioned segmental FRP-confined and unconfined crumb rubber concrete columns", Journal of Composites for Construction, 21, 4, 04017006.

[28] Bowland, A. G. (2011)"Comparison and analysis of the strength, stiffness, and damping characteristics of concrete with rubber, latex, and cqrbonate additives", PhD thesis, Virginia Polytechnic Institute and State University, Blacksburg, VA.

[29] Chopra, A. K. (1995). "Dynamics of Structures", New Jersey: Prentice Hall.

Industry-Academia Forum on Advances in Structural Engineering(2018)

第八届结构工程新进展论坛简介

本届论坛主题：可持续结构与材料

会议时间：2018年9月7日～9日

会议地点：中国　上海　同济大学

主办单位

 同济大学《建筑钢结构进展》编辑部

 中国建筑工业出版社

 香港理工大学《结构工程进展》编委会

承办单位

同济大学土木工程学院建筑工程系

➢ 论坛介绍

"结构工程新进展论坛"自 2006 年首次举办以来,十余年间已经打造成为行业内一个颇有影响的交流平台。论坛旨在促进我国结构工程界对学术成果和工程经验的总结及交流,汇集国内外结构工程各方面的最新科研信息,提高专业学术水平,推动我国建筑行业科技发展。

论坛原则上以两年一个主题的形式轮流出现,前七届的主题分别为:
- **新型结构材料与体系**(第一届 2006,北京)
- **结构防灾、监测与控制**(第二届 2008,大连)
- **钢结构研究和应用的新进展**(第三届 2009,上海)
- **混凝土结构与材料新进展**(第四届 2010,南京)
- **钢结构**(第五届 2012,深圳)
- **结构抗震、减震技术与设计方法**(第六届 2014,合肥)
- **工业建筑及特种结构**(第七届 2016,西安)

论坛有三大特点:

其一,每一届论坛都会选择一个在结构工程领域广受关注的主题,请国内外顶尖专家作全面、深度的阐述;

其二,"论坛文集"为特邀报告人对演讲内容进行深度展开或延伸而成,充分表达其学术成果;

其三,本次论坛计入注册结构工程师继续教育选修课学时(20 学时)。

"结构工程新进展论坛"已作为结构工程领域重要的学术会议在国内外产生了重要影响,历届论坛都吸引了众多专家学者、工程设计人员、青年学生等参会。本届论坛由同济大学土木工程学院建筑工程系承办。

➢ 本届论坛主题

可持续结构与材料

细分为如下 10 个主要议题:
1. 再生混凝土材料及结构;
2. 海砂海水混凝土材料及结构;
3. 基于复合材料的可持续结构;
4. 基于自然材料(木、竹、自然纤维等)的可持续结构;
5. 高性能结构钢及高性能钢结构;
6. 高性能混凝土及高性能混凝土结构;
7. 绿色建造(预制装配、可拆装技术、3D 打印、BIM 技术);
8. 碳足迹及生命周期评价;
9. 可持续新型结构(组合结构、混合结构)抗灾理论与实践;
10. 结构可持续性设计及评价。

➤ 本届论坛特邀报告人

董石麟　院士（浙江大学）
何满潮　院士（深部岩土力学与地下工程国家重点实验室，同济大学）
周绪红　院士（重庆大学）
聂建国　院士（清华大学）
肖绪文　院士（同济大学，中国建筑股份有限公司）
岳清瑞　院士（中冶建筑研究总院）
滕锦光　院士（南方科技大学，香港理工大学）
Surendra P. Shah 院士（Northwestern University）

以下按专家姓氏拼音字母顺序：

Jorge de Brito 教授（Instituto Superior Técnico）
Luc Courard 教授（University of Liège）
曹万林　教授（北京工业大学）
丁洁民　教高（同济大学建筑设计研究院（集团）有限公司）
段文会　教授（Monash University）
方　秦　教授（陆军工程大学）
顾祥林　教授（同济大学）
龚　剑　教高（上海建工集团）
韩林海　教授（清华大学）
何敏娟　教授（同济大学）
吕西林　教授（同济大学）
李　杰　教授（同济大学）
李国强　教授（同济大学）
李秋胜　教授（香港城市大学）
李　恒　教授（香港理工大学）
李柱国　教授（Yamaguchi University）
刘伟庆　教授（南京工业大学）
刘加平　教授（东南大学）
牛荻涛　教授（西安建筑科技大学）
童乐为　教授（同济大学）
吴　波　教授（华南理工大学）
王元丰　教授（北京交通大学）
王翔宇　教授（Curtin University）
王发洲　教授（武汉理工大学）
肖　岩　教授（浙江大学-UIUC）
肖建庄　教授（同济大学）
赵宪忠　教授（同济大学）
张立海　教授（University of Melbourne）
张其林　教授（同济大学）
周　健　教高（上海现代建筑设计（集团）有限公司）
诸葛燕　教授（University of South Australia）

➤ **本届论坛组织机构**

    指导委员会：
    主　任：沈元勤　滕锦光　李国强
    委　员：韩林海　李宏男　吴智深　徐正安　任伟新　苏三庆　史庆轩　赵梦梅
    组织委员会：
    主　任：肖建庄
    委　员：陈　隽　陈素文　王　伟　宋晓滨　刘玉姝　李　征　刘婷婷　钱爱民
           罗佳明　周翌晖　孙晓蕾

第八届论坛特邀报告论文作者简介

(按本书论文顺序)

李 杰 工学博士,同济大学特聘教授,上海防灾救灾研究所所长。兼任国际结构安全性与可靠性协会(IASSAR)主席,中国振动工程学会副理事长、随机振动专业委员会主任,中国建筑学会结构计算理论专业委员会主任,国际核心期刊"Structural Safety""International Journal of Damage Mechanics"编委等学术职务。

李教授长期在结构工程与地震工程领域从事研究工作,在随机动力学、随机损伤力学、工程结构可靠性与生命线工程研究中取得了具有国际声望的研究成果。1998 年获国家杰出青年科学基金,1999 年入选教育部"长江学者奖励计划"首批特聘教授。2013 年,因在随机动力学与生命线工程可靠性方面的学术成就、被丹麦王国奥尔堡大学授予荣誉博士学位。2014 年,因在概率密度演化理论与大规模基础设施系统可靠性方面的学术成就、被美国土木工程师学会(ASCE)授予领域最高学术成就奖——Freudenthal 奖章。发表学术论文 400 余篇(SCI 收录 150 余篇),出版学术专著 4 部,研究论著被他人引用 11000 余次。以第一完成人获得国家自然科学二等奖(2016)、国家科技进步三等奖(1997)、部省级科技奖励一等奖 5 项。

Professor **Surendra P. Shah** is a Walter P. Murphy Emeritus Professor of Civil Engineering and was the founding director of the pioneering National Science Foundation Science and Technology Center for Advanced Cement-Based Materials. His current research interests include: fracture, fiber-reinforced composites, nondestructive evaluation, transport properties, processing, rheology, nano-technology, and use of solid waste materials. He has co-authored two books: *Fiber Reinforced Cement Based Composites* and *Fracture Mechanics of Concrete*. He has published more than 500 journal articles and edited more than 20 books. He is past editor of RILEM's journal, Materials and Structures. Professor Shah is a member of the US National Academy of Engineering. He is also a foreign member of the Chinese Academy of Engineering and the Indian Academy of Engineering, and the only civil engineer who is a member of these three academies. He has received many awards, including the Swedish Concrete Award, American Concrete Institute's Anderson Award, RILEM Gold Medal, ASTM Thompson Award, American Society of Civil Engineer's Charles Pankow Award, and Engineering News Record News Maker Award. He was named one of the ten most influential people in concrete by Concrete Construction Magazine. He has been awarded an honorary membership in American Concrete Institute and RILEM (based in Paris).

In 2007-2008 he spent time at the Indian Institute of Technology, Mumbai as an Honorary Professor under the auspices of a Fulbright grant and in 2014 he received a Fulbright Senior Lecturer award to spend five months at IIT Madras.

Besides teaching at Northwestern University, Professor Shah has taught at the University of Illinois Chicago and served as a visiting professor at MIT, University of Sydney, Denmark Technical University, University of Singapore, Darmstadt University, Laboratoire Central des Ponts et Chaussees, Paris, and University of Houston.

Currently, he is a member of the Institute of Advanced Studies at Hong Kong University of Science and Technology and anhonorary professor at Tongji University, Hong Kong Polytechnic University, Dalian Maritime University, Nanjing Technical University, South East University, Nanjing, and a Distinguished Professor at Indian Institute of Technology, Madras, India.

Professor **Wenhui Duan** graduated from Tianjin University (China) in engineering mechanics with B. Eng. and M. Eng. in 1997 and 2002, respectively. He received his Ph. D. from the Department of Civil Engineering, the National University of Singapore (NUS), Singapore in April 2006. Professor Duan joined the Department of Civil Engineering at Monash University 2008. He was appointed as an ARC Future Fellow in February 2013 and the Director of ARC Nanocomm Hub in 2016.

Professor Duan conducted interdisciplinary research on nanocomposites and nanomechanics in the context of Civil Enginering. By adding advanced nanomaterials such as 2D materails (for example, carbon nanotubes, graphene oxide, BN, MoS2 etc) into conventional engineering materials such as epoxy, Portland cement, and geopolymer, Professor Duan and his team have developed novel nanocomposites with high mechanical performances and reduced environmental impact. His research studies have resulted in more than 150 publications. Professor Duan's current h-index stands at 31 (Google Scholar). In the last ten years, Professor Duan was awarded several ARC and industry grants on nanocomposites and nanomechanics totalling $AUD 20 million including the most recent ARC IRTH Hub on nanoscience-based construction materials manufacturing.

Jorge de Brito is a Full Professor at the Department of Civil Engineering, Architecture and Georresources of Instituto Superior Técnico (IST), University of Lisbon, Portugal.

He is the Head of the CERIS research centre 2017-2018, with over 200 PhD researches and 300 PhD students. He is the Director of The Eco-Construction and Rehabilitation Doctoral Program at IST.

He is Editor-in-Chief of the Journal of Building Engineering (Elsevier).

He is co-coordinator of the CIB W80 Working Group, on Service Life Prediction.

He is the coordinator of the CIB W86 Working Group, on Building Pathology.

He is alsomember of the following scientific international commissions: TC RAC (RILEM); W115 (CIB); WC7 (IABSE).

His main research areas are Sustainable construction (green concrete and mortars), Building and bridge management systems, Service life prediction and life cycle assessment.

He has authored 6 books and over 400 papers in scientific peer-reviewed journals.

Luc Courard is Professor of Building Materials at the University of Liège in Belgium and President of the Department of Architecture, Geology, Environment and Constructions (200 people). After completing his PhD work on concrete surface characterization in the late 1990's, he went to Laval University (Quebec, Canada) for a postdoctoral fellowship devoted to surface preparation of concrete prior to repair. Most of his research activities are still dedicated today to concrete surface characterization, new repair materials, supplementary cementitious materials and use of by-products in concrete technology. Dr. Courard is a member ACI, RILEM and the Belgian Group of Concrete. He authored or co-authored more than 250 peer-reviewed papers and contributed to 4 books and 14 chapters. (https://orbi.uliege.be/ph-search? uid=U026276)

李柱国 教授，2000年3月获得名古屋大学工学博士学位。现为日本山口大学工学部建筑系教授。李教授主要从事新拌混凝土流变学及其工作性预测设计技术；火灾后混凝土性能恢复促进技术；地聚合物新材料以及建筑材料环境影响评价方法的研究。完整地建构了新拌混凝土流变学理论体系；提出了混凝土环境性能评价和设计方法；关于地聚合物添加剂和实用化技术就以第一发明人申报日本专利12项。已发表学术论文247篇，著书6册，4次获得日本学术成果奖。尤其是2005年获得了日本建筑学会奖励奖，它是40岁以下最优秀研究成果奖；2016年获得日本建筑学会奖，也是旅居日本的外国人在建筑材料研究方向首次获得日本建筑学会最高成就奖。

李　恒 博士现为香港理工大学建筑及房地产系讲座教授。在此之前，李博士曾任教于上海同济大学，澳大利亚悉尼大学（University of Sydney），James Cook大学，Monash大学，并且在澳大利亚的两家工程公司工作过。李博士的研究领域为信息技术在项目管理中的应用。他的研究成果，包括一些开发的工程管理软件，不但在许多工程项目中使用，而且在许多国际杂志中发表，从而被同行们所肯定。到目前为止，李教授已出版了3本专著，发表300多篇国际杂志论文。同时，他也是10个国际杂志的编委或主编。

韩林海 现任清华大学教授,土木水利学院副院长。首批国家"百千万人才工程"人选、国家杰出青年基金获得者、教育部长江学者特聘教授、国务院特殊津贴专家、清华大学"百人计划"入选者、英国皇家工程院杰出访问学者、多本领域著名国际期刊副主编或编委。近四年连续入选 Elsevier 发布的中国高被引学者榜单;入选上海软科与 Elsevier 发布的 2016 年"全球土木工程学科高被引学者"榜单。根据中国科学文献计量评价研究中心发布的《中国高被引图书年报》(2017 年版),其《钢管混凝土结构》专著位居 1949—2011 年出版时间段内建筑结构学科高被引榜单第 2 名。在领域主流国际会议上做特邀报告多次。研究成果为多部国家或行业标准采纳,并在典型工程(如北京中国尊、北京奥林匹克塔、广州新电视塔和四川干海子特大桥等)中应用;获国家教育部科技进步一等奖等奖励多次。

何敏娟 教授长期从事以木结构为中心的研究工作,在木结构试验、理论分析、数值模拟及可靠度分析等方面积累了丰富的研究经验。何敏娟教授先后主持了国家自然科学基金项目、科技部绿色建筑及建筑工业化重点专项课题、国家科技支撑计划课题、国际合作研究项目等;是我国《木结构设计规范》GB 50005、《胶合木结构技术规范》GB/T 50708、《多高层木结构建筑技术标准》GB/T 51226 和《装配式木结构建筑技术规范》GB/T 51233 等九部木结构国家规范标准的主要编制成员;多次在世界木结构工程大会(World Conference on Timber Engineering)、国际薄壳和空间结构会议(International Associations for Shell and Spatial Structures)等高水平国际学术会议上做木结构方面的学术报告;并以第一作者或通讯作者在《Journal of Structural Engineering ASCE》、《Construction and Building Materials》、《建筑结构学报》等国内外高水平期刊上发表 200 多篇学术论文,在国内外木结构研究领域具有较大影响力。

Xiangyu Wang is Curtin-Woodside Chair Professor for Oil, Gas and LNG Construction and Project Management in Australia. He received his Ph. D (2006) from Purdue University, U. S. A, in the Division of construction engineering and management. He was offered a personal chair Professor with 100% time research in July 2011 by the School of Built Environment at Curtin University. Since March 2012, he has taken the role of directing Australasian Joint Research Centre for Building Information Modelling. As a remarkable recognition for his excellent performance, Prof Wang was nominated by Curtin as the Curtin-Woodside Chair for Oil, Gas and LNG Construction and Project Management since July 2013. Prof.

Wang's research interests focus mainly on BIM, visualization and Information Technology in Construction Engineering and Management. He is on the Board of Directors and country representatives of International Society of Computing in Civil and Building Engineering (ISCCBE) and International Association of Automation and Robotics in Construction (IAARC), two most highly regarded academic societies in Automation in Construction. He was the editorial board member for ASCE Journal of Construction Engineering and Management (ERA-A*) for numerous years and Journal of Information Technology in Construction (ERA-A). He is the editor/co-editor of four conference proceedings and the guest editor of Journal of Automation in Construction" (ERA-A), "Journal of Information Technology in Construction" (ERA-A)", on various topics in Construction Engineering and Management. He chaired three international conferences specialized in the area of Construction IT and is chairing the Global Lean Construction Conference 2015. He is the Editor-in-Chief of Springer Journal "Visualization in Engineering". Prof Wang established and led a reputable industry-focused collaborative research alliance in Australia that strives to improve productivity in LNG construction and maintenance by synergizing theory with practice. Impact of his work is highlighted by his establishment and leadership of a reputable industry alliance where over 40 industry partners are significantly contributing. The alliance is well positioned to lead the research, innovation, and industry practices towards improved productivity in Australia's energy industry.

王元丰 北京交通大学土建学院教授。中国城市科学研究会可持续土木工程专业委员会主任委员，曾任中国公路学会理事。主要从事混凝土材料与结构徐变、阻尼，绿色建筑及装配式建筑，工程可持续生命周期量化评价等方面的研究。由科学出版社出版学术专著3部，发表论文230余篇，其中被SCI检索60余篇，在 Nature 上发表评论文章2篇，是以"土木工程可持续发展面临的挑战和应对技术路径"为主题的第559次香山科学会议发起人和执行主席及第599次和S42次香山科学会议的执行主席。在全球最大的中文MOOC平台"清华大学学堂在线"主讲通识课程《技术创新简史》。出版长篇小说2部、诗集1部，在《人民日报》和新加坡《联合早报》等国内外主流媒体上发表评论文章180余篇。

肖岩博士 南京工业大学土木工程学院"千人计划"特聘教授、博士生导师,现任院长,兼美国南加州大学教授。1982年天津大学毕业,1986、1989年日本九州大学获工学硕士、工学博士学位。在日本青木建设、美国加州大学圣迭戈校从事研究和教学工作后,受聘于南加州大学,历任助理教授、副教授和教授(tenured)。2001年被聘为教育部"长江学者"、湖南大学特聘教授,并获国家自然科学基金委杰出青年基金(海外)。2010年入选中组部"千人计划"、湖南省"百人计划"。任 ASCE Journal of Structural Engineering,Journal of Bridge Engineering 副主编;Journal of Constructional Steel Research,Int. Journal of Advances in Structural Engineering 编委;国际组合结构学会理事;自然灾害学报副主任委员,建筑结构学报编委。美国混凝土学会 ACI、土木工程师学会 ASCE Fellow。研究兴趣包括结构在抗极端荷载下的性能分析和实验及设计方法,高性能及新型绿色结构材料等。

肖建庄 1968年出生于山东省沂南县,1997年博士毕业于同济大学。现任同济大学土木工程学院建筑工程系主任、教授、博导,上海高峰学科讲座教授;德国洪堡学者、国家杰出青年科学基金获得者。长期从事高性能混凝土、再生混凝土材料与结构基础研究。兼任国际 RILEM 再生混凝土结构行为与创新技术委员会主席、中国再生混凝土专业委员会主任委员、中国可持续土木工程研究专业委员会副主任委员兼秘书长、中国建筑废物资源化专业委员会副主任委员、中国高强与高性能混凝土专业委员会副主任委员。出版学术专著4本,英文专著1本,发表 SCI 论文100余篇;培养博士、硕士80余名。

张其林 同济大学土木工程学院建筑工程系空间结构研究室主任、教授。1982年毕业于东南大学土木工程系,1985及88年在同济大学分别获硕士和博士学位。1994—1996年获洪堡基金资助在 T. U. Braunschweig 工作,1999年在 T. U. Deflt 任高级研究员一年。长期从事金属结构与空间结构的教学、科研及工程实践。担任国家标准《铝合金结构技术标准》主编。目前担任中国钢协和中国金属结构协会钢结构专家委员会委员、中国土木工程学会空间结构委员会副主任委员、上海市金属结构协会和中国钢协空间结构分会专家委员会副主任委员、上海市空间结构工程技术中心技术委员会主任等。

Dr. Lihai Zhang is an academic staff in the Department of Infrastructure Engineering at The University of Melbourne. He received his Master of Engineering degree from the National University of Singapore with Research Scholarship in 1995. Following a successful ten year career in industry, he went on the Melbourne International Research Scholarship to further his PhD study at the University of Melbourne and graduated in 2009. He has been employed at The University of Melbourne since obtaining his PhD.

Dr. Zhang is best known for his expertise in numerically modelling of fluid flow, mass transfer, and reactive transport in deformed porous media as well as the development of advanced stochastic analysis methods. This gives him a leading advantage and a unique angle to solving outstanding engineering problems in both Civil Engineering (Life-cycle performance assessment of bridges) and Biomedical Engineering fields (Orthopaedic Biomechanics & Mechano-Biology). Dr. Zhang has been successful in making use of strong and sustained external collaborations both within Australia, and internationally, to seek cash funding from both public and industrial sources (e. g. ARC, NHMRC and CRC) along with supports from industrial collaborators (e. g. China Aerospace Science & Industry Corporation, and Johnson & Johnson Medical) to produce a coherent body of knowledge that has been disseminated in high impact publications.

Dr. Zhang has published more than 90 international journal papers, conference papers and book chapters. He has been awarded more than 20 National and International Competitive Grants with a total of over AU $8 million. He the Founding Director of Melbourne-Shenzhen Rehabilitation Research Centre and Editorial Board Member of International Journal of Applied Mechanics (World Scientific). He is also an Adjunct Associate Professor at Monash University, and a Visiting Professor in Earthquake Engineering Research & Test Centre at Guangzhou University, China.

Dr. **Yan Zhuge** is a Professor in Structural Engineering at School of Natural & Built Environments, University of South Australia, Australia. Yan has lectured in several Australian universities for more than 20 years. She has a BEng (Hons) in Civil Engineering and a Masters in Structural Engineering from Beijing, China, a PhD in Structural Engineering from Queensland University of Technology (QUT), Australia. Professor Zhuge's main research interests include green concrete materials, composite materials, utilisation of waste to construction materials, fibre composite structures (sandwich panels) and seismic retrofitting of unreinforced masonry and concrete structures using fibre reinforced polymer (FRP). She has published more than 150 SCI technical papers in the referred international journals and conferences and has been invited as a keynote speaker at several international conferences. Yan has successfully supervised many PhD graduates. Yan has won several Australian and Queensland government awards and fellowships and attracted funding from Australian Research council and industry. She is the executive committee member of Concrete Institute of Australia (CIA).